LINEAR OPERATORS AND ILL-POSED PROBLEMS

LINEAR OPERATORS AND ILL-POSED PROBLEMS

M. M. Lavrent'ev and L. Ya. Savel'ev
Institute of Mathematics
Siberian Division of the Russian Academy of Sciences
Novosibirsk, Russia

Translated from Russian by
Nauka Publishers, Moscow

CONSULTANTS BUREAU • NEW YORK, LONDON, AND MOSCOW
NAUKA PUBLISHERS • MOSCOW

Library of Congress Cataloging-in-Publication Data

Lavrent'ev, M. M. (Mikhail Mikhaĭlovich)
  [Lineĭnye operatory i nekorrektnye zadachi. English]
  Linear operators and ill-posed problems / M.M. Lavrent'ev and L.
Ya. Savel'ev ; translated from Russian by Nauka Publishers.
    p. cm.
  Includes bibliographical references (p.   -   ) and index.
  ISBN 0-306-11035-0
  1. Numerical analysis--Improperly posed problems. 2. Linear
operators. 3. Boundary value problems. I. Savel'ev, L. I͡A. (Lev
I͡Akovlevich) II. Title.
QA297.L3813 1995
515'.7246--dc20   95-22545
                                                                 CIP

This translation is published under an agreement with the
Russian Authors' Society (RAO)

ISBN 0-306-11035-0

©1995 Consultants Bureau, New York
A Division of Plenum Publishing Corporation
233 Spring Street, New York, N.Y. 10013

10 9 8 7 6 5 4 3 2 1

All rights reserved

No part of this book may be reproduced, stored in a retrieval system, or transmitted in any form or
by any means, electronic, mechanical, photocopying, microfilming, recording, or otherwise, without
written permission from the Publisher

Printed in the United States of America

Preface

This book has its origin in a course of lectures on ill-posed problems and functional analysis given by the authors in the Department of Mathematics at the Novosibirsk State University. It essentially consists of two parts: Linear Operators – a mathematical background; and Ill-Posed Problems, which is devoted to ill-posed problem theory and its applications.

Part One, Linear Operators, is divided into three chapters: Differentiation, Integration, and Linear Operators. Chapter 1 expounds on the differential calculus in normed spaces. It provides the reader with the necessary information about normed spaces and linear mappings. Chapter 2 is devoted to the integration of functions of an abstract variable. The general definition of the integral presented in this chapter allows one to derive delta functions that are used in solving ill-posed problem. Also it formulates the general theorems on change of variables and operations in the integrand function. Chapter 3 covers some elements of the theory of linear operators in normed spaces and describes some special spaces and classes of operators. It considers various types of generalized functions and introduces Fourier-Stieltjes, Fourier-Plancherel, and Radon's transforms. An ample portion is devoted to the spectral theory of operators.

While serving as a prerequisite for reading Part Two, Part One is also significant in its own right. The wide scope of the material covered makes it a useful reference guide to certain areas in functional analysis.

The main content of the book is presented in Part Two, Ill-Posed Problems. It consists of five chapters. Chapter 4, Classical Problems, describes some classical well-posed problems which reduce to Fredholm's equations of the second kind. Chapter 5, Ill-Posed Problems, is devoted to ill-conditioned problems of mathematical physics related to integral continuation of functions. Chapter 6, Operator and Integral Equations, presents the fundamentals of the theory of operator equations and treats two classes of integral equations. Chapter 7, Evolution Equations, deals mainly with the improperly posed Cauchy problems for equations of evolution. Chapter 8, Problems of Integral Geometry, discusses the results relevant to integral geometry.

The book concludes with two short chapters appended to the main text by its revising editor, A.L. Bukhgeim. This supplement is an extension of Part Two and contains new results recently obtained in ill-posed problem theory. Specifically it treats tomographic applications in relationship with the theory of functions of a complex variable, discusses logarithmic convexity and uniqueness in the inverse problems, and presents inversion formulas for integral equations of the first kind.

We gratefully acknowledge the assistance of A.L. Bukhgeim and E.Yu. Derevtsov in the preparation and editing of the Russian manuscript.

The work has been supported by the Russian Fundamental Research Fund, Grant No. 93-011-1753.

Novosibirsk, 1991
M.M. Lavrent'ev
L.Ya. Savel'ev

Contents

Part One. Linear Operators

Chapter 1. Differentiation 1
  1.1. Normed Spaces 1
     1.1,1. Norm 1
     1.1.2. Banach Spaces 4
     1.1.3. Linear Operators 4
     1.1.4. Operator Algebra 6
     1.1.5. Small Mappings 8
  1.2. Definition of Differential 10
     1.2.1. Increment of a Function 10
     1.2.2. Differential 11
     1.2.3. Derivative 14
     1.2.4. Partial Differentials 16
     1.2.5. Partial Derivatives 17
  1.3. Rules for Differentiation 19
     1.3.1. Differentiation of a Linear Combination 19
     1.3.2. Differentiation of a Product 19
     1.3.3. The Chain Rule 21
     1.3.4. Lagrange's Theorem 23
     1.3.5. Total Differentials 27
  1.4. Solution of Functional Equations 32
     1.4.1. Illustrative Examples and Definitions 32
     1.4.2. The Implicit Function Theorem 36
     1.4.3. Theorem of Inverse Function 37

1.4.4.	The Fixed Point Theorem	39
1.4.5.	Sum of Operator Progression	40

1.5. Taylor's Formula 41
   1.5.1. Sequential Derivatives 42
   1.5.2. Maclaurin's Formula 44
   1.5.3. Taylor's Formula 45
   1.5.4. Symmetry of Derivatives 46
   1.5.5. Particular Case 47

1.6. Local Minima 49
   1.6.1. Definitions 49
   1.6.2. Unconditional Local Minima 50
   1.6.3. Conditional Local Minima 51
   1.6.4. The Lagrange Method 53
   1.6.5. Smooth Surfaces 54

Chapter 2. Integration 60

2.1. Measures 60
   2.1.1. Simple Sets and Simple Functions 61
   2.1.2. Measures on Simple Sets 62

2.2. Integral Sums 63
   2.2.1. Definition 63
   2.2.2. Order Properties 65

2.3. Convergence Almost Everywhere 66
   2.3.1. The Concept of "Almost Everywhere" 66
   2.3.2. Continuity and Countable Additivity of Measures 67
   2.3.3. Convergence of Monotonic Sequences 68

2.4. Integrals 70
   2.4.1. Definition of Measurable Functions 71
   2.4.2. Approximating Sequences 72
   2.4.3. Integrable Functions 74
   2.4.4. Definition of Integral 76
   2.4.5. Order Properties of Integrals 79

2.5. Limit Theorems 80
   2.5.1. Levy's Theorem 80
   2.5.2. Fatou's Theorem 81
   2.5.3. Integrability Criterion 84
   2.5.4. Lebesgue's Theorem 86
   2.5.5. Fourier Transform 87

2.6. Measurable Functions 88
   2.6.1. Measurability and Integrability 88

Contents

2.6.2.	Sequences of Measurable Functions	88
2.6.3.	Measurable and Integrable Sets	89

2.7. Fubini and Tonelli's Theorems 90
    2.7.1. Product of Measures 91
    2.7.2. Double and Iterated Integrals 92
    2.7.3. Fubini's Theorem 93
    2.7.4. Tonelli's Theorem 97
    2.7.5. Rules for Operations Under the Integral Sign 99

2.8. Indefinite Integrals 108
    2.8.1. The Radon-Nikodym Theorem 108
    2.8.2. Theorem on the Change of Variables in Integration 109

Chapter 3. Linear Operators 112

3.1. Hilbert Spaces 112
    3.1.1. Euclidean Spaces 112
    3.1.2. Geometry of Euclidean Spaces 115
    3.1.3. Orthogonal Projection 116
    3.1.4. Linear Functionals 119
    3.1.5. Hilbert Space $\mathscr{L}^2$ 122

3.2. Fourier Series 127
    3.2.1. Orthonormal Bases 127
    3.2.2. The Bessel Inequality 128
    3.2.3. Fourier Series Expansion 130
    3.2.4. The Parseval Equation 132
    3.2.5. The Riesz-Fisher Theorem 132

3.3. Spaces of Functions 133
    3.3.1. Metric Spaces 133
    3.3.2. Smooth Functions 138
    3.3.3. Lebesgue Spaces 143
    3.3.4. Distributions 147
    3.3.5. Sobolev Spaces 155

3.4. Fourier Transforms 157
    3.4.1. Fourier Transforms of Functions of Rapid Decrease 157
    3.4.2. Transformations of Tempered Distributions 159
    3.4.3. The Fourier-Plancherel Transform 160
    3.4.4. The Fourier-Stieltjes Transform 161
    3.4.5. The Radon Transform 162

3.5. Bounded Linear Operators 163
    3.5.1. Extension of Functionals 163
    3.5.2. Uniform Boundedness of Operators 164
    3.5.3. Inverting Operators 166

>   3.5.4. Closedness of Operator Graphs 168
>   3.5.5. Weak Compactness 169
> 3.6. Compact Linear Operators 170
>   3.6.1. Examples of Compact Operators 170
>   3.6.2. Properties of Compact Operators 172
>   3.6.3. Adjoint Operators 173
>   3.6.4. Fredholm Operators 174
>   3.6.5. Fredholm's Theorems 177
> 3.7. Self-Adjoint Operators 179
>   3.7.1. Banach Space Adjoints 179
>   3.7.2. Hilbert Space Adjoints 180
>   3.7.3. Hermitian and Normal Operators 182
>   3.7.4. Unitary Operators 183
>   3.7.5. Positive Operators 183
> 3.8. Operator Spectra 184
>   3.8.1. Classification of Spectra 184
>   3.8.2. Spectrum of a Closed Operator 189
>   3.8.3. Spectrum of a Bounded Operator 191
>   3.8.4. Spectrum of a Compact Operator 192
>   3.8.5. Spectrum of a Self-Adjoint Operator 193
> 3.9. The Spectral Theorem 200
>   3.9.1. Projection-Valued Measures 200
>   3.9.2. Integrals of Bounded Functions 205
>   3.9.3. Integrals of Unbounded Functions 212
>   3.9.4. The Spectral Theorem 215
>   3.9.5. Operator-Valued Functions 219
> 3.10. Operator-Valued Exponential 220
>   3.10.1. Problem Formulation 221
>   3.10.2. Operator Semigroup 222
>   3.10.3. The Laplace Transform 223
>   3.10.4. Stone's Theorem 224
>   3.10.5. Evolution Equations 225

Part Two. Ill-Posed Problems

Chapter 4. Classical Problems 229

> 4.1. Mathematical Description of Physical Laws 229
> 4.2. Classification of Differential Equations of the Second Order 234
> 4.3. Elliptic Equations 236
> 4.4. Hyperbolic and Parabolic Equations 241
> 4.5. Concept of Well-Posedness 243

Contents   xi

Chapter 5. Ill-Posed Problems 245
  5.1. Ill-Posed Cauchy Problems 245
  5.2. Analytic Continuation and Interior Problems 248
  5.3. Physical Problems Leading to Ill-Posed Formulations 250
  5.4. Reduction of Ill-Posed Problems to Integral Equations 251
  5.5. Well-Posedness of Integral Equations 254
  5.6. Two Ill-Posed Problems 255
  5.7. Tomography 256

Chapter 6. Operator and Integral Equations 260
  6.1. Fredholm Equations of the Second Kind 260
  6.2. Equations of the First Kind 264
  6.3. Well-Posedness in the Sense of Tychonoff 266
  6.4. Regularization 269
  6.5. Equations with Bounded Operators 274
  6.6. Equations with Weak Singularities 276
  6.7. Volterra Equations for Functions of One Variable 277
  6.8. Operator Volterra Equations 280

Chapter 7. Evolution Equations 283
  7.1. Cauchy Problems and Operator Semigroups 283
  7.2. Equations in Hilbert Spaces 285
  7.3. Equations with Variable Operator 288
  7.4. Equations of the Second Order 290
  7.5. Well-Posed and Ill-Posed Cauchy Problems 292

Chapter 8. Problems of Integral Geometry 293
  8.1. Determining a Function from Its Spherical Averages 293
  8.2. General Problems of Plane Integral Geometry 298
  8.3. General Problems of Integral Geometry in Space 306
  8.4. Volterra-Type Problems of Integral Geometry with
       Motion-Group Invariant Manifolds 319

Supplement. Inversion Formulas in Inverse Problems
(A.L. Bukhgeim)

Introduction 323

Chapter I. Tomography and the Theory of A-Analytic Functions 324

   1.1. Radon's Problem: a Complex Function Approach 324

   1.2. A-Analytic Functions 332
      1.2.1. The Cauchy Formula 332
      1.2.2. Limit Values of Cauchy-Type Integrals 337
      1.2.3. Illustrative Examples 341

   1.3. A-Harmonic Functions 347
      1.3.1. Green's Formulas 347
      1.3.2. The Dirichlet Problem 356

   1.4. The Radon Problem on a Ring with Weight 360

   1.5. Problems 364

Chapter 2. Reconstruction of Phonon Spectra from the Heat Capacity Data 367

   2.1. Problem Statement and Theoretical Background 367
   2.2. Inversion Formulas 370
   2.3. Problems 374
   Reference Guide 376

References 377

Subject Index 379

LINEAR OPERATORS AND ILL-POSED PROBLEMS

Part One

Linear Operators

The first two chapters expound the fundamentals of differential calculus for normed spaces and the integral calculus for functions of abstract variables. Chapter 3 covers topics on linear operator theory.

Chapter 1. DIFFERENTIATION

This chapter discusses the fundamentals of differential calculus for normed spaces. At the outset the spaces proper are considered.

1.1. Normed Spaces

A normed space $(E, +, \cdot, \| \|)$ consists of a vector space $(E, +, \cdot)$ over the (real or complex) field $\mathbb{F}$ of scalars and a norm $\| \|$. The elements of the set E are referred to as *points* or *vectors* of the space $(E, +, \cdot, \cdot, \| \|)$, which will be also denoted simply by E. The properties of scalars (numbers) and vectors are assumed to be known.

1.1.1. Norm

A norm measures the distance from a given point x to the point 0. The norm of a vector may be thought of as its length.

DEFINITION. *A norm over E is defined as a semiadditive, absolutely homogeneous, real-valued function* $\| \|$ *on E:*

$$\| x + y \| \leq \| x \| + \| y \|$$

and

$$\| \alpha x \| = | \alpha | \cdot \| x \|$$

for all x and y in E and $\alpha \in \mathbb{F}$.

If $\| x \| > 0$ and $x \neq 0$, then we say that the norm $\| \ \|$ separates points in E. This type of norm is usually considered, while those defined above are called seminorms. It would be natural to refer to spaces endowed with norms separating the points as separable, but we will omit this modifier in most situations. Henceforth, unless otherwise noted, by a *normed space* we mean a *separable normed space*.

In any case we consider nondegenerate normed spaces where there are vectors with nonzero norm.

The distance between points x and y in space E is measured by the number $\rho(x, y) = \| x - y \|$.

If a norm does not separate points, then $\rho(x, y) = 0$ for $x \neq y$.

The sets of points

$$B(x, r) = \{ y : \| x - y \| < r \}$$

and

$$\overline{B}(x, r) = \{ y : \| x - y \| \leq r \}$$

are called open and closed balls of radius $r > 0$ centered on $x \in E$. Any set $V(x) \subseteq E$ containing some ball $B(x, r)$ is called the neighborhood of point x.

EXAMPLE 1. Given an n-dimensional space $\mathbb{F}^n$ ($= \mathbb{R}^n$ or $\mathbb{C}^n$) with the points $x = (\xi_1, ..., \xi_n)$, the norms can be defined by the equalities

$$\| x \|_2^2 = | \xi_1 | + ... + | \xi_n |,$$

$$\| x \|_2 = | \xi_1 |^2 + ... + | \xi_n |^2,$$

$$\| x \|_\infty = \max\{ | \xi_1 |, ..., | \xi_n | \}.$$

Product of Normed Spaces

Consider the normed spaces $(E_1, +, \cdot, \| \ \|)$ and $(E_2, +, \cdot, \| \ \|)$ and the product $(E, +, \cdot)$ of vector spaces $(E_1, +, \cdot)$ and $(E_2, +, \cdot)$. Elements of the set $E = E_1 \times E_2$ are ordered pairs $x = (x_1, x_2)$ of the vectors $x_1 \in E_1$ and $x_2 \in E_2$. The sum of the vectors and multiplication by a scalar are defined in terms of components as

$$x + y = (x_1, x_2) + (y_1, y_2) = (x_1 + x_2, y_1 + y_2)$$

1.1.2. Banach Spaces

The norm can be used to define convergence for sequences of points (vectors) in a normed space.

DEFINITION 1. *We will say that a sequence of points $x_n \in E$ converges in itself, symbolically $x_n \to$, if $\| x_n - x_m \| \to 0$ as $m \to \infty$ and $n \to \infty$.* Such a sequence is referred to as *a fundamental or Cauchy sequence*.

DEFINITION 2. *A sequence of points $x_n \in E$ is said to converge to a point $x \in E$, written $x_n \to x$, if $\| x_n - x \| \to 0$.*

Because the norms of differences measure distances between points, the sense of these definitions should be obvious.

Separability of space ensures that the limit is unique.

In view of the semiadditive property of the norm, convergence at a point implies convergence in itself. The converse is not always true. Normed spaces in which every sequence of points converging in itself tends to some point of the space are called *Banach spaces*. The Cauchy general principle of convergence holds in such spaces.

By way of example, the spaces $E = \mathbb{F}^n$ and $E = \mathcal{B}(T)$ described in Section 1.1.1. are Banach spaces, whereas the subspace P of the polynomials over $\mathcal{B}([0, 1])$ is not.

1.1.3. Linear Operators

Consider normed spaces E and F.

DEFINITION. *An additive homogeneous map $L : E \to F$ is linear:*

$$L(u + v) = L(u) + L(v),$$

$$L(cu) = cL(u) \qquad (c \in \mathbb{F}; u, v \in E)$$

From the homogeneity of L it follows that $L(0) = 0$.

In agreement with the general definition, the continuity of L at a point u implies that for every $\varepsilon > 0$ there exists $\delta > 0$ such that

$$\| L(u + v) - L(u) \| \leq \varepsilon, \quad (\| v \| \leq \delta).$$

Since, for a linear map L,

$$L(u + v) - L(u) = L(v)$$

for all u, v in E, the continuity of L at any point u is equivalent to L being continuous at 0. From this equality it should be evident that the continuity of this linear map is uniform.

EXAMPLE. If $E = \mathbb{F}^n$ and $F = \mathbb{F}^m$, then every linear mapping L of

1.1. Normed Spaces

and
$$\alpha x = \alpha(x_1, x_2) = (\alpha x_1, \alpha x_2)$$

for any $x = (x_1, x_2)$ and $y = (y_1, y_2)$ in E and $\alpha \in \mathbb{F}$.

There are different ways to define a norm for $(E, +, \cdot)$, but the stan norm is assumed to be the largest of the vector norms, namely,

$$\|x\| = \|(x_1, x_2)\| = \max\{\|x_1\|, \|x_2\|\}$$

for any $x = (x_1, x_2)$ in E. The properties of semiadditivity and abs homogeneity for such a norm function $\|\ \| : E \to R$ are readily veri as

$$\|x_i + y_i\| \leq \|x_i\| + \|y_i\| \leq \|x\| + \|y\|,$$
$$\|x + y\| \leq \|x\| + \|y\|;$$
$$\max\{\|\alpha x_i\|\} = \max\{|\alpha| \cdot \|x_i\|\} = |\alpha| \max\{\|x_i\|\},$$
$$\|\alpha x\| = |\alpha| \cdot \|x\|$$

for any $x = (x_i)$ and $y = (y_i)$ in E, $i = 1, 2$, and $\alpha \in \mathbb{F}$.

The normed space $(E, +, \cdot, \|\ \|)$ is *the product of the norme* $(E_1, +, \cdot, \|\ \|)$ and $(E_2, +, \cdot, \|\ \|)$. For more than two norme $(E_i, +, \cdot, \|\ \|)$, the product is defined in a similar manner. Elements $E = \prod E_i$ are the families $x = (x_i)$ of the vectors x_i in E_i. The multiplication by a scalar are defined in terms of components

$$x + y = (x_i) + (y_i) = (x_i + y_i),$$
$$\alpha x = \alpha(x_i) = (\alpha x_i)$$

for all $x = (x_i)$ and $y = (y_i)$ in E and $\alpha \in \mathbb{F}$.

The norm is taken to be the maximum vector norm,

$$\|x\| = \max\{\|x_i\|\}$$

for any $x = (x_i)$ in E $(i = 1, ..., n)$.

If $E_i = \mathbb{F}$ $(i = 1, ..., n)$, then the norm $\|\ \|$ is equal to th of Example 1.

EXAMPLE 2. Consider a vector space $\mathcal{B}(T, A)$ of scalar fi bounded on the subset A of T. A norm for $\mathcal{B}(T, A)$ is defin

$$\|x\| = \sup\{|x(t)| : t \in A\}.$$

If $A = T$, this norm separates the points in $\mathcal{B}(T, A)$; if A not.

For $A = T$ we will write $\mathcal{B}(T)$ in place of $\mathcal{B}(T, T)$.

1.1. Normed Spaces

these finite-dimensional spaces is continuous under the norms $\|\ \|_1$, $\|\ \|_2$, and $\|\ \|_\infty$. This assertion is easily verified using the matrix algebra of the linear maps $\mathbb{F}^n \to \mathbb{F}^m$ and the inequalities $\|x\|_2 \le \|x\|_1 \le c\|x\|_\infty$.

A linear map $L: E \to F$ is said to be *bounded* if it is bounded on the unit ball $\overline{B}(0, 1)$ in E, namely, for some $c > 0$

$$\|L(v)\| \le c \quad [v \in \overline{B}(0,1)].$$

Boundedness of a linear map is equivalent to its continuity. This fact can be readily verified by comparing the definitions and resorting to the linearity of compression or extension, transforming any balls with center 0 into a unit ball. This equivalence allows one to say *bounded linear map* instead of *continuous linear map*.

Linear maps are called *linear operators*. The modifier linear may be omitted where there is no risk of misunderstanding. Also, the terms *continuous operators* and *bounded operators* are interchangeable.

EXAMPLE. **Projectors and embeddings.** Let $E = \Pi E_j$ be the product of normed spaces E_j ($j = 1, ..., n$). The operator $p_j : E \to E_j$ mapping $x = (x_1, ..., x_n) \in E$ onto $x_j \in E_j$ is called a projector of E onto E_j. Clearly, this is a linear operator.

Specifically, if $n = 2$ and $E_1 = E_2 = R$, then p_1 and p_2 are the projectors onto the 1st and 2nd coordinate axes.

In some circumstances it is more convenient to project $E = \Pi E_j$ not onto a cofactor E_j but rather on a subspace D_j of E naturally connected with E_j and composed of $x = (x_1, ..., x_n) \in E$ in which all coordinates are zeros except possibly the jth. The operator $r_j : E_j \to E$ mapping $x_j \in E_j$ onto $x = (0, ..., 0, x_j, 0, ..., 0) \in E$ is called a *natural embedding* of E_j into E. Clearly, this is a linear operator mapping E_j onto D_j one-to-one.

The operators p_j and r_j are bounded. Indeed, from the definitions it follows that for all x in E and x_j in E_j

$$\|p_j(x)\| = \|x_j\| \le \|x\|,$$
$$\|r_j(x_j)\| = \|x_j\|.$$

These inequalities suggest also that

$$\|p_j\| = 1 \text{ and } \|r_j\| = 1; \quad j = 1, ..., n.$$

The composite operator $r_j p_j$ projects E onto D_j and is a bounded linear operator with norm 1.

1.1.4. Operator Algebra

Linear operators may be added, multiplied by a scalar, and applied in succession, preserving the linearity and continuity of the operators if available.

Let us consider normed spaces E, F, G and the linear operators $A : E \to F$ and $B : F \to G$. The composition $BA : E \to G$ is defined by

$$BA(u) = B(A(u))$$

for any u in E. It is also called the *product* of A and B. One can readily verify that BA is a linear operator. If A and B are continuous, then BA is also continuous.

Some mathematicians prefer writing Au in place of $A(u)$, a very convenient notation for linear operators.

In finite-dimensional spaces, to an operator algebra there correspond operations with matrices. In particular, to the product of operators there corresponds the product of their matrices in selected bases.

For a bounded linear operator, a norm may be defined as follows.

DEFINITION. *A norm of the bounded linear operator* $A : E \to F$ *is the number*

$$\|A\| = \sup\{\|Au\| : \|u\| \leq 1\}.$$

Here $\|\ \|$ represents three different norms, that of the operator A, of the vector $Au \in F$, and of the vector $u \in E$.

EXAMPLE. Let $E = F = \mathbb{R}$ and A be a linear function. Then

$$\|A\| = \sup\{|Au| : |u| \leq 1\} = |A(1)|.$$

The norm of A equals the absolute value of the coefficient $A(1)$. It measures the angular separation from the horizontal axis.

CAUCHY INEQUALITY. $\|Au\| \leq \|A\| \cdot \|u\|$.

PROOF. The case of $\|u\| = 0$ is trivial as both sides of the inequality are zero. If $\|u\| \neq 0$, then

$$\|u\|^{-1} \|A(u)\| = \|A(\|u\|^{-1} u)\| \leq \|A\|$$

for $\|\ \|u\|^{-1} \cdot u\| = \|u\|^{-1} \cdot \|u\| = 1$, and the Cauchy inequality holds again. ∎

COROLLARY. $\|BA\| \leq \|B\| \cdot \|A\|$.

1.1. Normed Spaces

PROOF. Applying the Cauchy inequality in succession yields

$$\|BA(u)\| = \|B(Au)\| \leq \|B\| \cdot \|Au\| \leq \|B\| \cdot \|A\| \cdot \|u\|$$

Hence,

$$\|BA(u)\| \leq \|B\| \cdot \|A\| \quad (\|u\| \leq 1)$$

and we see that the desired inequality is satisfied. ∎

Consider the normed space $\mathcal{B} = \mathcal{B}(E, F)$ of bounded linear operators $L : E \to F$ having the ordinary sum, product by a scalar, and an operator norm. Note that the space $\mathcal{B}$ is separable if F is separable (while E may not). Indeed, let $L \neq 0$; then $L(a) \neq 0$ and $\|La\| > 0$ for some a in E. Choose an $\alpha > 0$ such that $\|\alpha a\| \leq 1$. Then

$$\|L\| \geq \|L(\alpha a)\| = \alpha \|La\| > 0.$$

THEOREM. *If F is Banach, then $\mathcal{B} = \mathcal{B}(E, F)$ is also Banach.*

PROOF. Let a sequence of operators $L_n \in \mathcal{B}$ converge in itself, i.e., be a Cauchy sequence. This means that for any $\varepsilon > 0$ there is a number l such that $\|L_n - L_m\| \leq \varepsilon$ for $n, m \geq l$. Then, by the Cauchy inequality,

$$\|L_n u - L_m u\| = \|(L_n - L_m)u\| \leq \|L_n - L_m\| \cdot \|u\| \leq \varepsilon \cdot \|u\|$$

for $n, m \geq l$ and u in E. Hence, the sequence $L_n u \in F$ converges in itself. If F is Banach, it converges to some point y in F. If F is separable, which is implicitly assumed, then such a point y is unique and a mapping $L : E \to F$ with $L(u) = y$ ($u \in E$), is defined.

From the linearity of the limit it follows that the mapping L is also linear. From the above inequalities, semi-additivity of the norm, and the Cauchy inequality it follows that L is bounded on a unit ball:

$$\|L(u)\| - \|L_m(u)\| \leq \|L(u) - L_m(u)\| \leq \varepsilon \cdot \|u\|,$$

$$\|L(u)\| \leq \|L_m(u)\| + \varepsilon \cdot \|u\| \leq (\|L_m\| + \varepsilon) \cdot \|u\|$$

$$\leq \|L_m\| + \varepsilon \quad (m = l, \ \|u\| \leq 1, \ u \in E)$$

for $L(u) = \lim L_n(u)$. Finally, from the inequalities written at the beginning of the proof it follows that $\|L - L_m\| \leq \varepsilon$ for $m \geq l$. Hence, $L_n \to L$. ∎

If $E = F = G$, then augmenting the sum and the product by a scalar with the product of operators converts $\mathcal{B} = \mathcal{B}(E, F)$ into a Banach operator algebra. By virtue of the Cauchy inequality

$$\|BA\| \leq \|B\| \cdot \|A\|$$

for $A, B \in \mathcal{B}$. This is an essential distinction of the operator norm from the absolute value of a number.

1.1.5. Small Mappings

Such mappings describe errors incurred by an approximation.

Let us consider normed spaces E, G, and H, a neighborhood $V = V(0) \subseteq E$ of a point $0 \in E$, the mappings $g : V \to G$ and $h : V \to H$, and the functions $\varphi : V \to \mathbb{R}$ and $\psi : V \to \mathbb{R}$.

The function φ is said to be *locally small compared to* ψ, written $\varphi = o(\psi)$, if for every $\varepsilon > 0$ there exists $\delta > 0$ such that

$$|\varphi(v)| \leq \varepsilon \cdot |\psi(v)| \quad (\|v\| \leq \delta, \ v \in V).$$

The function φ is said to be *locally bounded compared to* ψ, written $\varphi = O(\psi)$, if there exists $\alpha > 0$ and $\delta > 0$ such that

$$|\varphi(v)| \leq \alpha \cdot |\psi(v)| \quad (\|v\| \leq \delta, \ v \in V).$$

If $|\psi(v)| = 1$, $v \in V$, we will say respectively that φ is *locally small* or *locally bounded* and write $\varphi = o(1)$ or $\varphi = O(1)$. Moreover, if $|\psi(v)| = \|v\|$, $(v \in V)$, we will briefly say that φ is *comparatively small* or *bounded* and write $\varphi(v) = o(v)$ or $\varphi(v) = O(v)$. Also, we will say simply *small* instead of *locally small*.

The concepts of comparative smallness and boundedness are carried over from the numerical functions φ and ψ to the mappings g and h with the aid of norms induced in the spaces G and H.

A mapping g is said to be locally small (bounded) compared to h, written $g = o(h)$ [$g = O(h)$], if $\varphi = \|g\|$ is locally small (bounded) relative to $\psi = \|h\|$.

By definition, *local smallness* of g means that for every $\varepsilon > 0$ there exists $\delta > 0$ such that

$$\|g(v)\| \leq \varepsilon \quad (\|v\| \leq \delta, \ v \in V),$$

that is, $g(v) \to 0$ as $v \to 0$. *Local boundedness* implies that there exist

1.1. Normed Spaces

$\alpha > 0$ and $\delta > 0$ such that

$$\|g(v)\| \leq \alpha \quad (\|v\| \leq \delta, \ v \in V).$$

Comparative smallness of g implies that for any $\varepsilon > 0$ there exists $\delta > 0$ such that

$$\|g(v)\| \leq \varepsilon \|v\| \quad (\|v\| \leq \delta, \ v \in V),$$

that is, $\|v\|^{-1} g(v) \to 0$ as $v \to 0$, $\|v\| \neq 0$, and $g(0) = 0$.

Comparative boundedness means that there are $\alpha > 0$ and $\delta > 0$ such that

$$\|g(v)\| \leq \alpha \|v\| \quad (\|v\| \leq \delta, \ v \in V).$$

From the definitions, it follows that the local smallness (boundedness) of g is equivalent to the smallness (boundedness) of g with respect to a constant mapping $h = c$. Therefore, the fact that g is locally small (bounded) may be written as $g = o(c)$ [$g = O(c)$]. We will conditionally write $o(1)$ and $O(1)$ for mappings in place of $o(c)$ and $O(c)$. In much the same way, the comparative smallness and boundedness of g may be represented without any risk of ambiguity as $g(v) = o(v)$ and $g(v) = O(v)$.

Clearly, if $g = o(h)$, then $g = O(h)$, but the opposite is not true: if $g(v) = O(v)$, then $g = o(1)$.

Comparison of these definitions suggests that the boundedness of a linear operator means its local boundedness.

EXAMPLE. **Scale of infinitesimals.** Let $E = G = H = V = \mathbb{F}$ and $e^n(v) = v^n$, $n = 0, 1, 2$, etc.; then $g = e^n$ is small for $n \geq 1$, comparatively small for $n \geq 2$, and the function $g = e^{n+1}$ is small compared with $h = e^n$ for $n \geq 0$.

A linear combination of comparatively small (bounded) maps is also comparatively small (bounded). A composition of a comparatively bounded mapping and a comparatively small one is itself comparatively small. It is straightforward to verify that

$$o(h) + o(h) = o(h),$$

$$O(h) + O(h) = O(h),$$

$$\alpha \cdot o(h) = o(h), \quad \alpha \cdot O(h) = O(h) \text{ for } (\alpha \in \mathbb{F});$$

$$o(h) \circ O(h) = O(h) \circ o(h) = o(h).$$

1.2. Definition of Differential

The differential of a function is the main linear part of its increment. The comparative smallness of the difference between the function's increment and the differential allows the function to be approximated by the differential in the neighborhood of a given point, however close.

1.2.1. Increment of a Function

Let E, F, and G be normed spaces with a set $U \subseteq E$, a function $f: U \to F$, and a point $u \in U$. We will assume that U is open, that is, for every point $u \in U$ the set U contains a ball $B(u, r)$ of radius $r > 0$ centered on u. Let a parallel transfer of E carrying u at 0 convert U into the set

$$V = U - u = \{t - u \mid t \in U\}.$$

The *increment in a function* $f: U \to F$ at *a point* $u \in U$ is called the function $\Delta fu : V \to F$ with values

$$\Delta fu(v) = f(u + v) - f(u)$$

for $v \in V$.

Two properties of the increment are worth mentioning. Let g be a function from U to F; then it is easy to verify that

$$\Delta(f + g)u = \Delta fu + \Delta gu$$

and

$$\Delta(c \cdot f)u = c \cdot \Delta fu.$$

Now, let $Y \subseteq F$ be an open set, $g: Y \to G$, $y \in Y$, and $Z = Y - y = \{x - y \mid x \in Y\}$. Suppose that $f(U) \subseteq Y$. In this case we are in a position to define the composite function $g \circ f: U \to G$ and the increment in this function $\Delta(g \circ f) u : V \to G$ at a point $u \in U$. This increment equals the composition of the increments $\Delta f u$ and $\Delta g y$ for $y = f(u)$. Indeed, for $v \in V$,

$$\Delta gy(\Delta fu(v)) = g(f(u + \Delta fu(v))) - g(f(u)) = g(f(u + v)) - g(f(u));$$

hence, for $y = f(u)$,

$$\Delta gy \circ \Delta fu = \Delta(g \circ f)u.$$

Note that the increment in a function at a point is identically zero if and only if the function is constant.

1.2. Definition of Differential

Hence

$$d_1 fu(\|v\|^{-1} v) - d_2 fu(\|v\|^{-1} v) \to 0 \quad (v \to 0, \ v \neq 0),$$

$$\|d_1 fu - d_2 fu\| = 0$$

and $d_1 fu = d_2 fu$. ∎

Thus, the differential dfu and the remainder rfu are uniquely defined. Note that in the proof use is made of the separability of the space F, which ensures uniqueness of the limit.

The differential dfu is often termed a *strong* differential, or *Frechet's differential*.

If a function $f: U \to F$ is differentiable at each point $u \in U$, then we will simply say that f *is differentiable*. In this case there exists a differential df of f. The differential maps U into the space $\mathcal{B} = \mathcal{B}(E, F)$ of bounded linear operators from E into F. The image of df at $u \in U$ is the differential $dfu \in \mathcal{B}$. It should be stressed that the operator df is, as a rule, nonlinear, unlike dfu.

Denote by $D_u = D_u(U, F)$ and $D = D(U, F)$ the sets of all functions $f: U \to F$ differentiable at a given point u and at each point of U. The differential dfu will be denoted also $d_u f$. In addition to df, with the differential dfu there are associated the operators $d: D \to \mathcal{F}(U, \mathcal{B})$ and $d_u: D_u \to \mathcal{B}$. The operator d maps $f \in D$ onto $df \in \mathcal{F}(U, \mathcal{B})$, while d_u maps $f \in D_u$ onto $d_u f = dfu \in \mathcal{B}$. Looking a little bit ahead, we note that both operators are linear, namely, for $f, g \in D$ and $f, g \in D_u$,

$$d(f + g) = df + dg, \quad d_u(f + g) = d_u f + d_u g;$$

In addition since f is differentiable at u, it follows that f is continuous at u:

$$\Delta fu(v) = dfu \cdot v + rfu(v) \to 0 \quad \text{as} \quad v \to 0, \ v \neq 0.$$

A few simple examples follow.

EXAMPLE 1. Given $E = F = U = \mathbb{R}$, $u = 0$, and $f(t) = 2v + v^2$, then, for $v \in \mathbb{R}$,

$$dfu \cdot v = 2v \quad \text{and} \quad rfu(v) = v^2.$$

EXAMPLE 2. Given $E = F = \mathbb{R}$, $U = \mathbb{R}$, $u = 2$, and $f(t) = 2 - 2t + t^2$ then, for $t, v \in \mathbb{R}$,

$$\Delta fu(v) = 2v + v^2, \quad dfu \cdot v = 2v, \quad \text{and} \quad rfu(v) = v^2.$$

EXAMPLE 3. Given $E = F = \mathbb{R}$, $U = \mathbb{R}$, $u \in \mathbb{R}$, and $f(t) = e^t$ then for t, $v \in \mathbb{R}$,

$$\Delta f u(v) = e^{u+v} - e^u = e^u(e^v - 1),$$
$$dfu \cdot v = e^u \cdot v, \quad \text{and} \quad rfu(v) = e^u(e^v - 1 - v).$$

EXAMPLE 4. Given $E = \mathbb{R}^2$, $F = \mathbb{R}$, $U = \mathbb{R}^2$, $u = (2, 0)$, and $f(t) = 2 - 2t_1 + t_1^2 + t_2^2$, then, for $(t = (t_1, t_2) \in \mathbb{R}^2)$, $v = \begin{pmatrix} v_1 \\ v_2 \end{pmatrix} \in \mathbb{R}^2$,

$$\Delta f u(v) = 2v_1 + v_1^2 + v_2^2,$$
$$dfu \cdot v = (2, 0) \cdot v = (2, 0) \begin{pmatrix} v_1 \\ v_2 \end{pmatrix} = 2 \cdot v_1 + 0 \cdot v_2 = 2v_1,$$
$$rfu(v) = v_1^2 + v_2^2, \quad \text{and} \quad |rfu(v)| = |v|^2.$$

EXAMPLE 5. Given $E = F = \mathbb{R}^2$, $U = \mathbb{R}^2$, $u = (u_1, u_2) \in \mathbb{R}^2$, $f(t) = (t_1 \cdot \cos t_2, t_1 \cdot \sin t_2)$, then, for $(t = (t_1, t_2) \in \mathbb{R}^2)$ and $v = (v_1, v_2)^T \in \mathbb{R}^2$,

$$\Delta f u(v) = \begin{array}{l} ((u_1 + v_1) \cdot \cos(u_2 + v_2) - u_1 \cdot \cos u_2, \\ (u_1 + v_1) \cdot \sin(u_2 + v_2) - u_1 \cdot \sin u_2), \end{array}$$

$$dfu \cdot v = \begin{bmatrix} \cos u_2 & -u_1 \cdot \sin u_2 \\ \sin u_2 & u_1 \cdot \cos u_2 \end{bmatrix} \cdot v$$

1.2.3. Derivative

Let $E = \mathbb{R}$. Each linear function $l : \mathbb{R} \to F$ is defined by its coefficient $l(1) \in F$:

$$l(t) = l(1) \cdot t \quad \text{for all} \quad t \in \mathbb{R}.$$

The function is bounded on the interval $[-1, 1]$ by the number $\| l(1) \|$, that is, for $|t| \leq 1$,

$$\| l(t) \| = \| l(1) \| \cdot | t | \leq \| l(1) \|.$$

Therefore, for $E = \mathbb{R}$, the set of bounded linear functions coincides with the set of all linear functions, $\mathcal{B}(\mathbb{R}, F) = \mathcal{L}(\mathbb{R}, F)$.

In particular, the differential dfu of a function $f : U \to F$ at $u \in U \subseteq \mathbb{R}$ is defined by its coefficient

$$f'(u) = dfu(1) \in F.$$

This coefficient is called the *derivative of f at u.*

1.2. Definition of Differential

PROPOSITION. *A function f is differentiable at a point u if and only if there exists a limit of the ratio $v^{-1} \cdot \Delta fu(v)$ as $v \to 0$, $v \neq 0$.*

PROOF. When this limit does exist we have

$$f'(u) = \lim(v^{-1} \cdot \Delta fu(v)).$$

If f is differentiable at u, then

$$v^{-1}\Delta fu(v) = v^{-1} \cdot dfu(v) + v^{-1} \cdot rfu(v) = f'(u) + v^{-1} \cdot rfu(v) \to f'(u).$$

If $v^{-1} \cdot \Delta fu(v) \to c \in F$ as v tends to zero ($v \neq 0$), then

$$v^{-1} \cdot (\Delta fu(v) - c \cdot v) = v^{-1} \cdot \Delta fu(v) - c \to 0 \text{ as } v \to 0, \ v \neq 0,$$

hence,

$$dfu(v) = c \cdot v,$$
$$rfu(v) = \Delta fu(v) - c \cdot v,$$
$$f'(u) = dfu(1) = c, \text{ as required.} \blacksquare$$

Thus, the definition of derivative of a vector-valued function of a real-valued variable as a coefficient of the differential is equivalent to the ordinary definition of derivative as the limit of an incremental ratio of the function over its argument. The definition of a derivative as the coefficient is advantageous in that it naturally generalizes to vector functions of a vector variable over spaces with finite or denumerable bases. The derivative is then defined as the matrix of the differential in the selected bases.

EXAMPLE. Let $U = E = \mathbb{R}$, $F = \mathbb{R}^2$, and for $t \in \mathbb{R}$, $f: U \to F$ is defined as

$$f(t) = \begin{bmatrix} \cos t \\ \sin t \end{bmatrix}.$$

Then, for $u, v \in \mathbb{R}$,

$$dfu(v) = \begin{bmatrix} -\sin u \\ \cos u \end{bmatrix} \cdot v,$$

and the derivative

$$f'(u) = \begin{bmatrix} -\sin u \\ \cos u \end{bmatrix}$$

is a 2×1 matrix.

1.2.4. Partial Differentials

Let us consider the products $F = \Pi F_i$ and $E = \Pi E_j$ of normed spaces F_i and E_j with $i = 1, ..., m$, and $j = 1, ..., n$, the rectangle $U = \Pi U_j$ with the open sides $U_j \subseteq E_j$, a mapping $f: U \to F$, a point $u = (u_j) \in U$, and the sets $V = U - u$, and $V_j = U_j - u_j$, $j = 1, ..., n$.

Given the translation $Tu : E \to E$, projectors $q_i : F \to F_i$ and $p_j : E \to E_j$, and natural embeddings $s_i : F_i \to F$ and $r_j : E_j \to E$, we define the particular functions $f_{ij}u : V_j \to F_i$ with values

$$f_{ij}u(v_j) = q_i f(u + r_j v_j) \text{ for all } v_j \in V_j.$$

These functions describe the variation of f in a space F_i when the argument t varies in a space E_j. The variation of the argument is counted from the point u, namely,

$$t = Tu(r_j v_j) = u + r_j v_j,$$

$$t_j = u_j + v_j, \ t_l = u_l, \text{ for } l \neq j \text{ and } v_j \in V_j.$$

The partial functions $f_{ij}u$ are the composites of the function f, translation Tu, projectors q_i, and embeddings r_j:

$$f_{ij}u = q_i \cdot f \cdot Tu \cdot r_j.$$

We denote conditionally the increment in the function $f_{ij}u$ at a point $o_j \in E_j$ by $\Delta f_j u$, rather than $\Delta f_{ij} u \, o_j$, so that for all $v_j \in V_j$,

$$\Delta f_{ij} u(v_j) = f_{ij}u(v_j) - f_{ij}u(o_j) = q_i(f(u + r_j v_j) - f(u))$$

$$= q_i \cdot \Delta fu \cdot r_j(v_j).$$

Hence,

$$\Delta f_{ij} u = q_i \cdot \Delta fu \cdot r_j : V_j \to F_i.$$

We will call $\Delta f_{ij} u$ the *partial increments of f at u*. It is convenient to associate with these increments the composites

$$\Delta_{ij} fu = s_i \cdot \Delta f_{ij} u \cdot p_j = s_i q_i \cdot \Delta fu \cdot r_j p_j : V \to F$$

defined on $V \subseteq E$ with values in F for any i and j. Such composites are suitable for algebraic operations. Note that $s_i q_i : F \to F$ and $r_j p_j : E \to E$ project F and E onto subspaces composed of vectors having all their coordinates equal to zero except, possibly, the ith and jth.

1.2. Definition of Differential

The differentials $df_{ij}u$ of functions f_{ij} at points o_j are called the *partial differentials of f at u*. It is easy to verify that

$$df_{ij}u = q_i \cdot dfu \cdot r_j : E_j \to F_i.$$

It is convenient to associate these differentials with the composites

$$d_{ij}fu = s_i \cdot df_{ij}u \cdot p_j = s_i q_i \cdot dfu \cdot r_j p_j : E \to F$$

defined over E and having the image in F for any i and j. Such compositions are amenable to algebraic operations. We will conditionally call $d_{ij}fu$ also the *partial differentials of f at u*.

Note that the sums $\sum_i s_i q_i$ and $\sum_i r_j p_j$ are identity functions over F and E, viz.,

$$\sum_i s_i q_i = 1_F \text{ and } \sum_i r_j p_j = 1_E.$$

Therefore, in view of the linearity of the operators under consideration,

$$dfu = 1_F \cdot dfu \cdot 1_E = \sum_j s_i q_i \cdot dfu \cdot \sum_j r_j p_j$$
$$= \sum_i \sum_j s_i q_i \cdot dfu \cdot r_j p_j = \sum_{ij} d_{ij}fu.$$

The differential dfu is called the *total differential*. The equality

$$dfu = \sum d_{ij}fu$$

reads "the total differential is the sum of the partial differentials".

REMARK. All expressions for the differentials have been written assuming that all the considered differentials exist.

1.2.5. Partial Derivatives

Let $E_1 = \ldots = E_n = \mathbb{F}$, and $E = \mathbb{F}^n$ be a real or complex n-dimensional space. Then the differentials of partial functions may be expressed by the derivatives of these functions:

$$df_{ij}u(v_j) = f'_{ij}(u) \cdot v_j \quad (v_j \in V_j).$$

Here $f'_{ij}(u) \in F_i$ is a vector, and $v_j \in \mathbb{F}$ is a number.

It is natural to associate with the matrix of partial functions $f_{ij}u$ the

Chapter 1. Differentiation

$m \times n$ matrix

$$f'(u) = (f'_{ij}(u))$$

called the *Jacobi matrix* and having as its elements the derivatives of the partial functions $f_{ij}u$ ($i = 1, ..., m$; $j = 1, ..., n$).

Writing $dfu(v)$ as a column-matrix with elements $q_i \cdot dfu(v)$ and noting that in view of the linearity of the differential

$$q_i \cdot dfu(v) = \sum_j q_i \cdot dfu(r_j v_j) = \sum_j df_{ij}u(v_j) = \sum_j f'_{ij}(u) \cdot v_j$$

we obtain the matrix equality

$$dfu(v) = f'(u) \cdot v \quad (v \in E).$$

Therefore, it would be natural to call $f'(u)$ the *derivative of f at u*. For $m = n = 1$, in agreement with the previous definition $f'(u)$ will be a 1×1 matrix, i.e., a vector in F.

In the general case when $E_j \neq \mathbb{F}$ and the derivatives $f'_{ij}(u)$ are linear operators and $v_j \in E_j$ are vectors, the differential dfu may be expressed in matrix form through $f'(u)$ using a multiplicative representation of values of the linear operators.

EXAMPLE. Let $F = \mathbb{R}^3$, $E = U = \mathbb{R}^2$ and for $t = (t_1, t_2)$ the function $f : U \to F$ is defined by

$$f(t) = \begin{bmatrix} t_1 \cdot \sin t_2 \\ t_2 \cdot \cos t_1 \\ t_1 t_2 \end{bmatrix}.$$

Then

$$f'(u) = \begin{bmatrix} f'_{11}(u) & f'_{12}(u) \\ f'_{21}(u) & f'_{22}(u) \\ f'_{31}(u) & f'_{32}(u) \end{bmatrix} = \begin{bmatrix} \sin u_2 & u_1 \cos u_2 \\ -u_2 \sin u_1 & \cos u_1 \\ u_2 & u_1 \end{bmatrix}$$

Thus, the derivative is a 3×2 matrix.

REMARK. In order to emphasize the central role of linear approximation we have chosen the differential to serve as a central concept. The derivative has been defined with its aid. These concepts are often mixed, and a differential is referred to as a derivative. We will do so as this simplifies the exposition and enables us to avoid repeating the formulas.

1.3. Rules for Differentiation

Thus, $f'(u)$ will denote a derivative in some situations and the differential of f at u in others, the connotation being evident from the context.

Common designations of derivative are primes, overhead dots, and the symbols D, ∂, d/dt, and $\partial/\partial t$. The fractions are normally written upright for convenience of handling the derivatives.

1.3. Rules for Differentiation

Appropriate application of some simple rules of differentiation yields the differentials of many functions.

1.3.1. Differentiation of a Linear Combination

Let E and F be normed spaces, an open set $U \subseteq E$, a point u in U, a set $V = U - u$, functions $f: U \to F$ and $g: U \to F$, and numbers $a \in \mathbb{F}$ and $b \in \mathbb{F}$.

PROPOSITION. *If f and g are differentiable at u, then the linar combination $af + bg$ is also differentiable at u and*

$$(af + bg)'(u) = a \cdot f'(u) + b \cdot g'(u).$$

PROOF. Indeed,

$$\Delta(af + bg)u = a \cdot \Delta fu + b \cdot \Delta gu = [a \cdot f'(u) + b \cdot g'(u)] \cdot v + [a \cdot rfu(v) + b \cdot rgu(v)]$$

for any v in V. ■

Thus, differentiation is a linear operation. The derivative D_u is a linear operator in the vector space $\mathscr{D}_u$ of functions differentiable at u, i.e.,

$$D_u \in \mathscr{L}(\mathscr{D}_u, F).$$

1.3.2. Differentiation of a Product

Consider normed spaces E, F, G and H, an open set $U \subseteq E$, a point $u \in U$, a set $V = U - u$, and functions $f: U \to F$ and $g: U \to G$.

The product $F \times G \to H$ is a continuous bilinear function $(x, y) \to x \cdot y$, $x \in F$, $y \in G$, $x \cdot y \in H$, defined on the product $F \times G$ and having values in H. The bilinearity of the product implies that for every $a \in F$ and $b \in G$, the functions $y \to a \cdot y$ $(y \in G)$ and $x \to x \cdot b$ $(x \in F)$ are linear. Continuity of the product is equivalent to its boundedness on a unit ball in $F \times G$, namely,

$$\|x \cdot y\| \leq c \quad (\|x\| \leq 1, \|y\| \leq 1, x \in F, y \in G).$$

Along with the product of vectors $x \in F$ and $y \in G$ we define the product $f \cdot g : U \to H$ of functions $f : U \to F$ and $g : U \to G$ as the product of their values $x = f(t) \in F$ and $y = g(t) \in G$,

$$(f \cdot g)(t) = f(t) \cdot g(t) \quad (t \in U).$$

In writing out products the dot is commonly omitted.

PROPOSITION. *If f and g are differentiable at u, the product $f \cdot g$ is differentiable at u and*

$$(f \cdot g)'(u) = f'(u) \cdot g(u) + f(u) \cdot g'(u).$$

PROOF. For a point v in V,

$$\Delta(f \cdot g)u(v) = f(u+v) \cdot g(u+v) - f(u) \cdot g(u) = l(v) + r(v),$$

$$l(v) = f'(u)v \cdot g(u) + f(u) \cdot g'(u)v,$$

$$r(v) = f(u) \cdot rgu(v) + f'(u)v \cdot (g'(u)v + rgu(v))$$

$$+ rfu(v) \cdot (g(u) + g'(u)v + rgu(v)).$$

The function $l = f'(u) \cdot g(u) + f(u) \cdot g'(u)$ with values $l(v)$ is continuous and linear. The function r with values $r(v)$ is comparatively small. Therefore, the product $f \cdot g$ is differentiable at u and the equality to be proved is true. ∎

REMARK. In the equality enunciating the rule for differentiation of a product, $f'(u)$ is a linear mapping $E \to F$, and $g(u)$ is the constant of $E \to G$. The product $f'(u) \cdot g(u)$ is a mapping $E \to H$ such that for every v in E

$$(f'(u) \cdot g(u))v = f'(u)v \cdot g(u) \in H.$$

Similarly, $f(u)$ is the constant of $E \to F$, $g'(u) : E \to G$, and $f(u) \cdot g'(u) : E \to H$, so for every v in E,

$$(f(u) \cdot g'(u))v = f(u) \cdot g'(u)v \in H.$$

Therefore, $(fg)'(u) : E \to H$ and

$$(fg)'(u)v = f'(u)v \cdot g(u) + f(u) \cdot g'(u)v \quad (v \in E).$$

The multiplicative representation has been used throughout to avoid many extra parentheses for linear operators.

1.3. Rules for Differentiation

1.3.3. The Chain Rule

Consider some normed spaces E, F and G; an open set $U \subseteq E$, a point u in U, a set $V = U - u$, an open set $Y \subseteq F$, a point y in Y, a set $Z = Y - y$, and functions $f: U \to F$, $g: Y \to G$.

We shall assume that $f(U) \subseteq Y$ and $f(u) = y$, and will consider a composite function $g \circ f: U \to G$ with value $(g \circ f)(u) = g(f(u)) = g(y)$ at u.

THEOREM. *If f is differentiable at u and g differentiable at $y = f(u)$, then the composite function $g \circ f$ is differentiable at u and*

$$(g \circ f)'(u) = g'(f(u)) \circ f'(u).$$

PROOF. It is an easy matter to verify that for any v in V

$$\Delta(g \circ f)u(v) = \Delta gy(\Delta fu(v)) = g'(y) \cdot \Delta fu(v) + rgy(\Delta fu(v))$$
$$= g'(y)(f'(u)v + rfu(v)) + rgy(\Delta fu(v)) = l(v) + r(v),$$

with

$$l(v) = g'(y) \cdot f'(u)v$$

and

$$r(v) = g'(y) \cdot rfu(v) + rgy(\Delta fu(v)).$$

The function $l = g'(y) \cdot f'(u)$ with values $l(v)$ is continuous and linear. The function r with values $r(v)$ is relatively small. Therefore, the composite function $g \circ f$ is differentiable at u and the equality to be proved holds. ■

REMARK. In the identity representing the chain rule, which is sometimes misleadingly called the "function of a function" rule, $f'(u)$ is a linear mapping $E \to F$, and $g'(y)$ is a linear mapping $F \to G$. The composite $g'(y) \circ f'(u)$ is a linear mapping $E \to G$. If we resort to the multiplicative manner of representation for a composite of linear operators and call this composite a product, as is commonly done, then the chain rule will be given by the equality

$$(g \circ f)'(u) = g'(f(u)) \cdot f'(u),$$

or in words: the derivative of a composite function is the product of the derivatives of the composants at the respective points.

Another representation of this efficient rule is

$$d(g \circ f)u = dg(f(u)) \cdot dfu.$$

Three illustrative examples follow.

EXAMPLE 1. Given $E = \mathbb{R}^2$, $F = \mathbb{R}^3$, $G = \mathbb{R}^2$, $U = E$, $Y = F$,

$$f(t) = [t_1 \cdot \sin t_2, \ t_2 \cdot \cos t_1, \ t_1 t_2] \quad (t = [t_1, t_2] \in \mathbb{R}^2),$$

$$g(x) = [x_1 x_2 x_3, \ x_3^2] \quad (x = [x_1, x_2, x_3] \in \mathbb{R}^3),$$

$$(g \circ f)(t) = [t_1^2 t_2^2 \cdot \cos t_1 \cdot \sin t_2, \ t_1^2, \ t_2^2],$$

$$f'(u) = \begin{bmatrix} \sin u_2 & u_1 \cdot \cos u_2 \\ -u_2 \cdot \sin u_1 & \cos u_1 \\ u_2 & u_1 \end{bmatrix} \quad (u = [u_1, u_2] \in \mathbb{R}^2),$$

and

$$g'(y) = \begin{bmatrix} y_2 y_3 & y_1 y_3 & y_1 y_2 \\ 0 & 0 & 2 y_3 \end{bmatrix} \quad (y = [y_1, y_2, y_3] \in \mathbb{R}^3),$$

then

$$(g \circ f)'(u) = g'(f(u)) \cdot f'(u)$$

$$= \begin{bmatrix} u_1 u_2^2 \cdot \cos u_1 & u_1^2 u_2 \cdot \sin u_2 & u_1 u_2 \cdot \cos u_1 \cdot \sin u_2 \\ 0 & 0 & 2 u_1 u_2 \end{bmatrix}$$

$$\times \begin{bmatrix} \sin u_2 & u_1 \cdot \cos u_2 \\ -u_2 \cdot \sin u_1 & \cos u_1 \\ u_2 & u_1 \end{bmatrix}.$$

EXAMPLE 2. Given $E = \mathbb{R}$, $F = \mathbb{R}^2$, $G = \mathbb{R}$, $U =]-2^{-1}, 2^{-1}[$, $Y = B(0, 1)$,

$$f(t) = [t, (2^{-1} - t^2)^{1/2}] \quad (t \in]-2^{-1}, 2^{-1}[),$$

$$g(x) = (1 - x_1^2 - x_2^2)^{1/2} \quad (x = [x_1, x_2] \in B(0,1)),$$

$$(g \circ f)(t) = 2^{-1/2},$$

$$f'(u) = \begin{bmatrix} 1 \\ -u(2^{-1} - u^2)^{-1/2} \end{bmatrix} \quad (u \in]-2^{-1}, 2^{-1}[),$$

and

$$g'(y) = -(1 - y_1^2 - y_2^2)^{-1/2} [y_1, y_2] \quad (y = [y_1, y_2] \in B(0, 1)),$$

1.3. Rules for Differentiation

then

$$(g \circ f)'(u) = g'(f(u)) \cdot f'(u)$$

$$= -2^{1/2}[u, (2^{-1} - u^2)^{1/2}] \begin{bmatrix} 1 \\ -u(2^{-1} - u^2)^{-1/2} \end{bmatrix} = -2^{1/2}(u - u) = 0.$$

EXAMPLE 3. Given $E = F = G = \mathbb{R}^2$, $U = E$, $Y = F$,

$$f(t) = [t_1 \cdot \cos t_2, \; t_1 \cdot \sin t_2] \quad (t = [t_1, t_2] \in \mathbb{R}^2),$$
$$g(x) = [x_1^2 + x_2^2, \; x_1 - x_2] \quad (x = [x_1, x_2] \in \mathbb{R}^2),$$
$$(g \circ f)(t) = [t_1^2, \; t_1 \cdot (\cos t_2 - \sin t_2)],$$

$$f'(u) = \begin{bmatrix} \cos u_2 & -u_1 \cdot \sin u_2 \\ \sin u_2 & u_1 \cdot \cos u_2 \end{bmatrix} \quad (u = [u_1, u_2] \in \mathbb{R}^2),$$

and

$$g'(y) = \begin{bmatrix} 2y_1 & 2y_2 \\ 1 & -1 \end{bmatrix} \quad (y = [y_1, y_2] \in \mathbb{R}^2),$$

then

$$(g \circ f)'(u) = \begin{bmatrix} 2u_1 \cdot \cos u_2 & 2u_1 \cdot \sin u_2 \\ 1 & -1 \end{bmatrix} \begin{bmatrix} \cos u_2 & -u_1 \cdot \sin u_2 \\ \sin u_2 & u_1 \cdot \cos u_2 \end{bmatrix}$$

$$= \begin{bmatrix} 2u_1 & 0 \\ \cos u_2 - \sin u_2 & -u_1(\sin u_2 + \cos u_2) \end{bmatrix}.$$

1.3.4. Lagrange's Theorem

This theorem may be deemed the principal theorem of differential calculus. It establishes a connection between the differential and variation of a function.

Let there be given an open interval $U \subseteq \mathbb{R}$, a normed space F, and differentiable functions $h: U \to F$ and $\varphi: U \to \mathbb{R}$.

THEOREM (Lagrange). *Let t and u be points in U and $t > u$, then from* $\|h'(u)\| < \varphi'(u)$ *follows*

$$\|h(t) - h(u)\| < \varphi(t) - \varphi(u).$$

PROOF. Assume that the theorem is not true and, hence, there exist points $t_0 > u_0$ in U for which

$$\|h(t_0) - h(u_0)\| \geq \varphi(t_0) - \varphi(u_0).$$

Let for a positive integer n and points $t_n > u_n$ in U

$$\|h(t_n) - h(u_n)\| \geq \varphi(t_n) - \varphi(u_n); \tag{1}$$

then for $c_n = 2^{-1}(u_n + t_n)$ there holds at least one of the inequalities

$$\|h(t_n) - h(c_n)\| \geq \varphi(t_n) - \varphi(c_n),$$
$$\|h(c_n) - h(u_n)\| \geq \varphi(c_n) - \varphi(u_n).$$

If the first is true we take $u_{n+1} = c_n$ and $t_{n+1} = t_n$, otherwise we take $u_{n+1} = u_n$ and $t_{n+1} = c_n$; in any case,

$$\|h(t_{n+1}) - h(u_{n+1})\| \geq \varphi(t_{n+1}) - \varphi(u_{n+1}).$$

By induction it follows that there exists a contracting sequence of intervals $[u_n, t_n]$ at the ends of which inequality (1) is true. This sequence contracts to a point c in U.

We will prove that there exists a c-converging sequence of points $s_n \in U$ for which

$$\|(s_n - c)^{-1}(h(s_n) - h(c))\| \geq (s_n - c)^{-1}(\varphi(s_n) - \varphi(c)). \tag{2}$$

From (1) and the continuity of functions h and φ it follows that for each n at least one of the inequalities

$$\|h(t_n) - h(c)\| \geq \varphi(t_n) - \varphi(c), \tag{3}$$
$$\|h(c) - h(u_n)\| \geq \varphi(c) - \varphi(u_n)$$

holds true with three possibilities

$$u_n < c < t_n, \quad u_n = c < t, \quad \text{and} \quad u_n < c = t_n.$$

If $u_n < c < t_n$ and the first of the inequalities in (3) is true, then we take $s_n = t_n$. If $u_n < c < t_n$ and the first of the inequalities in (3) does not hold, we take $s_n = u_n$. If $u_n = c < t_n$, then from (1) follows the first inequality in (3) and we take $s_n = t_n$. Finally, if $u_n < c = t_n$, then the second inequality in (3) is true and we may take $s_n = u_n$. Inequality (2) holds under all circumstances. Since $u_n \to c$ and $t_n \to c$, then $s_n \to c$.

1.3. Rules for Differentiation

From (2) it follows that

$$\|h'(c)\| \geq \varphi'(c),$$

which is contrary to the condition of the theorem. ∎

Consider a few corollaries of this theorem.

COROLLARY 1. *If* $\|h'(u)\| \leq \varphi'(u)$ $(u \in U)$, *then*

$$\|h(t) - h(u)\| \leq \varphi(t) - \varphi(u) \quad (t \geq u;\ tu \in U).$$

PROOF. If $u = t$, then the inequality to be proved is equivalent to $0 = 0$.

For every number $n > 0$ we take a function

$$\varphi_n(t) = \varphi(t) + n^{-1}t : U \to R \quad (t \in U)$$

which is differentiable and

$$\varphi'_n(u) = \varphi'(u) + n^{-1} > \varphi'(u) \geq |h'(u)| \quad (u \in U).$$

By Lagrange's theorem it follows that for $t > u$,

$$|h(t) - h(u)| < \varphi_n(t) - \varphi_n(u) = \varphi(t) - \varphi(u) + n^{-1}(t - u).$$

and the stated inequality is proved. ∎

COROLLARY 2. *Given* $c > 0$, u *and* t *in* U *and* $t \geq u$, *if* $\|h'(u)\| \leq c$, *then* $\|h(t) - h(u)\| \leq c(t - u)$.

PROOF. Consider a function $\varphi: U \to \mathbb{R}$ with values $\varphi(t) = ct$ $(t \in U)$. It is differentiable and $\varphi'(u) = c$ for u in U. The required inequality follows from Corollary 1. ∎

These propositions and rules for differentiation lead to other ones.

Consider normed spaces F and G; an open set $Y \subseteq F$ and a function $g : Y \to G$.

Let $a, b \in F$. The set

$$[a, b] = \{x = a + t(b-a) : t \in [0, 1]\}$$

will be called a *closed interval* with end points a and b. The set

$$]a, b[= \{x = a + t(b-a) : t \in]0, 1[\}$$

will be called an *open interval* with end points a and b.

We take $[a, b] \subseteq Y$ $(a \neq b)$, $c \geq 0$ and assume that g is continuous at each point of the interval $[a, b]$ and differentiable at each point of the open interval $]a, b[$.

COROLLARY 3. *If* $\|g(y)\| \leq c$ $(y \in \,]a, b\,[\,)$, *then*

$$\|g(b) - g(a)\| \leq c\|b - a\|.$$

For each $y \in \,]a, b\,[$ we choose a number $c(y) \geq 0$. As before we shall assume that g is continuous on $[a, b]$ and differentiable on $]a, b\,[$.

COROLLARY 4. *If*

$$\|g'(x) - g'(y)\| \leq c(y) \qquad (x, y \in \,]a, b[\,),$$

then

$$\|g(b) - g(a) - g'(y) \cdot (b - a)\| \leq c(y) \cdot \|b - a\| \qquad (y \in \,]a, b[\,).$$

An open set Y is said to be *connected* if it cannot be partitioned into two nonempty open subsets.

A set $X \subseteq F$ containing along with any two points a and b the closed interval $[a, b]$ is called a *convex* set. Every ball in F is a convex set.

Now we assume in addition that the open set $Y \subseteq F$ is convex and a function $g : Y \to G$ is differentiable. The Corollary 3 leads us immediately to

COROLLARY 5. *If*

$$\|g'(y)\| \leq c \qquad (y \in Y),$$

then

$$\|g(x) - g(y)\| \leq c \cdot \|x - y\| \qquad (x, y \in Y).$$

Consider the open set $Y \subseteq F$ connected and the function $g : Y \to G$ differentiable. Then we have

COROLLARY 6. *A function g is equal to a constant if and only if* $g'(y) = 0$ $(y \in Y)$.

Lagrange's theorem may be used also to prove a theorem on termwise differentiation of a sequence of functions.

Let F and G denote Banach spaces, $Y \subseteq F$ denote an open connected set with a point $a \in Y$, and $g_n : Y \to G$ denote a sequence of differentiable functions. We will say that the sequence g_n is *differentiable termwise* if it converges at every point to some differentiable function $g : Y \to G$ and the sequence of derivatives g'_n converges at each point to the derivative g'.

1.3. Rules for Differentiation

THEOREM. *If a sequence of derivatives g' converges uniformly and the sequence of points $g_n(a)$ is converging, then the sequence of functions g_n is termwise differentiable and converges uniformly on each bounded subset of Y.*

1.3.5. Total Differentials

Let us consider the products $F = \Pi F_i$ and $E = \Pi E_j$ of normed spaces F_i and E_j ($i = 1, ..., m; j = 1, ..., n$), a rectangle $U = \Pi U_j$ with open sides $U_j \subseteq E_j$, a mapping $f: U \to F$, a point $u = (u_j) \in U$, a set $V = U - u$, and sets $V_j = U_j - u_j$ ($j = 1, ..., n$).

In Section 1.2.4 we defined the partial differentials

$$d_{ij} f u = s_i \cdot d f_{ij} u \cdot p_j = s_i q_i \cdot d f u \cdot r_j p_j$$

assuming the existence of the differentials $df_{ij} u$ of partial functions $f_{ij} u$ and the differential dfu of function f. It is natural to expect that under certain conditions the fact that the partial differentials exist means that the total differential also exists, and vice versa.

If the total differential dfu exists, the partial differentials $df_{ij} u$ exist too and by the theorem of differentiation of a composite function we have

$$d f_{ij} u = d(q_i \cdot f \cdot Tu \cdot r_j) o_j = q_i \cdot d(f \cdot Tu \cdot r_j) o_j$$
$$= q_i \cdot d f u \cdot r_j.$$

(From the definition of a differential it follows that the differential of a bounded linear operator at a point is equal to this operator itself.)

There is evidence that the converse is not true — the partial differentials can exist while the total differential will not. Additional propositions are needed.

Now we come to the functional terminology and notation and will, conventionally, call the differentials derivatives and denote them by primes. If a function f is differentiable at a point u, then by definition the derivative $f'(u)$ is a bounded linear operator, $f'(u) \in \mathcal{B} = \mathcal{B}(E, F)$. If a function f is differentiable at every point u of the domain of definition $U \subseteq E$ the derivative f' with values $f'(u)$ maps U into $\mathcal{B}$. Since both E and $\mathcal{B}$ are normed spaces we may say of the continuity of the derivative $f': U \to \mathcal{B}$. By definition, it is continuous at a point u if for any $\varepsilon > 0$ there exists $\delta > 0$ such that

$$\|f'(u + v) - f'(u)\| \leq \varepsilon \quad (\|v\| \leq \delta, \ v \in V).$$

We shall also refer to the partial differentials $d_{ij} f u$ as partial derivatives and denote them by $f'_{ij}(u)$. Then

$$f'_{ij}(u) = s_i \cdot df_{ij} u \cdot p_j \in \mathcal{B},$$

and, if these partial functions are differentiable, the partial derivatives f'_{ij} also map U on $\mathcal{B}$, like the full derivative f', and can be continuous as well.

Functions having continuous derivatives will be said to be continuously differentiable, or smooth.

THEOREM. *The total derivative f' exists and is continuous if all the partial derivatives f'_{ij} exist and are continuous.*

PROOF. (i) If f' exists and is continuous, then, as has been demonstrated, the partial functions are differentiable and

$$df_{ij} u = q_i \cdot f'(u) \cdot r_j.$$

Consequently, all the partial derivatives

$$f'_{ij}(u) = s_i \cdot df_{ij} u \cdot p_j = s_i q_i \cdot f'(u) \cdot r_j p_j$$

exist.

One can easily verify that $\|s_i q_i\| \leq 1$ and $\|r_j p_j\| \leq 1$. Now using the Cauchy inequality we obtain

$$\|f'_{ij}(u+v) - f'_{ij}(u)\| = \|s_i g_i \cdot (f'(u+v) - f'(u)) \cdot r_j p_j\|$$

$$\leq \|s_i q_i\| \cdot \|f'(u+v) - f'(u)\| \cdot \|r_j p_j\| \leq \|f'(u+v) - f'(u)\|.$$

Therefore, the continuity of all f'_{ij} follows from the continuity of f'.

(ii) The idea underlying the proof of the inverse statement is rather simple. If f' exists, then

$$f'(u) = \sum f'_{ij}(u).$$

Therefore, to prove that f' exists we need to show that the difference $r: V \to F$ defined by

$$r(v) = \Delta f u(v) - \sum f'_{ij}(u) \cdot v$$

is comparatively small when f'_{ij} are continuous. The proof is established with the aid of Lagrange's increment theorem.

1.3. Rules for Differentiation

The continuity of f' follows from $f' = \Sigma f'_{ij}$ and the continuity of f'_{ij}. ∎

REMARK. One cannot replace the partial derivatives of f with the derivatives of the partial functions. They may exist and be continuous even when f is not differentiable.

If $F_i = \mathbb{F}^{m(i)}$ and $E_j = \mathbb{F}^{n(j)}$, then $f'(u)$ may be written as a partitioned matrix whose entries will be $m(i) \times n(j)$ matrices. In particular for $m(i) = n(j) = 1$, we have an ordinary constant matrix.

Let us consider a few examples devoted to evaluation of derivatives.

EXAMPLE 1. Let $E_1 = E_2 = U_1 = U_2 = F = \mathbb{R}$ and

$$f_1(t) = \begin{cases} t_1 t_2 (t_1^2 + t_2^2)^{-1}, & \text{if } t = (t_1, t_2) \neq (0, 0) \\ 0, & \text{if } t = (t_1, t_2) = (0, 0). \end{cases}$$

The function f_1 is discontinuous and therefore not differentiable at $u = (0, 0)$ ($f_1(t) = 2^{-1}$ at $t_1 = t_2 \neq 0$). At the same time the derivatives of the partial functions f_{1j} do exist at $u_j = 0$ and

$$f'_{11}(0) = 0, \qquad f'_{12}(0) = 0,$$

with

$$f_{11}(t_1) = t_1 0(t_1^2 + 0)^{-1} = 0 \quad (t_1 \neq 0)$$
$$f_{12}(t_2) = 0 t_2(0 + t_2^2)^{-1} = 0 \quad (t_2 \neq 0).$$

If $u_2 \neq 0$, then the partial function f_{11} is the ratio of the differentiable functions g_1 and h_1 with values, for $t_1 \in \mathbb{R}$,

$$g_1(t_1) = t_1 u_2,$$
$$h_1(t_1) = t_1^2 + u_2^2 \neq 0.$$

Hence, f_{11} is differentiable and

$$df_{11} u(v_1) = -u_2(u_1^2 - u_2^2)(u_1^2 + u_2^2)^{-2} v_1 \quad (v_1 \in \mathbb{R}).$$

Similarly, if $u_1 \neq 0$, then the partial function f_{12} is the ratio of the differentiable functions g_2 and h_2 with values

$$g_2(t_2) = u_1 t_2,$$
$$h_2(t_2) = u_1^2 + t_2^2 \neq 0, \quad (t_2 \in \mathbb{R}).$$

Consequently, f_{12} is differentiable and

$$df_{12}u(v_2) = u_1(u_1^2 - u_2^2)(u_1^2 + u_2^2)^{-2} \cdot v_2 \quad (v_2 \in \mathbb{R}).$$

At each point $u \in \mathbb{R}^2$ there exist partial derivatives $f'_{11})(u)$ and $f'_{12})(u)$, i.e., f'_{11} and f'_{12}). Since for $u = (\alpha, 2\alpha)$ and $\alpha \neq 0$

$$f'_{11}(u) = (6 \cdot 5^{-2} \cdot \alpha^{-1}, 0) \quad \text{and} \quad f'_{12}(u) = (0, 6 \cdot 5^{-2} \cdot \alpha^{-1}),$$

the partial derivatives f'_{11} and f'_{12} have a discontinuity at $(0, 0)$. At the same time, for $u = (u_1, u_2) \neq (0, 0)$, the derivatives of the partial functions $f_{11} = f_{11}u$ and $f_{12} = f_{12}u$ with values

$$(g_1/h_1)'(u_1) = -u_2(u_1^2 - u_2^2)(u_1^2 + u_2^2)^{-1},$$

$$(g_2/h_2)'(u_2) = u_1(u_1^2 - u_2^2)(u_1^2 + u_2^2)^{-1}$$

are continuous with respect to their variables, namely, $f_{11} = g_1/h_1$ with respect to u_1, and $f_{12} = g_2/h_2$ with respect to u_2.

EXAMPLE 2. Let $Y = \,]0, \infty[\, \times \,]0, \pi/2[\, \times \,]0, \pi/2[\,$ and $X = \,]0, \infty[\, \times \,]0, \infty[\, \times \,]0, \infty[\, \subseteq \mathbb{R}^3$ and let functions $g: Y \to \mathbb{R}^3$ and $f: X \to \mathbb{R}^3$ be defined by $x = g(y)$ and $y = f(x)$ with

$$\begin{pmatrix} x_1 \\ x_2 \\ x_3 \end{pmatrix} = \begin{pmatrix} y_1 \cdot \cos y_2 \cdot \sin y_3 \\ y_1 \cdot \sin y_2 \cdot \sin y_3 \\ y_1 \cdot \cos y_3 \end{pmatrix}, \quad \begin{pmatrix} y_1 \\ y_2 \\ y_3 \end{pmatrix} = \begin{pmatrix} (x_1^2 + x_2^2 + x_3^2)^{1/2} \\ \arctan(x_1^{-1} x_2) \\ \arccos(x_3(x_1^2 + x_2^2 + x_3^2)^{-1/2}) \end{pmatrix}.$$

These functions express Cartesian coordinates about spherical and spherical coordinates about Cartesian; they are mutually inverse.

Using appropriate rules of differentiation we obtain

$$g'(y) = \begin{pmatrix} \cos y_2 \cdot \sin y_3 & -y_1 \cdot \sin y_2 \cdot \sin y_3 & y_1 \cdot \cos y_2 \cdot \cos y_3 \\ \sin y_2 \cdot \sin y_3 & y_1 \cdot \cos y_3 \cdot \sin y_3 & y_1 \cdot \sin y_2 \cdot \cos y_3 \\ \cos y_3 & 0 & -y_1 \sin y_3 \end{pmatrix}$$

$$f'(x) = \begin{pmatrix} x_1 r^{-1} & x_2 r^{-1} & x_3 r^{-1} \\ -x_2(x_1^2 + x_2^2)^{-1} & x_1(x_1^2 + x_2^2)^{-1} & 0 \\ x_1 x_3(x_1^2 + x_2^2)^{-1/2} r^{-2} & x_2 x_3(x_1^2 + x_2^2)^{-1/2} r^{-2} & -(x_1^2 + x_2^2)^{1/2} r^{-2} \end{pmatrix},$$

where $r = (x_1^2 + x_2^2 + x_3^2)^{-1/2}$.

1.3. Rules for Differentiation

Note that since $y = f(x)$, then

$$f'(x) = \begin{pmatrix} \cos y_2 \cdot \sin y_3 & \sin y_2 \cdot \sin y_3 & \cos y_3 \\ -y_1^{-1} \cdot \sin y_2 \cdot \sin^{-1} y_3 & y_1^{-1} \cdot \cos y_2 \cdot \sin^{-1} y_3 & 0 \\ y_1^{-1} \cdot \cos y_2 \cdot \cos y_3 & y_1^{-1} \cdot \sin y_2 \cdot \cos y_3 & -y_1^{-1} \cdot \sin y_3 \end{pmatrix},$$

$$g'(y) \cdot f'(x) = \begin{pmatrix} 1 & 0 & 0 \\ 0 & 1 & 0 \\ 0 & 0 & 1 \end{pmatrix}.$$

Like g and f the derivatives g' and f' are inverse to one another at the respective points.

EXAMPLE 3. Consider a differentiable function $f: \mathbb{R}^n \to \mathbb{R}^n$ and its derivative $f'(x) = (f'_{ij}(x))$ at a point $x \in \mathbb{R}^n$. The trace

$$\operatorname{tr} f'(x) = f'_{11}(x) + \ldots + f'_{nn}(x)$$

of the matrix $A = f'(x)$ is called the *divergence* of f at x and denoted by $\operatorname{div} f(x)$.

We wish to prove that for any invertible linear transformation $L: \mathbb{R}^n \to \mathbb{R}^n$, say $x = L(t)$,

$$\operatorname{div}(L^{-1} f L)(t) = \operatorname{div} f(x).$$

This identity is equivalent to

$$\operatorname{tr}(L^{-1} f L)'(t) = \operatorname{tr} f'(x).$$

Applying the chain rule and observing the linearity of L yields

$$B = (L^{-1} f L)'(t) = (L^{-1})'(f(L(t))) \cdot f'(L(t)) \cdot L'(t) = C^{-1} A C,$$

where C is the matrix of L in standard bases.

Let $\lambda \in \mathbb{C}$ and $\mathbf{I}$ denote an $n \times n$ unit matrix. Then

$$\det(\lambda \mathbf{I} - B) = \det(\lambda \mathbf{I} - C^{-1} A C) = \det(C^{-1}(\lambda \mathbf{I} - A) C)$$

$$= \det(\lambda \mathbf{I} - A),$$

$$\det(\lambda \mathbf{I} - A) = \lambda^n - \operatorname{tr} A \lambda^{n-1} + \ldots + (-1)^n \det A$$

$$= (\lambda - \lambda_1) \ldots (\lambda - \lambda_n)$$

$$= \lambda^n - (\lambda_1 + \ldots + \lambda_n) \lambda^{n-1} + \ldots + (-1)^n \lambda_1 \ldots \lambda_n,$$

$$\operatorname{tr} A = \lambda_1 + \ldots + \lambda_n,$$

$$\det A = \lambda_1 \ldots \lambda_n,$$

where $\lambda_1, ..., \lambda_n$ are the eigenvalues of A. Consequently, $\operatorname{tr} B = \operatorname{tr} A$.

Thus, the divergence does not change under invertible linear transformations.

1.4. Solution of Functional Equations

Equations with smooth functions can be differentiated to reduce them to linear equations and study the local properties of the solutions.

1.4.1. Illustrative Examples and Definitions

In analysis of the solutions we shall use a geometrical language whereby functions will be deemed sets of pairs and identified with graphs.

EXAMPLE 1. Given some numbers P, Q, c and a function $g : \mathbb{R} \times \mathbb{R} \to \mathbb{R}$ defined by

$$g(x, y) = Px + Qy + c \quad (x \in \mathbb{R}, \ y \in \mathbb{R}).$$

Consider the set

$$g^{-1}(0) = \{(x, y) : g(x, y) = 0\}$$

of all points $(x, y) \in \mathbb{R} \times \mathbb{R}$ for which $g(x, y) = 0$.

Let $(a, b) \in g^{-1}(0)$ that is $g(a, b) = 0$. Then given Q^{-1} exists, $Q \neq 0$, the set $g^{-1}(0)$ is the graph of function f with values

$$y = f(x) = b + Q^{-1}P(x - a) \quad (x \in \mathbb{R}).$$

We say that f is a solution of the equation $g(x, y) = 0$:

$$g(x, f(x)) = 0 \quad (x \in \mathbb{R}).$$

Clearly, this solution is unique and

$$f'(x) = Q^{-1}P.$$

If Q^{-1} does not exist ($Q = 0$), $P = 0$ and $c \neq 0$, then $g^{-1}(0) = O$ which implies that there are no points $(x, y) \in \mathbb{R} \times \mathbb{R}$ for which $g(x, y) = 0$. Hence, the equation at hand has no solution.

EXAMPLE 2. Let P, Q and c are matrices of order $m \times n, m \times m$, and $m \times 1$, respectively, and $g : \mathbb{R}^n \times \mathbb{R}^m \to \mathbb{R}^m$ be a function defined as

$$g(x, y) = Px + Qy + c \quad (x \in \mathbb{R}^n, \ y \in \mathbb{R}^m).$$

1.4. Solution of Functional Equations

Consider the set

$$g^{-1}(0) = \{(x, y): g(x, y) = 0\}$$

of all points $(x, y) \in \mathbb{R}^n \times \mathbb{R}^m$ for which $g(x, y) = 0$.
Let $(a, b) \in g^{-1}(0)$ implying $g(a, b) = 0$. Then

$$g(x, y) = g(x, y) - g(a, b) = P(x - a) + Q(y - b).$$

If Q^{-1} exists ($\det Q \neq 0$), the set $g^{-1}(0)$ is the graph of the map $f: \mathbb{R}^n \to \mathbb{R}^m$ for which

$$y = f(x) = b + Q^{-1}P(x - a) \qquad (x \in \mathbb{R}^n).$$

We say that f is a solution of the equation $g(x, y) = 0$:

$$g(x, f(x)) = 0 \qquad (x \in \mathbb{R}^n).$$

It is obvious that this solution is unique and

$$f'(x) = Q^{-1}P \qquad (x \in \mathbb{R}^n).$$

If Q^{-1} does not exist ($\det Q = 0$), then the considered system of linear equations can have no solution or many solutions.

EXAMPLE 3. Let $g: \mathbb{R} \times \mathbb{R} \to \mathbb{R}$ be a function given by

$$g(x, y) = x^2 + y^2 - 1 \qquad (x \in \mathbb{R}, y \in \mathbb{R}).$$

Consider the set

$$g^{-1}(0) = \{(x, y): g(x, y) = 0\}$$

of all points $(x, y) \in \mathbb{R} \times \mathbb{R}$ for which $g(x, y) = 0$.
Let

$$(a, b) = (3/5, 4/5) \in g^{-1}(0): g(a, b) = 0.$$

Notice that $g^{-1}(0)$ is not a graph of the function since $(0, -1) \in g^{-1}(0)$ and $(0, 1) \in g^{-1}(0)$. However, there exists a rectangle $A \times B$ with open sides $A = {]}0, 1{[}$ and $B = {]}0, 1{[}$ such that $(a, b) \in A \times B$ and

$$f = g^{-1}(0) \cap A \times B = \{(x, y): g(x, y) = 0, x \in A, y \in B\}$$

is the graph of the function on A:

$$y = f(x) = (1 - x^2)^{1/2} \qquad (0 < x < 1).$$

The function f is said to be the local solution of the equation $g(x, y) = 0$ in the neighborhood of point (a, b). In the plane $\mathbb{R} \times \mathbb{R}$, f is represented by a circular arc over the interval $A = \,]\,0, 1\,[$, which is confined in the square $A \times B = \,]\,0, 1\,[\, \times \,]\,0, 1\,[$ and passes through the point $(a, b) = (3/5, 4/5)$.

Consider Banach spaces E, F, G, open sets $X \subseteq E$, $Y \subseteq F$, a function $g: X \times Y \to G$, points $a \in X$, $b \in Y$ and their open neighborhoods $A \subseteq X$, $B \subseteq Y$, a point $(a, b) \subseteq X \times Y$ and its open neighborhood $A \times B$.

DEFINITION 1. *Every function* $f: A \to F$ *with image* $f(A) \subseteq B$ *for which* $g(x, f(x)) = 0$ $(x \in A)$ *is called a solution of the equation*

$$g(x, y) = 0, \qquad (1)$$

defined on A and contained in $A \times B$.

Among these local solutions of equation (1) we distinguish those satisfying the condition

$$f(a) = b. \qquad (2)$$

We say that they pass through the point (a, b).

For such a solution to exist it is necessary that equation (1) hold true at (a, b), that is,

$$g(a, b) = 0. \qquad (3)$$

A solution of Eq. (1) defined on A and contained in $A \times B$ is unique if and only if the set $g^{-1}(0) \cap A \times B \subseteq X \times Y$ is the graph of $f: A \to B$. It is often convenient to identify this function with its graph and write

$$f = g^{-1}(0) \cap A \times B.$$

When a solution f of Eq. (1) defined on A and contained in $A \times B$ is unique, equations (2) and (3) are equivalent.

We will say that *the equation $g(x, y) = 0$ has a unique solution f in the neighborhood of a point* (a, b) if there exists an open neighborhood $A \times B$ about (a, b), and the intersection of this neighborhood with $g^{-1}(0)$ is the graph of f, that is, provided that $f = g^{-1}(0) \cap A \times B$ is a function.

Suppose now that g is a smooth (continuously differentiable) function and for each point (x, y) in $X \times Y$ consider the partial functions $g(\cdot, y): X \to G$, $g(x, \cdot): Y \to G$, their derivatives at points (x, y)

$$P(x, y) = dg(\cdot, y)x \in \mathcal{B}(E, G),$$
$$Q(x, y) = dg(x, \cdot)y \in \mathcal{B}(F, G),$$

1.4. Solution of Functional Equations

and the functions

$$P: X \times Y \to \mathcal{B}(E, G) \quad \text{and} \quad Q: X \times Y \to \mathcal{B}(F, G)$$

with values $P(x, y)$ and $Q(x, y)$. By analogy with partial differentials it is straightforward to prove that P and Q are continuous.

The derivative of g is related to P and Q by the equality

$$g'(x, y) \cdot (u, v) = P(x, y) \cdot u + Q(x, y) \cdot v \quad (u \in E, v \in F)$$

that follows from the partial derivative theorem (here again we have used the multiplicative representation of linear operators). Therefore it is natural to associate with $g(x, y) = 0$ the linear equation

$$P(x, y) \cdot u + Q(x, y) \cdot v = 0. \tag{4}$$

If for the linear operator $Q(x, y) \in \mathcal{B}(F, Q)$ there exists the inverse operator $Q^{-1}(x, y) \in \mathcal{L}(G, F)$, then equation (4) has the unique solution

$$v = -Q^{-1}(x, y) \cdot P(x, y) \cdot u. \tag{5}$$

In this case equation (4) is said to be nonsingular with respect to v.

If $Q^{-1}(x, y)$ is bounded, then it is continuous and the solution v is a continuous function of u. We say that equation (4) is correctly resolvable for v, or well posed.

Since a smooth function can be locally approximated by its derivative at the given point, it is natural to expect that the correctness of equation (4) for the derivatives at $(x, y) = (a, b)$ ensures the existence and uniqueness of the smooth solution to Eq. (1) for the considered function in the neighborhood of (a, b). The solution (5) of equation (4) yields the derivative of this solution.

If $Q(a, b)$ maps F injectively on G and the inverse operator of $Q(a, b)$ is continuous, then $Q(a, b)$ is a homeomorphism of F on G. For $F = G = \mathbb{R}^m$, the operator $Q(a, b)$ is a homeomorphism if $\det Q(a, b) \neq 0$. The condition of $Q(a, b)$ being a homeomorphism secures the correctness of equation (4) at $(x, y) = (a, b)$.

DEFINITION 2. *We say that the equation $g(x, y) = 0$ with smooth function g such that the derivative $Q(a, b)$ of the partial function $g(a, \cdot)$ at b is a homeomorphism of F on G is nonsingular with respect to y at point (a, b).*

Note in retrospect that Examples 1 through 3 dealt with nonsingular and singular equations.

1.4.2. The Implicit Function Theorem

This theorem claims that a nonsingular functional equation has a unique solution. We formulate it with the aid of the notation and definitions introduced earlier.

THEOREM. *Let $g(x, y) = 0$ be an equation with smooth function g, which is true and nonsingular with respect to y at a point (a, b). It has a unique solution f in the neighborhood of (a, b). This solution f passes through this point, is smooth and*

$$f'(x) = -Q^{-1}(x, f(x)) \cdot P(x, f(x))$$

at every point x from the domain of definition of f.

The solution f is called an implicit function defined by the equation $g(x, y) = 0$.

PROOF. (i) The existence and uniqueness of f is proved by the fixed point theorem which will be proved in Section 1.4.4. The proof of existence of the solution bears on the equivalence of the equations $g(x, y) = 0$ and $h(x, y) = y$ for a function $h : X \times Y \to F$, defined by

$$h(x, y) = y - Q^{-1}(a, b) \cdot g(x, y) \quad (x \in X, \ y \in Y).$$

For every x the solution $y = f(x)$ of $h(x, y) = y$ is a fixed point of the function $h(x, \cdot) : Y \to F$, given by

$$h(x, \cdot)(y) = h(x, y) = y$$

for all y in Y. To prove that $h(x, \cdot)$ satisfies the theorem of fixed point we invoke the Lagrange theorem of increments.

Another theorem involved in the proof is the one of operator progression (Neumann series) which will be proved in Section 1.4.5.

(ii) To prove that f is smooth and obtain the expression for f' one should evaluate the relevant increments. This argument is rather long and therefore omitted. ∎

The implicit function theorem allows us to say that the equation $g(x, y) = 0$ with the stated properties is correctly resolvable for y at point (a, b) since in some neighborhood of this point there exists a unique smooth solution $y = f(x)$ of this equation. It should be emphasized that the theorem does not delineate an exact domain of definition of the implicit function f.

REMARK. One may change the roles of x and y in $g(x, y) = 0$ and solve this equation for x. Then P will play the role of Q. If $g(a, b) = 0$ and $P(a, b)$ is a homeomorphism from E onto G, then we say that $g(x, y) = 0$ is correctly resolvable for x at (a, b): in some neighborhood

1.4. Solution of Functional Equations

$A \times B$ there exists a unique smooth solution $x = h(y)$ of this equation, and for the derivative of $h : B \to A$ we have

$$h'(y) = -P^{-1}(h(y), y) \cdot Q(h(y)) \qquad (y \in B).$$

Formally, the procedure reduces to a permutation $x \leftrightarrow y$; no other argument is required.

For brevity we shall call an equation $g(x, y) = 0$ correctly resolvable for x or for y at (a, b) or well-posed at (a, b), implying that g is smooth once its correctness, or well-posedness, is defined.

1.4.3. Theorem of Inverse Function

This theorem states that a smooth function is invertible in the neighborhood of each point where the derivative has an inverse. Such an inverse function is also smooth and its derivative is the inverse of the derivative of the given function at the respective points. Since smooth functions may be locally approximated by their derivatives, such a result is intuitively obvious. On formal grounds the inverse function theorem is equivalent to the implicit function theorem – they are derivable from one another.

In the following consideration we assume given Banach spaces E and F, open sets $X \subseteq E$ and $Y \subseteq F$, points $a \in X$ and $b \in Y$ with open neighborhoods $A \subseteq X$ and $B \subseteq Y$, and functions $h : X \to F$, $f : A \to F$ with images $h(X) \subseteq Y$, $f(A) = B$.

If f is one-one, then it has the inverse $f^{-1} : B \to A$ with image $f^{-1}(B) = A$.

A function f is a diffeomorphism of A on B if (1) f is a bijection of A on B, and (2) both f and f^{-1} are smooth. Accordingly, a diffeomorphism may be treated as a one-one mutually smooth function.

THEOREM. *If h is a smooth function and its derivative $h'(a)$ is a homeomorphism of E on F, then there exists an open neighborhood A of point a and an open neighborhood B of point $b = h(a)$ such that the restriciton f of h on the set A is a diffeomorphism of A on B, and*

$$df^{-1}(f(x)) = (dfx)^{-1}$$

for all points x in A.

PROOF. To use the implicit function theorem we consider $g : X \times Y \to F$, defined by

$$g(x, y) = y - h(x) \qquad (x \in X, y \in Y).$$

The equality $y = h(x)$ is equivalent to $g(x, y) = 0$. To prove the inverse function theorem it suffices to demonstrate that g satisfies the conditions of the implicit function theorem and verify that there exists a solution f of $g(x, y) = 0$ possessing the desired properties. ∎

It should be emphasized that the theorem does not specify an exact domain of definition for the inverse function f^{-1} and merely denotes it by B.

Two following examples illustrate the concept of diffeomorphism.

EXAMPLE 1. Let $E = F = \mathbb{R}$, $X = Y = \mathbb{R}$, and $h(x) = x^2$ with x in $\mathbb{R}$. Since $h'(a) = 2a$, the linear function dha with coefficient $h'(a)$ is a homeomorphism of $\mathbb{R}$ on $\mathbb{R}$ at $a \neq 0$.

If $a > 0$, the restriction f_+ of h to the open neighborhood $A_+ = {]}0, \infty{[}$ of point a is the diffeomorphism of A_+ on the open neighborhood $B = {]}0, \infty{[}$ of the point $b = h(a) = a^2$. The inverse function $\varphi_+ = f_+^{-1} : {]}0, \infty{[} \to {]}0, \infty{[}$ has the values $\varphi_+(y) = y^{1/2}$ for $0 < y < \infty$, and for its derivatives we have

$$\varphi'_+(y) = 2^{-1} y^{-1/2} = (2x)^{-1} = (f'_+(x))^{-1}$$

for $y = x^2$ $(0 < y < \infty)$.

If $a < 0$, the restriction f_- of h to the open neighborhood $A_- = {]}-\infty, 0{[}$ of point a is a diffeomorphism of A_- in the open neighborhood $B = {]}0, \infty{[}$ of the point $b = h(a) = a^2$. The inverse function $\varphi_- = f_-^{-1}$ is defined by $\varphi_-(y) = -y^{1/2}$ $(0 < y < \infty)$. For its derivative we have

$$\varphi'_-(y) = 2^{-1} y^{-1/2} = (2x)^{-1} = (f'_-(x))^{-1}$$

for $y = x^2$ $(-\infty < x < 0)$ and the arithmetic value of the root $y^{1/2}$.

EXAMPLE 2. Let $Y = {]}0, \infty{[} \times {]}0, \pi/2{[}$ and $X = {]}0, \infty{[} \times {]}0, \infty{[} \subseteq \mathbb{R}^2$, and functions $g : Y \to \mathbb{R}^2$ and $f : X \to \mathbb{R}^2$ given by $x = g(y)$ and $y = f(x)$ for

$$\begin{pmatrix} x_1 \\ x_2 \end{pmatrix} = \begin{pmatrix} y_1 \cdot \cos y_2 \\ y_1 \cdot \sin y_2 \end{pmatrix} \quad \begin{pmatrix} y_1 \\ y_2 \end{pmatrix} = \begin{pmatrix} (x_1^2 + x_2^2)^{1/2} \\ \arctan(x_1^{-1} x_2) \end{pmatrix}.$$

These functions express Cartesian coordinates in terms of polar coordinates and conversely. The functions g and f are inverse of one another.

Using appropriate differentiation rules we have

$$g'(y) = \begin{pmatrix} \cos y_2 & -y_1 \cdot \sin y_2 \\ \sin y_2 & y_1 \cdot \cos y_2 \end{pmatrix}$$

$$f'(x) = \begin{pmatrix} x_1 (x_1^2 + x_2^2)^{-1/2} & x_2 (x_1^2 + x_2^2)^{-1/2} \\ -x_2 (x_1^2 + x_2^2)^{-1} & x_1 (x_1^2 + x_2^2)^{-1} \end{pmatrix}$$

1.4. Solution of Functional Equations

Since $y = f(x)$, then

$$g'(y) f'(x) = \begin{pmatrix} \cos y_2 & -y_1 \cdot \sin y_2 \\ \sin y_2 & y_1 \cdot \cos y_2 \end{pmatrix} \begin{pmatrix} \cos y_2 & \sin y_2 \\ -y_1^{-1} \cdot \sin y_2 & y_1^{-1} \cdot \cos y_2 \end{pmatrix}$$

$$= \begin{pmatrix} 1 & 0 \\ 0 & 1 \end{pmatrix}.$$

Like the functions g and f, the derivatives g' and f' are the inverses of each other at the respective points.

The function f is a diffeomorphism of X on Y and the function g is a diffeomorphism of Y on X.

1.4.4. The Fixed Point Theorem

We present here the simplest of the many theorems devoted to fixed points. It is often applied in solving equations.

Let there be given a Banach space E, and a nonempty closed set $C \subseteq E$ which includes together with each convergent sequence x_n its limit x.

A mapping $h : C \to C$ for which there exists a number $\alpha \in\ ]0, 1[$ such that

$$\|h(x) - h(y)\| \leq \alpha \|x - y\| \quad (x, y \in C)$$

is called *a compressing transformation* of the set C, or *a compression* for short. (As usual E is assumed to be separable.)

Clearly compressions are continuous or even uniformly continuous.

A point $c \in C$ for which $h(c) = c$ is a fixed point. It is a solution of the equation $h(x) = x$.

The fixed point theorem claims that when h is a compression of a closed set in a Banach space, this equation has a unique solution $x = c$.

THEOREM. *For every compression h of a closed set C of a Banach space, there exists a unique fixed point.*

PROOF. **Existence.** We take $x_0 \in C$ and consider a sequence of points x_n in C defined by

$$x_{n+1} = h(x_n) \quad (n \geq 0).$$

Since h is a compression, this sequence converges. Indeed,

$$\|x_2 - x_1\| = \|h(x_1) - h(x_0)\| \leq \alpha \|x_1 - x_0\|,$$
$$\|x_{n+1} - x_n\| = \|h(x_n) - h(x_{n-1})\| \leq \alpha \|x_n - x_{n-1}\|, \quad (n \geq 1).$$

Consequently for each index m and $n \geq 0$,

$$\|x_{n+1} - x_n\| \leq \alpha^n \|x_1 - x_0\|,$$

$$\|x_{n+m} - x_n\| \leq \sum_{0 < k < m} \|x_{n+k+1} - x_{n+k}\|$$

$$\leq \sum_{0 \leq k < m} \alpha^{n+k} \|x_1 - x_0\| \leq \alpha^n (1 - \alpha)^{-1} \|x_1 - x_0\|.$$

By the definition of compression $\alpha \in]0, 1[$. Hence, $\alpha^n \to 0$ as $n \to \infty$.

Recalling that the considered space is complete we have that the sequence of points x_n converges to some point c of this space. Since C is closed, this c is in C. From $x_{n+1} = h(x_n)$ and the continuity of h it follows that $c = h(c)$.

Uniqueness. Suppose that there exists another fixed point $b \in C$ for h. Since h is a compression, then

$$\|c - b\| = \|h(c) - h(b)\| \leq \alpha \|c - b\|,$$

which contradicts to the condition $\alpha \in]0, 1[$ if $c \neq b$. ∎

REMARK. For $\alpha = 1$ the theorem does not hold. Indeed, under the identical mapping $h(x) = x$ of space E each point is fixed, and the translation $h(x) = x + a$ ($a \neq 0$) leaves no point fixed. The theorem is not true also for open sets: for $h(x) = 2^{-1}x$ the set $C = \{x : x \neq 0\}$ has no fixed point.

1.4.5. Sum of Operator Progression

Take a Banach space F and a Banach space $\mathcal{B} = \mathcal{B}(F, F)$ of bounded linear operators transforming F into F (see Section 1.1.4).

Let I denote the identity transformation of a set F given by

$$\mathrm{I}(x) = \mathrm{I} \cdot x = x \quad (x \in F).$$

We shall write the values of transformations from $\mathcal{B}$ and compositions of these transformations in product form.

For every transformation $T \in \mathcal{B}$, $\mathrm{I} - T$ will denote its difference with the identity function, and $(\mathrm{I} - T)^{-1}$ will denote the inverse of this difference. If $\|T\| < 1$, such a transformation exists and is the sum of a series constituted by powers T^n. This sum is defined by analogy with numerical series as

$$\sum_{n \geq 0} T^n = \lim_{n \to \infty} \sum_{k < n} T^k.$$

By definition $T^0 = \mathrm{I}$.

1.5. Taylor's Formula

THEOREM. *If $\|T\| < 1$, then $I - T$ is a linear homeomorphism from F onto F and*

$$(I - T)^{-1} = \sum_{n \geq 0} T^n.$$

PROOF. From the Cauchy inequality it follows that

$$\|T^n\| \leq \|T\|^n \quad (n \geq 0).$$

Therefore, if $\|T\| < 1$, then a sequence of $T^n \in \mathcal{B}$ has the sum

$$S = \sum_{n \geq 0} T^n \in \mathcal{B}.$$

This follows from the Cauchy criterion and the inequalities

$$\left\| \sum_{0 \leq k < m} T^{n+k} \right\| \leq \sum_{0 \leq k < m} \|T^{n+k}\| \leq \sum_{0 \leq k < m} \|T\|^{n+k} \quad (m > 0).$$

For here

$$(I - T) \sum_{k < n} T^k = \sum_{k < n} T^k (I - T) = I - T^n \quad (n > 0)$$

and the sequence of points T^n converges to 0 in $\mathcal{B}$, then

$$(I - T)S = S(I - T) = I.$$

Hence, $I - T$ and S are inverse of one another and one-one maps from F onto F. Their linearity and continuity follows from the definitions. ∎

COROLLARY. $\|(I - T)^{-1} - I - T\| \leq \|T\|^2 (1 - \|T\|)^{-1}$ $(\|T\| < 1)$.
Indeed, from the theorem we have

$$\left\|(I - T)^{-1} - I - T\right\| = \left\| \sum_{n \geq 2} T^n \right\| \leq \sum_{n \geq 2} \|T\|^n = \|T\|^2 (1 - \|T\|)^{-1}$$

for all T in $\mathcal{B}$, such that $\|T\| < 1$.

1.5. Taylor's Formula

This formula describes approximation of functions by polynomials obtained by successive differentiation.

1.5.1. Sequential Derivatives

We shall consider a function f from an open interval $\mathscr{J} \subseteq \mathbb{R}$ into a normed space F.

A continuous function f is said to be 0 times differentiated. Its 0-th derivative, $f^{(0)}$, is the function f itself.

Let n be a positive integer. By definition, an n times differentiable function f whose n-th derivative, $f^{(n)}$, is differentiable is deemed $n + 1$ times differentiable. Its $(n + 1)$-th derivative, $f^{(n+1)}$, is the derivative of the function $f^{(n)}$.

For any positive integer n, the concept of n times differentiable function and its n-th derivative is defined by induction on n to be

$$f^{(0)} = f, \quad f^{(n+1)} = (f^{(n)})'.$$

Functions claimed to be n times differentiable for any positive integer n are said to be *infinitely differentiable*.

We now prove two lemmas.

Let $t = u + v$ and u in $\mathscr{J}$. Choose an interval $X \subseteq \mathbb{R}$ such that $[0, 1] \subseteq X$ and $u + xv \in \mathscr{J}$ for all x in X.

Let φ be a function on X with values

$$\varphi(x) = u + xv \quad (x \in X),$$

f be an $n + 1$ times differentiable function from $\mathscr{J}$ into F, and $g = f \circ \varphi : X \to F$ be a composite function with values

$$g(x) = f(u + xv) \quad (x \in X).$$

LEMMA 1. *A function g is $n + 1$ times differentiable and*

$$g^{(k)}(x) = f^{(k)}(u + xv) \cdot v^k \quad (x \in X, \ 0 \leq k \leq n + 1).$$

PROOF. Since f is differentiable, then it is continuous. The function φ is also continuous. Hence, $g = f \circ \varphi$ is continuous, that is to say, it is 0 times differentiable and

$$g^{(0)}(x) = f^{(0)}(u + xv) \cdot v^0 \quad (x \in X).$$

Take $k \leq n$. If g is differentiable k times and

$$g^{(k)}(x) = f^{(k)}(u + xv) \cdot v^k \quad (x \in X), \tag{1}$$

then g is $k + 1$ times differentiable and

$$g^{(k+1)}(x) = f^{(k+1)}(u + xv) \cdot v^{k+1} \quad (x \in X). \tag{2}$$

1.5. Taylor's Formula

Indeed, using the differentiability of φ and $f^{(k)}$ and the chain rule of differentiation we obtain the differentiability of $f^{(k)} \circ \varphi$ and the identity

$$(f^{(k)} \circ \varphi)'(x) = f^{(k+1)}(\varphi(x)) \cdot \varphi'(x)$$
$$= f^{(k+1)}(u + xv) \cdot v \quad (x \in X).$$

This identity and Eq. (1) entail Eq. (2).

By induction from the above argument it follows that Lemma 1 is true. ∎

LEMMA 2. *A function* ψ *with values*

$$\psi(x) = (n!)^{-1}(1 - x)^n \quad (x \in X)$$

is infinitely differentiable and

$$\psi^{(l)}(x) = (-1)^l ((n - l)!)^{-1}(1 - x)^{n-l} \quad (0 \leq l \leq n, \ x \in X),$$
$$\psi^{(m)}(x) = 0 \quad (m > n, \ x \in X).$$

PROOF. The continuity of ψ and the identity

$$\psi^{(0)}(x) = (-1)^0 ((n - 0)!)(1 - x)^{n-0} \quad (x \in X)$$

follow from definitions.

Take $l < n$. If ψ is l times differentiable and

$$\psi^{(l)}(x) = (-1)^l ((n - l)!)^{-1} (1 - x)^{n-l} \quad (x \in X)$$

then ψ is $l + 1$ times differentiable and

$$\psi^{(l+1)}(x) = (-1)^{l+1} ((n - l - 1)!)^{-1} (1 - x)^{n-l-1} \quad (x \in X).$$

By induction from the above argument it follows that ψ is n times differentiable and the statement of the lemma for the l-th derivative holds for $0 \leq l \leq n$.

In particular,

$$\psi^{(n)}(x) = (-1)^n,$$

whence it follows that ψ is infinitely differentiable and for any $m > n$ its m-th derivative is identically zero. ∎

1.5.2. Maclaurin's Formula

Consider an open interval $X \subseteq R$, containing the closed interval $[0, 1]$, a function $g : X \to F$, a positive integer n, and a positive number $\bar{c}$. Suppose that g is differentiable $n + 1$ times and

$$\left|g^{(n+1)}(x)\right| \leq \bar{c} \quad (x \in X).$$

Then we have the *Maclaurin formula*.

THEOREM.

$$g(1) = \sum_{0 \leq k \leq n} (k!)^{-1} g^{(k)}(0) + \bar{r}_n, \quad |\bar{r}_n| \leq \bar{c}((n+1)!)^{-1}.$$

PROOF. Suppose that h is a function on X with values

$$h(x) = \sum_{0 \leq k \leq n} (k!)^{-1} g^{(k)}(x) (1 - x)^k \quad (x \in X).$$

The proof reduces to evaluation of $h'(x)$ and application of the Lagrange theorem.

Since h is the sum of products of the differentiable functions $g^{(k)}$ and $(-1)^{n-k}\psi^{(n-k)}$ for $0 \leq k \leq n$, it is differentiable and

$$h'(x) = \sum_{0 \leq k \leq n} (-1)^{n-k} (g^{(k)} \psi^{(n-k)})'(x)$$

$$= \sum_{0 \leq k \leq n} (-1)^{n-k} (g^{(k+1)}(x) \psi^{(n-k)}(x) + g^{(k)}(x) \psi^{(n-k+1)}(x))$$

$$= g^{(n+1)}(x) \psi(x) + (-1)^n g(x) \psi^{(n+1)}(x) = g^{(n+1)}(x) \psi(x) \quad (x \in X).$$

Consider a function $\bar{h}$ over X with values

$$\bar{h}(x) = -\bar{c}((n+1)!)^{-1}(1 - x)^{n+1} \quad (x \in X).$$

It is differentiable and

$$\bar{h}'(x) = \bar{c}(n!)^{-1}(1 - x)^n \quad (x \in X).$$

Hence,

$$|h'(x)| \leq \bar{h}'(x) \quad (x \in \,]0,1[).$$

By Lagrange's theorem we have

$$|h(1) - h(0)| \leq \bar{h}(1) - \bar{h}(0).$$

1.5. Taylor's Formula

From this inequality and the identities

$$h(1) - h(0) = g(1) - \sum_{0 \leq k \leq n} (k!)^{-1} g^{(k)}(0) = \bar{r}_n,$$

$$\bar{h}(1) - \bar{h}(0) = \bar{c}((n+1)!)^{-1}$$

follows the assertion proved. ∎

1.5.3. Taylor's Formula

Consider a normed space F, an open interval $U \subseteq \mathbb{R}$, a point $u \in U$, a set $V = U - u$, a function $f: U \to F$, an integer $n \geq 0$, and a constant $c > 0$.

THEOREM. *If a function f is differentiable $n+1$ times and*

$$\left\| f^{(n+1)}(t) \right\| \leq c \quad (t \in U),$$

then

$$f(u-v) = \sum_{0 \leq k \leq n} (k!)^{-1} \cdot f^{(k)}(u) \cdot v^k + r_n f u(v),$$

$$\left\| r_n f u(v) \right\| \leq c \cdot ((n+1)!)^{-1} \cdot |v|^{n+1} \quad (v \in V).$$

PROOF. At $v = 0$ this assertion is true and reduces to the identities

$$f(u) = f^{(0)}(u) \quad \text{and} \quad r_n f u(0) = 0.$$

We take $v \neq 0$ and the open interval $X = v^{-1}V = \{ v^{-1} \cdot z : z \in V \}$. Since $0 \in V$ and $v \in V$, then $[0, 1] \subseteq X$. By Lemma 1 a function $g: X \to F$ with values $g(x) = f(u + xv)$, $x \in X$, $u + xv \in U$, is $n + 1$ times differentiable and for $0 \leq k \leq n+1$ and any x in X

$$g^{(k)}(x) = f^{(k)}(u + xv) \cdot v^k.$$

Hence,

$$\left\| g^{(n+1)}(x) \right\| \leq c|v|^{n+1} = \bar{c} \quad (x \in X).$$

Consequently, for g the Maclaurin formula holds. Since

$$f(u + v) = g(1),$$
$$f^{(k)}(u) \cdot v^k = g^{(k)}(0), \quad \text{and} \quad \bar{c} = c|v|^{n+1}$$

this formula establishes the theorem. ∎

REMARK. The theorem has been proved for vector-valued functions of a real-valued variable. It holds also for functions of vector-valued variable with an appropriate definition of sequential derivatives as polylinear operators. In multiplicative representation, Taylor's formula for vector-variable functions has the same form as for functions of numerical variable, except for the absolute value of the increment in the argument must be replaced with its norm in the estimate of the remainder.

1.5.4. Symmetry of Derivatives

If a function of several variables is differentiable, its repeated derivatives are symmetric: the order of variables with respect to which differentiation is performed is immaterial.

Consider product $E = E_1 \times E_2$ of normed spaces E_1 and E_2, a normed space F, a rectangle $U = U_1 \times U_2$ with open sides $U_1 \subseteq E_1$ and $U_2 \subseteq E_2$, a function $f: U \to E$, a point $u \in U$, and a set $V = U - u$.

We shall assume that f is twice differentiable. This implies that there exist the derivative $f': U \to \mathcal{B}$ of f and the derivative $f'': U \to \mathcal{B}^2$ of the function f'. Here $\mathcal{B} = \mathcal{B}(E, F)$ is the normed space of continuous linear mappings $E \to F$. The normed space $\mathcal{B}^2 = \mathcal{B}(E, \mathcal{B})$ of continuous linear mappings $E \to \mathcal{B}$ is identified with the space of continuous bilinear maps $E \times E \to F$.

Take the partial differentials $d_j f$ of the function f defined by

$$d_j f u = df u \cdot r_j p_j \quad (j = 1, 2; \; u \in U).$$

By the rules of differentiation

$$d(d_j f) u = d(df) u \cdot r_j p_j \quad (j = 1, 2; \; u \in U),$$

whence

$$d(df) = d_1(d_1 f) + d_2(d_1 f) + d_1(d_2 f) + d_2(d_2 f).$$

The iterated differential $d(df) u: E \to \mathcal{B}$ defines the bilinear mapping $d^2 fu: E \times E \to F$ by the equality

$$d^2 fu(x, y) = d(df) u \cdot x \cdot y \quad (x \in E, \; y \in E).$$

THEOREM. *A bilinear mapping $d^2 fu$ is symmetric.*

PROOF. We must prove that for all x and y in E,

$$d^2 fu(x, y) = d(df) u \cdot x \cdot y = d(df) u \cdot y \cdot x = d^2 fu(y, x).$$

This can be achieved with the readily verifiable identity

$$\Delta(\Delta f) u \cdot v \cdot w = \Delta(\Delta f) u \cdot w \cdot v \quad (v \in V, \; w \in V),$$

1.5. Taylor's Formula

differentiation rules and Lagrange's theorem which allows us to estimate the desired increments. ∎

From this theorem we obtain

COROLLARY. $f''_{12}(u) = f''_{21}(u)$.

PROOF.

Indeed, applying the theorem we obtain

$$f''_{12}(u) \cdot v \cdot w = d_1(d_2 f) u \cdot w \cdot v = d(df) u \cdot (0, w) \cdot (v, 0)$$
$$= d(df) u \cdot (v, 0) \cdot (0, w) = d_2(d_1 f) u \cdot v \cdot w = f''_{21}(u) \cdot v \cdot w$$

for all v and w in V, as claimed. ∎

In differentiable form this identity can be written

$$d_1(d_2 f) u = d_2(d_1 f) u.$$

REMARK. If $f''_{12}(u) \neq f''_{21}(u)$ for some point u in U, then the function f is not twice differentiable over U.

1.5.5. Particular Case

Consider normed spaces E and F, an open set $U \subseteq E$, a function $f: U \to F$, a point $u \in U$, and the set $V = U - u$, a ball $B(u, \delta) \subseteq U$, $\delta > 0$, and assume that f is twice differentiable.

PROPOSITION. *Given some* $\gamma \in \,]0, \delta[$ *we have*

$$f(u+v) = f(u) + dfu(v) + 2^{-1}d^2fu(v, v) + r_2 fu(v),$$

where

$$\|v\|^{-2} r_2 fu(v) \to 0 \text{ as } v \to 0 \quad (0 < \|v\| < \gamma).$$

PROOF. Consider a function $\bar{g}$ defined by

$$\bar{g}(x, y) = f(u + x) - f(u) - dfu(x) - 2^{-1}d^2fu(x, y)$$

for all x, y in $B(0, \delta)$.

The differentiability of f implies the differentiability of a composite function h_1 defined by $h_1(x) = f(u + x)$, $x \in B(0, \delta)$, and also the identity

$$dh_1 s = df(u + s) \quad (s \in B(0, \delta)).$$

The continuous linear function $h_2 = dfu$ is differentiable with

$$dh_2 s = dfu \quad (s \in B(0, \delta)).$$

Now we take a function h with values
$$h(x, y) = h_1(x) - f(u) - h_2(x) \quad (x, y \in B(0, \delta)).$$
Its partial differentials d_1h and d_2h are defined by
$$d_1h(s, t) = (df(u + s) - dfu) \cdot p_1, \quad d_2h(s, t) = 0$$
$$p_1(x, y) = x$$
for all x, y in E and s, t in $B(0, \delta)$.

From the differentiability of df it follows that this mapping is continuous, and hence d_1h is also continuous. If the partial differentials d_1h and d_2h exist and are continuous, then the full differential dh exists and we have
$$dh(s, t)(x, y) = (df(u + s) - dfu)(x)$$
for $s, t \in B(0, \delta)$ and $x, y \in E$.

The mapping $b = d^2 fu$ is differentiable and
$$db(s, t)(x, y) = d^2 fu(x, t) - d^2 fu(s, y)$$
for $x, t \in B(0, \delta)$ and $x, y \in E$.

Consequently, the function $\bar{g} = h - 2^{-1}b$ is differentiable and
$$d\bar{g}(s, t)(x, y) = (df(u + s) - dfu)(x) - 2^{-1}(d^2 fu(x, t) + d^2 fu(s, y)).$$

Consider a function g with values
$$g(v) = \bar{g}(v, v) = r_2 fu(v) \quad (v \in B(0, \delta)).$$
It is the composite of function $\bar{g}$ and function φ with values
$$\varphi(v) = (v, v) \quad (v \in B(0, \delta)).$$
Using the chain rule of differentiation, the identity for $d\bar{g}(s, t)(x, y)$ and the symmetry theorem we obtain
$$dgs(x) = d\bar{g}(s, s)(x, x) = (df(u + s) - dfu)(s) - 2^{-1}(d^2 fu(x, s)$$
$$+ d^2 fu(s, x)) = (df(u + s) - dfu)(x) - d(df)usx$$
$$= (df(u + s) - dfu - d(df)us)x$$
and
$$dgs = r(df)us$$
for all s in $B(0, \delta)$ and x in E.

1.6. Local Minima

Since the mapping df is differentiable, for any $\varepsilon > 0$ there exists a $\gamma \in \,]0, \delta[$ such that

$$\|dgv\| = \|r(df)uv\| \leq \varepsilon \|v\| \qquad (\|v\| \leq \gamma).$$

By Corollary 5 of Lagrange's theorem we obtain from the above derivations

$$\|r_2 fu(v)\| = \|g(v)\| = \|g(v) - g(0)\| \leq \varepsilon \|v\|^2 \qquad (\|v\| \leq \gamma). \blacksquare$$

In the equivalent multiplicative form the identity proved above may be rewritten as

$$f(u + v) = f(u) + f'(u) \cdot v + 2^{-1} f''(u) \cdot v^2 + r_2 fu(v),$$

where

$$\|v\|^{-2} \cdot \|r_2 fu(v)\| \to 0 \text{ as } v \to 0 \qquad (0 < \|v\| < \gamma,\ v \in V).$$

REMARK. This equality is a particular case of Taylor's formula with remainder in Peano's form. A similar argument using the induction principle proves the general case.

By decomposing the permutations into transpositions and applying induction it is straightforward to prove that the derivatives of functions of several variables are symmetric.

1.6. Local Minima

The parabolic approximation by Taylor's series is a useful tool in deriving simple conditions for local minima and maxima of real functions of vector-valued variable.

1.6.1. Definitions

Consider a normed space E, an open set $U \subseteq E$, a point $u \in U$, and a function $f: U \to \mathbb{R}$.

DEFINITION: *A function f is said to have a local minimum at u if there exists a neighborhood $V \subseteq U - u$ of point 0 such that*

$$\Delta fu(v) \geq 0 \qquad (v \in V).$$

A function f is said to have a strict local minimum at u if there exists a neighborhood $V \subseteq U - u$ of point 0 such that

$$\Delta fu(v) > 0 \qquad (v \in V,\ v \neq 0)$$

REMARK. Inverting these inequalities defines a local maximum and a strict local maximum of f at u, but these extrema would coincide with the local minimum and strict local minimum of $g = -f$ at u. Therefore we will speak about minima only.

EXAMPLE. Let $E = U = \mathbb{R}$, $u = 0$, and $f(t) = t^2$ for $t \in \mathbb{R}$. This parabola has a strict local minimum at 0.

Note that $f'(0) = 0$ and $f''(0) = 2 > 0$. By approximating twice differentiable functions by parabolas one may derive for these functions similar conditions of local minima.

Suppose that f is twice differentiable. We will say that the quadratic form $f''(u)$ is positive and write $f''(u) \geq 0$ whenever

$$f''(u) \cdot v^2 \geq 0 \quad (v \in E).$$

We will call $f''(u)$ strictly positive and write $f''(u) > 0$ whenever

$$f''(u) \cdot v^2 \geq \alpha \cdot \|v\|^2 \quad (v \in E)$$

for some $\alpha > 0$. If space E is finite-dimensional, then the property of a quadratic form being strictly positive is equivalent to this form being positive and nondegenerate. (This follows from the compactness of spheres in finite-dimensional normed spaces.)

1.6.2. Unconditional Local Minima

Given a twice differentiable function $f: U \to \mathbb{R}$, we formulate the necessary conditions for a local minimum and sufficient conditions for a strict local minimum.

PROPOSITION 1. *If f has a local minimum at u, then $f'(u) = 0$ and $f''(u) \geq 0$.*

PROOF. If f has a local minimum at u, then

$$\Delta f u(v) = f'(u) \cdot v + 2^{-1} \cdot f''(u) \cdot v^2 + r_2 f u(v) \geq 0 \quad (v \in V)$$

in some neighborhood $V \subseteq U - u$ of point 0. Taking a ball $\overline{B}(0, \delta) \subseteq V$, $\delta > 0$, and observing that $r_2 f u(v) = o(\|v\|^2)$ we obtain first the equality $f'(u) = 0$, and later the inequality $f''(u) \geq 0$. ∎

REMARK. Let $E = U = \mathbb{R}$, $u = 0$, and $f(t) = t^3$ for $t \in \mathbb{R}$. The function f has no local minimum at 0 notwithstanding that $f'(0) = 0$ and $f''(0) \geq 0$. Hence, the conditions of Proposition 1 are necessary but not sufficient.

PROPOSITION 2. *If $f'(u) = 0$ and $f''(u) > 0$, then f has a strict local minimum at u.*

1.6. Local Minima

PROOF.
Indeed, under the said conditions

$$\Delta fu(v) = 2^{-1} \cdot f''(u) \cdot v^2 + r_2 fu(v) \geq (\alpha - 2^{-1}\alpha) \cdot \|v\|^2$$
$$= 2^{-1}\alpha \cdot \|v\|^2 > 0$$

for some $\alpha > 0$ and all $v \neq 0$ from some ball $B(0, \delta) \subseteq U - u$. ∎

REMARK. Let $E = U = \mathbb{R}$, $u = 0$, and $f(t) = t^4$ for $t \in \mathbb{R}$. The function f has a strict local minimum at 0 although $f''(0) = 0$. Hence, the conditions of Proposition 2 are sufficient but not necessary.

1.6.3. Conditional Local Minima

Consider Banach spaces E, F, G; open sets $X \subseteq E$, $Y \subseteq F$; points $a \in X$, $b \in Y$; a smooth function $g : X \times Y \to G$, the equation $g(x, y) = 0$, which is true and nondegenerate at a point (a, b), that is, correct at this point; g satisfies the conditions of the implicit function theorem, and a function $\varphi : X \times Y \to \mathbb{R}$.

The local minima of the contraction of φ to the set $g^{-1}(0) = \{(x, y) : g(x, y) = 0\}$ will be called its local minima over $g^{-1}(0)$ or its conditional local minima subject to $g = 0$.

DEFINITION. *A function φ is said to have a conditional local minimum at a point (a, b) when $g = 0$, if there exists an open neighborhood $A \times B \subseteq X \times Y$ of the point (a, b) such that*

$$\varphi(x, y) \geq \varphi(a, b)$$

for all $(x, y) \in g^{-1}(0) \cap A \times B$.

A function φ is said to have a conditional strict local minimum at a point (a, b) when $g = 0$ provided there exists an open neighborhood $A \times B \subseteq X \times Y$ of the point (a, b) such that

$$\varphi(x, y) > \varphi(a, b)$$

for all $(x, y) \in g^{-1}(0) \cap A \times B$ except $(x, y) = (a, b)$. (Putting $(x, y) \in g^{-1}(0) \cap A \times B$ implies that $x \in A, y \in B$ and $g(x, y) = 0$.)

By the implicit function theorem there exists an open neighborhood $A \times B \subseteq X \times T$ of (a, b) such that the set $g^{-1}(0) \cap A \times B$ is a graph of some smooth function $f: A \to B$.

Let $\psi : A \to \mathbb{R}$ be a function with values

$$\psi(x) = \varphi(x, f(x)) \qquad (x \in A).$$

From the above definition if follows that given $g = 0$ the function φ has a conditional (strict) local minimum at (a, b) if and only if ψ has a (strict) local minimum at a. Therefore, instead of looking for local minima of φ one could evaluate unconditional local minima of ψ making use of the conditions of Section 1.6.2. However, this approach often leads to cumbersome formulations. An easier way would be to derive special tests for conditional local minima.

First of all we formulate the general necessary condition for a local minimum.

LEMMA. *If a function $h : A \to \mathbb{R}$ is differentiable at a point $a \in A$ and has a local minimum at this point, then $h'(a) = 0$.*

PROOF. If h is differentiable and has a local minimum at a, then

$$\Delta ha(v) = h'(a) \cdot v + rha(v) \geq 0 \qquad (v \in A - a),$$

so the proposition of $h'(a) \neq 0$ would contradict the relative smallness of the remainder term rha. ∎

Take partial functions $\varphi(\cdot, y)$, $\varphi(x, \cdot)$, $g(\cdot, y)$, $g(x, \cdot)$ and their derivatives for $x \in A$ and $y \in B$:

$$p(x, y) = d\varphi(\cdot, y) x, \qquad q(x, y) = d\varphi(x, \cdot) y;$$
$$P(x, y) = dg(\cdot, y) x, \qquad Q(x, y) = dg(x, \cdot) y.$$

Applying the chain rule to the function ψ with values $\psi(x) = \varphi(x, f(x))$ yields

$$\psi'(x) = p(x, f(x)) + q(x, f(x)) \cdot f'(x),$$

which in view of the implicit function theorem is rewritten as

$$\psi'(x) = p(x, f(x)) + q(x, f(x)) Q^{-1}(x, f(x)) P(x, f(x)).$$

Relation between the conditional local minimum of φ and the local minimum of ψ paves the way for a straightforward proof of the following result.

PROPOSITION. *If a function φ has a conditional local minimum at a point (a, b) when $g = 0$, then*

$$p(a, b) - q(a, b) \cdot Q^{-1}(a, b) \cdot P(a, b) = 0.$$

PROOF. If φ has a conditional local minimum at (a, b) when $g = 0$, then ψ has a local minimum at a. Since φ and g are differentiable so is ψ. By the lemma, $\psi'(a) = 0$. From this equality and the identity for ψ' at $x = a$ and $f(x) = b$ we obtain the assertion of the proposition. ∎

1.6. Local Minima

1.6.4. The Lagrange Method

In seeking for conditional local minima it is convenient to use a method devised by Lagrange.

We retain the notation and propositions of Section 1.6.3. Consider a continuous linear function $l : G \to \mathbb{R}$ and a function $\lambda = \varphi + l \circ g : A \times B \to \mathbb{R}$ with values

$$\lambda(x, y) = \varphi(x, y) + l \cdot g(x, y) \quad (x \in A, \; y \in B).$$

We call λ the Lagrange function for φ, g and l. (Note the multiplicative representation for the operator l in the last equality.) The Lagrange function is differentiable along with g, and l, and the derivatives of its partial functions $\lambda(\cdot, y)$ and $\lambda(x, \cdot)$ are

$$d\lambda(\cdot, y)x = p(x, y) + l \cdot P(x, y),$$
$$d\lambda(x, \cdot)y = q(x, y) + l \cdot Q(x, y).$$

Note that if

$$l = -q(a, b) \cdot Q^{-1}(a, b),$$

then

$$d\lambda(\cdot, b)a = \psi'(a)$$

and

$$d\lambda(a, \cdot)b = 0.$$

Consequently, for the total differential $d\lambda(a, b)$ of λ at (a, b) we have

$$d\lambda(a, b) \cdot (\Delta x, \Delta y) = \psi'(a) \cdot \Delta x \quad (\Delta x \in E, \; \Delta y \in F).$$

Therefore, the equalities

$$d\lambda(a, b) = 0 \quad \text{and} \quad \psi'(a) = 0$$

are equivalent.

Thus, to check whether or not a function φ has a conditional local minimum at (a, b) when $g = 0$, one may use the Lagrange function λ for φ, g, and $l = -q(a, b) \cdot Q^{-1}(a, b)$ in place of the function ψ.

In view of the proposition of Section 1.6.3 this argument leads us to the following result.

COROLLARY. *If a function φ has a conditional local minimum at a point (a, b) when $g = 0$, then there exists a linear function $l \in \mathcal{B}(G, \mathbb{R})$*

such that

$$\varphi'(a, b) + l \cdot g'(a, b) = 0.$$

In other words, for φ to have a conditional local minimum at (a, b) when $g = 0$ we need that the derivative of Lagrange function λ for φ, g and some l be zero.

To verify whether φ does have a conditional local minimum at (a, b) when $g = 0$ a further effort is needed. It is customary to use the properties of the given φ and g to effect this check.

EXAMPLE. Let $X = E = \mathbb{R}^2$, $Y = F = G = \mathbb{R}$, and for x_1, x_2 and y in $\mathbb{R}$

$$g(x, y) = x_1^2 + x_2^2 + y^2 - 1,$$

$$\varphi(x, y) = x_1 + x_2 + y.$$

Let c be a number, and $\lambda = \varphi + cg$ be a function with values

$$\lambda(x, y) = (x_1 + x_2 + y) + c(x_1^2 + x_2^2 + y^2 - 1) \quad (x_1, x_2, y \in \mathbb{R}).$$

Since

$$\lambda'(x, y) = (1 + 2cx_1, 1 + 2cx_2, 1 + 2cy),$$

then

$$\lambda'(x, y) = (0, 0, 0) \quad \text{and} \quad g(x, y) = 0$$

if and only if $c = -\sqrt{3}/2$, $x_1 = x_2 = y = 1/\sqrt{3}$ or $c = \sqrt{3}/2$, $x_1, x_2 = y = -1/\sqrt{3}$. It is straightforward to verify that given $g = 0$ the function φ has a conditional local minimum of -1 at the point $(a, b) = -(1/3, 1/3, 1/3)$ and a maximum of 1 at the point $(1/3, 1/3, 1/3)$.

1.6.5. Smooth Surfaces

An equation $g(x, y) = 0$ with smooth function $g : X \times Y \to \mathbb{R}$ and $X \subseteq \mathbb{R}^n$, $Y \subseteq \mathbb{R}$ defines a smooth n-dimensional surface in $X \times Y \subseteq \mathbb{R}^{n+1}$. Therefore, when $g = 0$ the conditional local minima are local minima of real-valued functions on smooth surfaces. We describe such surfaces in more detail.

Definitions and Examples. Consider $E = \mathbb{R}^n$, $F = G = \mathbb{R} = \mathbb{R}^1$ and $E \times F = H = \mathbb{R}^{n+1}$, $n \geq 1$; an open set $W \subseteq \mathbb{R}^{n+1}$; a rectangle $X \times Y = Z \subseteq W$ with open sides $X \subseteq \mathbb{R}^n$ and $Y \subseteq \mathbb{R}$; points $z = (x_1, ..., x_n, y) \in Z$, $x = (x_1, ..., x_n) \in X$, $y \in Y$, $c = (a_1, ..., a_n, b) \in Z$, $a = (a_1, ..., a_n) \in X$, $b \in Y$; a smooth function $h : W \to \mathbb{R}$, and the restriction $g : X \times Y \to \mathbb{R}$ of function h into the rectangle $X \times Y = Z$.

1.6. Local Minima

If we identify $\mathbb{R}^n \times \mathbb{R}^1$ with $\mathbb{R}^{n+1}$ then the product $X \times Y$ of the rectangle $X \subseteq \mathbb{R}^n$ by the interval $Y \subseteq \mathbb{R}$ may be identified with the rectangle $Z \subseteq \mathbb{R}^{n+1}$, and the pairs (x, y) and (a, b) constituted by points $x, a \in \mathbb{R}^n$ and numbers $y, b \in \mathbb{R}$ may be identified with points $z, c \in \mathbb{R}^{n+1}$. The restriction of the function h of real-valued variables $x_1, \ldots, x_n, y$ to Z is identified with the function g of vector variable $x = (x_1, \ldots, x_n)$ and real-valued variable y on $X \times Y$. These natural identifications enable us to deem the equations

$$h(z) = 0 \quad \text{and} \quad g(x, y) = 0$$

equivalent for $z = (x, y)$.

The derivative

$$h'(w) = (h'_1(w), \ldots, h'_n(w), h'_{n+1}(w))$$

is a matrix of order 1 by $n + 1$ arrayed of the partial derivatives of h taken at a point $w \in W$ with respect to each of the $n + 1$ variables. The matrix $h'(w)$ has the largest rank 1 if and only if at least one of the constituent partial derivatives $h'_1(w), \ldots, h'_n(w), h'_{n+1}(w)$ is different from zero. We suppose that $h'_{n+1}(w) \neq 0$; this always can be arranged by relabeling the variables. Letting $a_1 = w_1, \ldots, a_n = w_n$, and $b = w_{n+1}$ we get $c = (a, b) = w$ with $a = (a_1, \ldots, a_n) \in \mathbb{R}^n$ and $b \in \mathbb{R}$. Now we may safely use the nomenclature outlined above.

By the implicit function theorem there exists an open rectangular neighborhood

$$A \times B = C \subseteq Z = X \times Y \quad (A \subseteq X, B \subseteq Y)$$

of the point $(a, b) = c$ such that the intersection

$$S = h^{-1}(0) \cap C$$

is a graph of some smooth function $f : A \to B$. It is natural to refer to this graph as a segment of the smooth surface defined by the equation $h(z) = 0$, and the union of these segments as a smooth surface.

If the matrix $h'(z)$ has rank 1 at each point $z \in S$, then we shall say that the function h is regular over the set S.

DEFINITION. *The set $S = \{z : h(z) = 0\}$ of the solutions $z \in W$ to the equation $h(z) = 0$ with a smooth function h from an open $W \subseteq \mathbb{R}^{n+1}$ into $\mathbb{R}$ and regular over S is called an n-dimensional smooth surface in W, which is defined by the equation $h(z) = 0$.*

REMARK. This definition describes only the simplest smooth surfaces. A general definition of a smooth surface is more involved.
A few examples to illustrate the definition follow.

EXAMPLE 1. Let $W = \mathbb{R}^2$ and $h : W \to \mathbb{R}$ be a function with values

$$h(z) = z_1^2 + z_2^2 - 1 \quad (z = (z_1, z_2) \in \mathbb{R}^2).$$

The matrix

$$h'(z) = (h_1'(z), h_2'(z)) = (2z_1, 2z_2)$$

has rank 1 for $z = (z_1, z_2) \neq (0, 0)$.

If we let $h(z) = 0$, we obtain an equation of circle S^1 in $\mathbb{R}^2$.

EXAMPLE 2. Let $W = \mathbb{R}^3$ and $h : W \to \mathbb{R}$ be a function defined by

$$h(z) = z_1^2 + z_2^2 + z_3^2 - 1 \quad (z = (z_1, z_2, z_3) \in \mathbb{R}^3).$$

The matrix

$$h'(z) = (2z_1, 2z_2, 2z_3)$$

has rank 1 for $z = (z_1, z_2, z_3) \neq (0, 0, 0)$.

If we let $h(z) = 0$ we get an equation of a sphere S^2 in $\mathbb{R}^3$.

EXAMPLE 3. Let $W = \mathbb{R}^{n+1}$ and $h : W \to \mathbb{R}$ be a function with values

$$h(z) = c_1 z_1 + \ldots + c_n z_n + c_{n+1} z_{n+1} - d$$

for $z = (z_1, \ldots, z_n, z_{n+1}) \in \mathbb{R}^{n+1}$ and $c_1, \ldots, c_n, c_{n+1}, d \in \mathbb{R}$.

The matrix

$$h'(z) = (c_1, \ldots, c_n, c_{n+1})$$

has rank 1 if at least one of the coefficients $c_1, \ldots, c_n, c_{n+1}$ is nonzero. In this case letting $h(z) = 0$ we get an equation of n-dimensional plane in $\mathbb{R}^{n+1}$, specifically at $n = 2$ an equation of a common plane in $\mathbb{R}^3$.

Tangents. Any smooth surface S may be approximated by some plane $TS(a, b)$ in a neighborhood of its every point (a, b). This plane is called a tangent to surface S at point (a, b).

Consider the neighborhood $A \times B \subseteq X \times Y$ of a point (a, b) on a smooth surface S in $X \times Y$, a piece of which $L = S \cap A \times B$ is defined by the equation $g(x, y) = 0$.

DEFINITION. *A tangent to a surface S at a point (a, b) is the plane*

1.6. Local Minima

$TS(a, b)$ defined by the equation

$$g'(a, b) \cdot (x - a, y - b) = 0. \tag{1}$$

This equation is equivalent to

$$P(a, b) \cdot (x - a) + Q(a, b) \cdot (y - b) = 0. \tag{2}$$

Since $Q(a, b)$ is a homeomorphism, equation (2) is equivalent to

$$y = b + Q^{-1}(a, b) \cdot P(a, b) \cdot (x - a). \tag{3}$$

In Section 2.2.5 below the implicit function theorem proves that equation (3) is equivalent to

$$y = f(a) + f'(a) \cdot (x - a). \tag{4}$$

All these equations are called the equations for a tangent to surface S at point (a, b).

EXAMPLE 1. For a tangent to a circle S^1 at a point $(a, b) \in S^1$, equations (2), (4) have the form

$$2a(x - a) + 2b(y - b) = 0, \quad y = b - b^{-1}a(x - a) \quad (b \neq 0).$$

EXAMPLE 2. For a tangent to a sphere S^2 at $(a, b) \in S^2$, equations (2), (4) have the form

$$2a_1(x_1 - a_1) + 2a_2(x_2 - a_2) + 2b(y - b) = 0,$$
$$y = b - b^{-1}(a_1(x_1 - a_1) + a_2(x_2 - a_2)).$$

Another corollary of the Section 1.6.3 proposition for a conditional local minimum of a differentiable function $\varphi : X \times Y \to \mathbb{R}$ can be formulated in geometrical terms.

COROLLARY. *If a function has a conditional local minimum at a point (a, b) of the smooth surface S, then*

$$\varphi'(a, b)(x - a, y - b) = 0 \quad ((x, y) \in TS(a, b)).$$

PROOF. Take a piece $L = g^{-1}(0)$ of surface S to which the point (a, b) belongs. For every point (x, y) of the tangent $TS(a, b)$ to S at (a, b),

$$y - b = f'(a) \cdot (x - a).$$

Recalling the expression for the differential of an implicit function

Chapter 1. Differentiation

and the equality of the proposition in Section 1.6.3 we have

$$\varphi'(a, b) \cdot (x - a, y - b) = p(a, b)(x - a) + q(a, b)(y - b)$$
$$= p(a, b)(x - a) + q(a, b) \cdot dfa(x - a)$$
$$= (p(a, b) + q(a, b) \cdot Q^{-1}(a, b) \cdot P(a, b))(x - a) = 0$$

for all $(x, y) \in TS(a, b)$. ∎

Directional Derivative. Consider normed spaces G and H, an open set $W \subseteq H$, a point $c \in W$, a ball $B(c, r) \subseteq W$ with $r > 0$, and a differentiable function $h : W \to G$. Any direction in H may be described by a vector $u \in H$ with norm $\|u\| = 1$. A function $\gamma : \mathbb{R} \to H$ with values

$$\gamma(t) = c + tu \quad (t \in \mathbb{R})$$

ties this direction with the point c, namely, γ describes a straight line in H, which passes through points $\gamma(0) = c$ and $\gamma(1) = c + u$ (parallel to the straight line $\gamma - c$ passing through 0 and u). We note that γ is differentiable and $\gamma'(0) = u$. Since $\gamma(\,]-r, r[\,) \subseteq B(c, r) \subseteq W$, then we have the composite function $h \circ \gamma : T \to G$ defined on the interval $T = \,]-r, r[$. This function is differentiable and

$$(h \circ \gamma)'(0) = h'(\gamma(0)) \cdot \gamma'(0) = h'(c) \cdot u.$$

The composite $h \circ \gamma$ and its derivative describe the variation of h in the direction u.

DEFINITION. *The derivative $(h \circ \gamma)'(0)$ of the function $h \circ \gamma$ at a point c is said to be the derivative of the function h at point c along the direction u, and denoted by $h'_u(c)$.*

Thus, by definition

$$h'_u(c) = h'(c) \cdot u,$$

and by the proposition of Section 1.2.3

$$h'_u(c) = (h \circ \gamma)'(0) = \lim_{t \to 0,\, t \neq 0} t^{-1}[h(c + tu) - h(c)].$$

REMARK. For any $z \in H$, $z \neq 0$, $u = \|z\|^{-1} \cdot z$, and $z = \|z\| \cdot u$,

$$h'(c) \cdot z = h'(c) \cdot (\|z\| \cdot u) = \|z\| \cdot h'(c) \cdot u = \|z\| \cdot h'_u(c).$$

Therefore, it would be natural to define the derivative $h'_z(c)$ of h at c along the direction $z \neq 0$ by the equality

$$h'_z(c) = \|z\|^{-1} \cdot h'(c) \cdot z.$$

1.6. Local Minima

In the expression for the directional derivative written in terms of the ratio limit, one can put z in place of u and drop the normalizing factor $\|z\|^{-1}$ – the limit will be the same.

Under certain conditions the derivatives $h'_z(c)$ define the *weak* derivative of h at c.

EXAMPLE. Let $G = \mathbb{R}$, $H = \mathbb{R}^n$ with $n \geq 1$, and the vector $u = e_j = r_j(1)$ defines the j-th coordinate direction in $\mathbb{R}^n$, $j = 1, \ldots, n$. Then

$$h'_u(c) = h'(c) \cdot u = h'(c) \cdot e_j = h'_j(c) \cdot r_j(1).$$

The derivatives along coordinate directions equal the partial derivatives.

Note that since

$$\varphi'_{(x, y) - (a, b)}(a, b) = \varphi'(a, b) \cdot (x - a, y - b),$$

the proved corollary of the proposition in Section 1.6.3 asserts that if a function φ has a conditional local minimum at (a, b) on a smooth surface S, then

$$\varphi'_{(x, y) - (a, b)}(a, b) = \varphi'(a, b) = 0$$

along each direction $(x, y) - (a, b)$ defined by the vector (x, y) in the tangent plane $TS(a, b)$.

Gradient. Consider a space $G = \mathbb{R}$, a space $H = \mathbb{R}^n$ with $n \geq 1$, an open set $W \subseteq \mathbb{R}^n$, a point $c \in W$, and a smooth function $h : W \to \mathbb{R}$.

The sphere

$$S^{n-1} = \{u : \|u\| = 1\} \subseteq \mathbb{R}^n$$

is compact and among the values

$$f(u) = h'_u(c) = h'(c) \cdot u \qquad (u \in S^{n-1})$$

of a continuous function $f : S^{n-1} \to \mathbb{R}$ there is the largest value

$$f(v) = h'_v(c) = \max\{h'_u(c) : \|u\| = 1\}.$$

The vector $v \in S^{n-1}$, at which this maximum is attained, is not hard to find. Indeed, since

$$h'(c) \cdot u = \Sigma h'_j(c) \cdot u_j \qquad (u = (u_j) \in \mathbb{R}^n),$$

then for $h'(c) \neq 0$ and $\|u\| = 1$, the largest value is achieved at

$$u_j = v_j = \|h'(c)\|^{-1} h'_j(c) \qquad (j = 1, \ldots, n).$$

In matrix form, $h'(c)$ is represented by a $1 \times n$ row, and the vector v — by an $n \times 1$ column:

$$h'(c) = (h'_1(c), \ldots, h'_n(c)), \qquad v = (v_1, \ldots, v_n)^T.$$

For the vector v, which defines the direction along which the derivative of a function h at a point c attains its largest value, we introduce a special term and notation which allow us to drop normalization and the condition $h(c) \neq 0$.

DEFINITION. *The vector*

$$\nabla h(c) = (h'(c))^T$$

is called the gradient of function h at point c, and denoted by grad $h(c)$ or $\nabla h(c)$.

This definition is valid for a function h differentiable at c.

Comparison of the definitions of the gradient and tangent to a plane shows that $\nabla g(a, b)$ is orthogonal to the plane $TS(a, b)$ tangent at a point (a, b) to a surface S given by the equation $g(x, y) = 0$. That is, the gradient is proportional to the normal to S at (a, b).

Using the concept of gradient one may formulate necessary conditions for local minima of differentiable function, which would be equivalent to that given in Sections 1.6.3 and 1.6.4 with $E = \mathbb{R}^n$, and $F = G = \mathbb{R}$.

If a function h has a local minimum at a point c, then $\nabla h(c) = 0$.

If a function φ has a conditional local minimum at a point (a, b) subject to $g(x, y) = 0$, then there exists a number l such than $\nabla \lambda(a, b) = 0$ for the Lagrange function $\lambda = \varphi + lg$.

These conditions are often used in search for extrema.

Chapter 2. INTEGRATION

This chapter discusses elements of integral calculus for functions of abstract variables. This subject is not much more complicated than the classical theory of the Lebesgue integral. In addition, the general definition of measure allows us to consider delta-functions that will be used in Part 2.

2.1. Measures

By a measure we mean an additive function on the algebra of simple sets.

2.1. Measures

2.1.1. Simple Sets and Simple Functions

Consider an abstract set U and an algebra $\mathcal{F} = \mathcal{F}(U)$ of all number-valued (real or complex) functions on U.

We introduce indicators on U that take values in the two-element set $\{0, 1\}$. Formally, a function

$$\text{ind } f : U \to \{0, 1\}$$

with values $f(x) = 1$ $(x \in X)$ and $f(x) = 0$ $(x \notin x)$ is called *indicator* of a set $X \subseteq U$, denoted by ind X or $\text{ind}_u X$. The class of all indicators on U will be denoted by $\mathcal{P} = \mathcal{P}(U)$.

We shall identify, when convenient, the function ind X with the set X, calling it a set and denoting by X. With this identification, operations with functions are carried over to sets: they can be added, subtracted, multiplied by other sets, numbers, and functions. (Note that these operations may result in functions as well as sets.)

EXAMPLES

1. The sum $X + Y$ of two sets X and Y is a set if and only if X and Y are disjoint.
2. The product XY of sets X and Y is always a set. It is equal to the minimum $X \vee Y$ of X and Y, i.e. their intersection.
3. The maximum $X \vee Y$ of the sets X and Y is their union.

The ordering $f \leq g \Leftrightarrow f(x) \leq g(x), x \in U$, for real-valued functions is extended to sets: for sets X and Y, the inequality $X \leq Y$ is equivalent to the inclusion $X \subseteq Y$.

The difference $Y - X$ of sets X and Y is a set if and only if $X \leq Y$.

Intersections of subalgebras of an algebra of functions $\mathcal{F}$ with the class $\mathcal{P}$ are called *algebras of sets*. In other words, a class $\mathcal{A} \subseteq \mathcal{P}$ forms an algebra of sets if and only if for any $A, B \in \mathcal{A}$:

1. $A + B \in \mathcal{A}$, if $A + B \in \mathcal{P}$ (i.e., $AB = 0$)
2. $A - B \in \mathcal{A}$, if $A - B \in \mathcal{P}$ (i.e., $B \leq \mathcal{A}$)
3. $AB \in \mathcal{A}$

EXAMPLES

1. Simple sets constituted by unions of intervals in $U = \mathbb{R}^m$ constitute an algebra of sets.
2. The class $\mathcal{P}$ forms an algebra of sets.
3. Let $A_1, A_2, ..., A_n, ...$ be pairwise disjoint sets ($A_i A_j = 0$ for $i \neq j$). Then the class $\mathcal{A}$ of all their finite sums forms an algebra of sets.

Consider an algebra of sets $\mathcal{R}$ in the set U under consideration. We will call the sets of this algebra *simple*. The abstract simple sets so defined need not possess any special properties; they should only belong to the chosen algebra.

Outstanding among the $\mathcal{R}$ algebras are those endowed with the identity

U, and those containing sets that form increasing sequences whose union is U. (Though U itself may not belong to an algebra.) Below, we deal, as a rule, with such algebras.

EXAMPLES

1. An algebra of simple sets in $\mathbb{R}^m$ constituted by unions of finite families of rectangles.
2. An algebra of all finite sets in U.
3. An algebra of all countable sets in U.
4. An algebra of all finite unions of given mutually disjoint sets $A_n \subseteq U$.
5. An algebra $\mathcal{P}$ of all sets in U.

Now that we have delineated simple sets we proceed to define simple functions.

DEFINITION. *We say simple functions of the linear combinations of simple sets.*

Simple functions form an algebra; we shall denote it by S.

A simple function may be represented by different linear combinations of simple sets:

$$f = \Sigma a_i X_i = \Sigma b_j Y_j$$

with different $a_i, b_j \in \mathbb{F}$ and $X_i, Y_j \in \mathcal{R}$. Among such representations we distinguish the one with sets X_i, pairwise disjoint and all the coefficients a_i, different and nonzero.

Because we agreed to identify sets with their indicators simple sets are also simple functions: $\mathcal{R} \subseteq S$.

Considering algebras of functions instead of algebras of sets simplifies operations with sets, because operations with sets may be replaced by more habitual operations with functions.

2.1.2. Measures on Simple Sets

A number-valued function μ defined on an algebra of sets $\mathcal{R}$ is called additive if

$$\mu(A + B) = \mu(A) + \mu(B) \quad (A, B, A + B \in \mathcal{R}).$$

DEFINITION. *By a measure we mean a positive additive function on an algebra of sets.*

EXAMPLES

1. Lebesgue's measure on an algebra of simple sets in $\mathbb{R}^m$ is the total volume of rectangles constituting this set.
2. The number of elements in a finite set.

2.2. Integral Sums

3. Let $p_n \geq 0$ and $\Sigma p_n = 1$. Then
$$\mu(A) = \Sigma p_n \quad (A = \Sigma A_n),$$
where the summation is over those p_n corresponding to mutually disjoint A_n constituting A.

The measure is monotonic and subtractive:
$$\mu(A) \leq \mu(B) \quad (A \leq B),$$
$$\mu(B - A) = \mu(B) - \mu(A) \quad (A \leq B).$$

This follows from its positiveness and additivity.

In what follows, sets of measure zero will play an important role.

DEFINITION. *A set $Z \subseteq U$ is called negligible or a set of measure zero if for any $\varepsilon > 0$ there exists a sequence of simple sets A_n which covers Z and has the measures of A_n summing up to less than ε.*

Let $\mathcal{N} = \mathcal{N}(\mu)$ be the class of all sets of measure zero for measure μ. Clearly such classes will be different for different measures. In particular, if $\mu(A) = 0$ for $A = O$ only, then $\mathcal{N}$ consists of an empty set alone, $\mathcal{N} = \{O\}$.

It is straightforward to verify that

(1) every subset of a set of measure zero is itself a set of measure zero;

(2) the union of any countable family of sets of measure zero is also a set of measure zero.

If $U = \cup U_n$ for some $U_n \in \mathcal{R}$, the algebra $\mathcal{R}$ and measure $\mu : \mathcal{R} \to \mathbb{R}$ are called *locally bounded*. If $\mathcal{N}(\mu) \subseteq \mathcal{R}$ and $\mu(Z) = O$ $(Z \in \mathcal{N}(\mu))$, then μ is called *complete*. We deal below as a rule with locally bounded, complete measures.

2.2. Integral Sums

We define the integral sum to be a linear extension of a measure to the algebra of simple functions.

2.2.1. Definition

A number-valued function φ defined on a vector space $\mathcal{A}$ is called a *functional*. A functional φ endowed with the property
$$\varphi(af + bg) = a\varphi(f) + b\varphi(g) \quad (a, b \in \mathbb{F}; f, g \in \mathcal{A})$$
is called *linear*.

We will extend the measure μ defined on the algebra of simple sets, $\mathcal{R}$, up to a certain linear functional S on the algebra of simple functions, S. Since S is a linear span of $\mathcal{R}$, this can be done easily by taking for the values of S the corresponding linear combinations of the values of μ.

Since the functional S is to continue the measure μ, we need

$$S(X) = \mu(X) \quad (X \in \mathcal{R}).$$

Also, since S is to be linear, the relation

$$S(\Sigma a_j x_j) = \Sigma a_j S(X_j) = \Sigma a_j \mu(X_j)$$

must be valid for all numbers a_j and simple sets X_j, $j = 1, ..., m$. Hence if a linear functional $S : \mathcal{S} \to \mathbb{F}$ extending the measure $\mu : \mathcal{R} \to \mathbb{F}$ exists, then for a simple function $f = \Sigma\, a_j X_j$ its value $S(f)$ must be defined by

$$S(f) = \Sigma a_j \mu(X_j).$$

Consider two pairs of families a_j, X_j and b_k, Y_k ($j = 1, ..., m$; $k = 1, ..., n$) of numbers and simple sets. It is straightforward to verify that from

$$\Sigma a_j X_j = \Sigma b_k Y_k$$

it follows

$$\Sigma a_j \mu(X_j) = \Sigma b_k \mu(Y_k).$$

This enables us to formulate

DEFINITION. *The integral sum of a simple function f is*

$$S(f) = \Sigma a_j \mu(X_j)$$

for an arbitrary pair of families $a_j \in \mathbb{F}$, $X_j \in \mathcal{R}$ representing f.

On an algebra $\mathcal{S}$ of simple functions a functional S with value $S(f)$ for a simple function f is an *integral sum*.

To show explicitly that the measure involved is μ we will write $S(f, \mu)$ instead of $S(f)$ and $S(\mu)$ or $S(\cdot, \mu)$ instead of S and call it the integral sum with respect to measure μ.

THEOREM. *The integral sum S is the unique linear functional on $\mathcal{S}$ which extends the measure μ.*

PROOF. (i) S extends μ. Indeed, for $X \in \mathcal{R}$, $S(X) = \mu(X)$.
(ii) The functional S is linear since

$$S(af + bg) = \Sigma a a_j \mu(X_j) + \Sigma b b_k \mu(Y_k) = aS(f) + bS(g)$$

for $a, a_j, b, b_k \in \mathbb{F}$; $X_j, Y_k \in \mathcal{R}$ and both

$$f = \Sigma a_j X_j \quad \text{and} \quad g = \Sigma b_k Y_k \in \mathcal{S}.$$

2.2. Integral Sums

(iii) S is the unique linear functional on S which extends μ.

Indeed, if T is a linear functional on S extending μ, then, for any simple function $f = \Sigma a_j X_j$

$$T(f) = T(\Sigma a_j X_j) = \Sigma a_j T(X_j) = \Sigma a_j \mu(X_j) = S(f). \blacksquare$$

2.2.2. Order Properties

Integral sums also possess some important order properties.
A linear functional φ having

$$\varphi(f) \geq 0 \quad (f \geq 0, \ f \in \mathcal{A})$$

on a vector space $\mathcal{A}$ is called *positive*.

We note that the positiveness of a linear functional does not imply that all its values are positive. Moreover, if $\varphi(f) > 0$, then, for $g = -f$, $\varphi(g) = \varphi(-f) = -\varphi(f) < 0$.

LEMMA. *The positive linear functional φ is increasing*:

$$\varphi(f) \geq \varphi(g) \quad (f \leq g; \ f, g \in \mathcal{A}).$$

PROOF. From the linearity and positiveness of φ it follows that

$$\varphi(g) - \varphi(f) = \varphi(g - f) \geq 0$$

whenever $g - f \geq 0$. $\blacksquare$

EXAMPLE. The value of a function at a given point is a positive linear functional on the algebra of functions.

PROPOSITION 1. *An integral sum is a positive linear functional.*

PROOF. Let $f = \Sigma a_j X_j$, where a_j are nonzero values of f, and X_j are their preimages. Then

$$S(f) = \Sigma a_j \mu(X_j) \geq 0$$

when $a_j \geq 0$ and $\mu(X_j) \geq 0$. $\blacksquare$

The positiveness of an integral sum implies its monotoneity (like for any linear functional).

PROPOSITION 2. *The integral sum is increasing*:

$$S(f) \leq S(g) \quad (f \leq g; \ f, g \in S).$$

From the monotoneity of integral sums, we obtain a useful inequality for the integral sum $S(f,\mu)$ of a real-valued simple function f, real numbers a and b, and a simple set A.

PROPOSITION 3. *If* $a \leq f(x) \leq b$ *for* $x \in A$, *and* $f(x) = 0$ *for* $x \notin A$, *then*

$$a \cdot \mu(A) \leq S(f,\mu) \leq b \cdot \mu(A).$$

PROOF. Since $a \cdot A \leq f \leq b \cdot A$, then $S(a \cdot A) \leq S(f) \leq S(b \cdot A)$. It suffices to note that

$$S(a \cdot A) = a \cdot \mu(A) \quad \text{and} \quad S(b \cdot A) = b \cdot \mu(A). \blacksquare$$

The following frequently used inequality is the *triangle inequality for integral sums*. It holds for any simple function f.

PROPOSITION 4. $|S(f)| \leq S(|f|)$.

PROOF. Let $f = \Sigma a_j X_j$, where a_j are nonzero values of f and X_j are their preimages. Then $|f| = \Sigma |a_j| X_j$ since $f(x) = a_j$ and $|f(x)| = |a_j|$ for $x \in X_j$. Consequently, $|S(f)| = |\Sigma a_j \cdot \mu(X_j)| \leq \Sigma |a_j| \cdot \mu(X_j) = S(|f|). \blacksquare$

Let f be a simple function, c a positive number, and A a simple set. Propositions 3 and 4 lead us to

PROPOSITION 5. *If* $|f(x)| \leq c$ *for* $x \in A$, *and* $f(x) = 0$ *for* $x \notin A$, *then*

$$|S(f,\mu)| \leq c\mu(A).$$

PROOF. By Propositions 3 and 4

$$|S(f,\mu)| \leq S(|f|,\mu) \leq c\mu(A). \blacksquare$$

2.3. Convergence Almost Everywhere

This convergence is central to the model of integral under consideration.

2.3.1. The Concept of "Almost Everywhere"

The term "almost everywhere" means "except for a set of measure zero". This term and its synonyms are widely used in the theory of integral. In each particular case it is specified which property holds almost everywhere and with respect to what measure.

Consider a measure $\mu: \mathcal{R} \to \mathbb{R}$ and number-valued functions f, g, f_n ($n = 1, 2, \ldots$) on a set U (not necessary simple). As a rule, we assume that μ is complete.

2.3. Convergence Almost Everywhere

DEFINITION. (i) *Functions f and g are equal almost everywhere with respect to measure μ if $\mu \{ x : f(x) \neq g(x) \} = 0$.*

(ii) *A sequence of functions f_n converges to f almost everywhere with respect to measure μ if $\mu \{ x : f_n(x) \not\to f(x) \} = 0$.*

Here $\{ x : f(x) \neq g(x) \}$ is the set of all points $x \in U$ for which the values of f and g are different, and $\{ x : f_n(x) \not\to f(x) \}$ is the set of all points $x \in U$ for which the sequence of numbers $f_n(x)$ does not converge to $f(x)$.

If f and g are equal almost everywhere with respect to measure μ we shall write $f = g(\mu)$, and say *functions are equivalent* (with respect to measure μ) instead of *functions are equal almost everywhere*.

We shall write $f_n \to f(\mu)$ to signify that a sequence of functions f_n converges to f almost everywhere with respect to measure μ.

The principal difference between convergence almost everywhere and convergence everywhere (at each point) is that the limit function of a sequence convergent almost everywhere may not be unique.

Consider arbitrary functions f_n, f, g, and a measure μ.

PROPOSITION. (i) *If $f_n \to f(\mu)$ and $f_n \to g(\mu)$, then $f = g(\mu)$.*
(ii) *If $f_n \to f(\mu)$ and $f = g(\mu)$, then $f_n \to g(\mu)$.*

PROOF. From the above definitions it follows that

(1) $\{ x | f(x) \neq g(x) \} \subseteq \{ x | f_n(x) \not\to f(x) \} \cup \{ x | f_n(x) \not\to g(x) \}$.

(2) $\{ x | f_n(x) \not\to g(x) \} \subseteq \{ x | f_n(x) \not\to f(x) \} \cup \{ x | f(x) \neq g(x) \}$.

Also, a subset of a set of measure zero and the union of two sets of measure zero are also sets of measure zero. ∎

EXAMPLES.
1. The Dirichlet function is equal to zero almost everywhere (with respect to a Lebesgue measure).
2. Every countable set is equal to zero almost everywhere (with respect to a Lebesgue measure).
3. The sequence $\{ f_n(x) \} = x^n$ ($0 \leq x \leq 1$) converges to zero and to a Dirichlet function almost everywhere (with respect to a Lebesgue measure).

2.3.2. Continuity and Countable Additivity of Measures

Measures continuous above at zero are distinguished among others by their good analytic properties, in particular, countable additivity. These measures will be used later to define the integral.

DEFINITION. *If $\mu(X_n) \to 0$ for every (almost) decreasing sequence of simple sets X_n which converges to zero (almost) everywhere, then such a measure μ is called continuous above at zero.*

We will omit the words "above at zero" and write briefly a "continuous measure" if there is no risk of misunderstanding.

DEFINITION. *If $\mu(X) = \Sigma\, \mu(X_n)$ for every sequence of pairwise disjoint simple sets X_n whose unity is an (almost) simple set X, then such a μ is called countably additive.*

DEFINITION. *If $\mu(X) \leq \Sigma\, \mu(X_n)$ for every sequence of simple sets X_n (almost) covering the simple set X, then such a measure μ is called countably semiadditive.*

All these properties of measure turn out to be equivalent: if one of them holds, so are the other two. A suitable one may be adopted in each case. It is possible to get three similar definitions using the convergence almost everywhere.

THEOREM. *The following statements are equivalent.*
(i) *The measure μ is continuous above at zero.*
(ii) *The measure μ is countably additive.*
(iii) *The measure μ is countably semiadditive.*

The proof of this theorem is simple but rather long.

EXAMPLES.

1. By definition, the Lebesgue measure dx of a simple set $A = \Sigma\, A_i$ constituted by mutually disjoint intervals $A_i = [a_i, b_i[\subseteq \mathbb{R}\ (a_i \leq b_i)$ is equal to total length of these intervals:

$$dx(A) = \Sigma\, (b_i - a_i).$$

It is an easy matter to prove that the Lebesgue measure is countably semiadditive.

2. By definition for the simple set A Stieltjes' measure dg is equal to the sum of increments of an increasing function g on the mutually disjoint intervals $A_i = [a_i, b_i[$ that constitute A:

$$dg(A) = \Sigma\, [g(b_i) - g(a_i)].$$

It is easily proved that if the function g is continuous on the left, then the Stieltjes measure is countably semiadditive.

3. By definition, the measure $\nu = \mu \times \lambda$ for a rectangle $C = B \times A$ with sides A and B belonging to the domains of definition of measure μ and λ is the product of these measures

$$\nu(C) = \mu(B) \times \lambda(A)\ [\mu \times \lambda\, (B \times A) = \mu(B) \times \lambda(A)].$$

The measure ν of a simple set A is the sum of the measures of the sets that constitute A.

2.3.3. Convergence of Monotonic Sequences

Monotonic sequences of simple functions f_n and their integral sums $S(f_n)$ are central to the definition of integral, which will be formulated in the forthcoming section. As usually, continuous measures are considered.

2.3. Convergence Almost Everywhere

If a sequence $\{S(f_n)\}$ is bounded, we will say that the sequence $\{f_n\}$ is integrally bounded.

LEMMA (Levy). (i) *Every integrally bounded monotonic sequence of simple functions f_n converges almost everywhere to a certain function f.*
(ii) *If function f is simple, then $S(f) = \lim S(f_n)$.*

PROOF. (i) The general case is readily reduced to considering increasing sequences of simple functions $f_n \geq 0$ and numbers $S(f_n)$.
Consider the sets

$$Z_n(m) = \{x: f_n(x) \geq m\}, \quad Z(m) = \bigcup_n Z_n(m), \quad Z = \bigcap_m Z(m).$$

Note that if a sequence $\{f_n(x)\}$ does not converge to $f(x)$, then it is unbounded and $x \in Z$. By the conditions of the lemma, $S(f_n) \leq c$ for some $c > 0$. Therefore, $\mu(Z_n(m)) \leq m^{-1} \cdot c$ for all $m = 1, 2, \ldots$, $n = 1, 2 \ldots$. Since μ is continuous, then $\mu(Z(m)) \leq m^{-1} \cdot c$ for $m = 1, 2 \ldots$. By definition, $Z \subseteq Z(m)$ for any m. Hence, $\mu(Z) = 0$.
Thus, the sequence $f_n(x)$ converges to some number $f(x)$ for almost all x. The sequence f_n converges almost everywhere to a function f equal to $f(x)$ for $x \notin Z$, and to 0 for $x \in Z$.

(ii) The general case reduces easily to considering the decreasing sequences of functions $f_n \geq 0$ together with the function $f = 0$.
Consider $\varepsilon > 0$ and sequences of the simple sets

$$X_n(\varepsilon) = \{x: f_n(x) \geq \varepsilon\}.$$

Using the linearity and positiveness of integral sums, we obtain

$$0 \leq S(f_n) \leq c[\mu(X_n(\varepsilon)) + \varepsilon]$$

for some $c > 0$ and every $\varepsilon > 0$. From the convergence of $f_n \to 0$ almost everywhere and the continuity of measure μ it follows that $\mu(X_n(\varepsilon)) \to 0$ as $n \to \infty$, or, in view of the above inequality, that $S(f_n) \to 0$, as claimed. ∎

The second assertion of Levy's lemma is often interpreted as *monotonic sequence of simple functions f_n, converging almost everywhere to a simple function f, may be integrated termwise*.

Consider two integrally bounded increasing sequences of simple functions f_n, g_n and sequences of their integral sums $S(f_n)$, $S(g_n)$. By Levy's lemma, $\lim f_n$ and $\lim g_n$ exist almost everywhere. In view of monotoneity and boundedness of the sequences of $S(f_n)$ and $S(g_n)$, $\lim S(f_n)$ and $\lim S(g_n)$ also exist.

PROPOSITION. *If $\lim f_n \leq \lim g_n$ almost everywhere, then $\lim S(f_n) \leq \lim S(g_n)$.*

Since the equality of the limits is equivalent to two opposite inequalities, this proposition entails the following result.

COROLLARY. *If* $\lim f_n = \lim g_n$ *almost everywhere, then* $\lim S(f_n) = \lim S(g_n)$.

Note that if the limit functions $f = \lim f_n$, $g = \lim g_n$ are equivalent to simple functions, then the assertion of the corollary follows directly from Levy's lemma.

2.4. Integrals

The notion of integral is rather complicated, therefore we will define it inductively for successively expanding classes of functions – first for sets identified with their indicators, then for their linear combinations, and, finally, for some functions approximated by such combinations.

Step 1. An algebra of simple sets, $\mathcal{R}$, and a continuous measure μ on it are selected. By definition, the integral of a simple set A is equal to its measure $\mu(A)$.

Step 2. An albegra $\mathcal{S}$ of simple functions which are linear combinations of simple sets is considered. The integral sum S is defined to be a linear extension of μ to $\mathcal{S}$. By definition, the integral of a simple function f is the integral sum $S(f)$.

Step 3. The algebra $\mathcal{M}$ of measurable functions which are the limits of almost everywhere convergent sequences of simple functions is considered. In $\mathcal{M}$, we choose the subspace $\mathcal{L}$ of integrable functions, which are the limits of certain special sequences of simple functions f_n, for which the sequences of integral sums $S(f_n)$ converge. By definition, the integral of the function $f = \lim f_n$ equals the number $\int f = \lim S(f_n)$. This integral is a continuous extension of the integral sum S to $\mathcal{L}$.

As a result, we arrive at the classical definition of integral as a limit of integral sums. The first two steps have been made. It remains to make the third more difficult step.

Now we summarize the notation for the following consideration:

U is abstract set;

$\mathcal{F}$ is algebra of all number-valued (real or complex) functions on U;

$\mathcal{R}$ is some algebra of subsets of U that are called simple sets and identified with their indicators, $U = \cup U_n$, $U_n \in \mathcal{R}$;

μ is measure on $\mathcal{R}$ (locally bounded, complete, and continuous from above at zero);

S is integral sum on $\mathcal{S}$, which extends measure μ (denoted also by $S(\mu)$ or $S(\cdot, \mu)$);

$\mathcal{S}$ is span of the class $\mathcal{R}$ of simple sets in $\mathcal{F}$, which consists of linear combinations of simple sets, called simple functions (they form a subalgebra of $\mathcal{F}$);

2.4. Integrals

$\mathcal{N}$ is class of sets of measure zero with respect to μ (denoted also by $\mathcal{N}(\mu)$). As a rule, it is assumed that $\mathcal{N}(\mu) \subseteq \mathcal{R}$ and μ is complete.

2.4.1. Definition of Measurable Functions

Integrable functions will be defined as the limits of almost everywhere convergent sequences of simple functions. Therefore, we consider first limits of such sequences.

DEFINITION. *Each function equal to the limit of an almost everywhere convergent sequence of simple functions is called measurable.*

By this definition, each function equivalent to a measurable one is measurable, too. Hence, each sequence of simple functions convergent almost everywhere defines a whole class of measurable functions equivalent to each other.

On the other hand, an entire class of sequences of simple functions may converge to each measurable function: if $f_n \to f$ and $g_n \to 0$, then $h_n = g_n + f_n \to f$.

Thus, the correspondence between almost everywhere convergent sequences of simple functions and measurable functions in not one–one.

Denote the set of all measurable functions by $\mathcal{M}$ or $\bar{\mathcal{S}}$. Obviously, $\mathcal{S} \subseteq \mathcal{M}$, that is each simple function is measurable. (It is the limit of the constant sequence with terms equal to it: if $f_n = f \in \mathcal{S}$ the $f_n \to f$ and does so everywhere.)

PROPOSITION. *Measurable functions form an algebra.*

PROOF. We need to prove that for measurable f and g and number c, the functions $f + g$, fg and cf are measurable.

Let f_n, g_n be simple functions and $f_n \to f$, $g_n \to g$. Then $f_n + g_n$, $f_n g_n$, cf_n are simple functions and so $f_n + g_n \to f + g$, $f_n g_n \to fg$, $cf_n \to cf$. Hence, $f + g$, fg, and cf are measurable. ∎

COROLLARY. *The function $f = g + ih$ is measurable if and only if its real and imaginary parts are measurable functions.*

REMARK. Different algebras of simple sets and different measures evolve different albegras of measurable functions.

In particular, if $\mathcal{R} = \{0, U\}$, then the albegra of measurable functions coincides with the algebra of simple functions, $\mathcal{M} = \mathcal{S}$. Measurable functions for such $\mathcal{R}$ are constants alone.

Sets are called measurable if their indicators are measurable. This corresponds to the convention to identify sets with their indicators. Since measurable functions form an algebra of functions, measurable sets form an algebra of sets. The class of all measurable sets will be denoted by $\overline{\mathcal{R}}$. Thus, by definition, $\overline{\mathcal{R}} = \overline{\mathcal{S}} \cap \mathcal{P} = \mathcal{M} \cap \mathcal{P}$. Obviously, simple sets are measurable: $\mathcal{R} = \mathcal{S} \cap \mathcal{P} \subseteq \overline{\mathcal{S}} \cap \mathcal{P} = \overline{\mathcal{R}}$. Each set equivalent to a measurable one is measurable, too.

2.4.2. Approximating Sequences

In order to define the integral of function f (to which a sequence of simple functions f_n converges almost everywhere) as the limit of the sequence of the integral sums $S(f_n)$, we need that $\lim S(f_n)$ exist and be the same for all such sequences of f_n.

Conditions to ensure the existence and uniqueness of the limit of the integral sums can be selected in different ways. Daniel's model is often used, in which the central role is played by increasing sequences of simple functions with bounded sequences of the integral sums. The advantage of this model is that it reduces to rather easy operations with increasing sequences. Also, it is easier to verify the boundedness of the sequences than their convergence, and for monotonic sequences, convergence and boundedness are equivalent.

Consider a modification of Daniel's model. First we describe the class of sequences having integrable functions as their limits. We introduce designations which will be used frequently. For every real number y we define its positive and negative parts by

$$y^+ = \max(0, y) = 2^{-1}(|y| + y),$$

$$y^- = \max(0, -y) = 2^{-1}(|y| - y).$$

Clearly, $y^+ \geq 0$, $y^- \geq 0$, $y = y^+ - y^-$ and $|y| = y^+ + y^-$.

For every real-valued function φ we define its positive and negative parts as

$$\varphi^+(x) = (\varphi(x))^+ = \max(0, \varphi(x)) = 2^{-1}(|\varphi(x)| + \varphi(x)),$$

$$\varphi^-(x) = (\varphi(x))^- = \max(0, -\varphi(x)) = 2^{-1}(|\varphi(x)| - \varphi(x)).$$

It is clear that $\varphi^+ \geq 0$, $\varphi^- \geq 0$, $\varphi = \varphi^+ - \varphi^-$ and $|\varphi| = \varphi^+ + \varphi^-$. (Here $|\varphi|$ denotes a function with values $|\varphi|(x) = |\varphi(x)|$.) If $\varphi^+(x) > 0$, then $\varphi^-(x) = 0$, and conversely.

Each number-valued function can be represented by a linear combination of positive-valued functions

$$f = g + ih = (g^+ - g^-) + i(h^+ - h^-)$$

where g and h are the real and imaginary parts of f. We will call positive

2.4. Integrals

and negative parts g^+, g^-, h^+, h^- of g, h the *positive components*, or, briefly, *components* of f.

Obviously, if f is simple, then its components g^+, g^-, h^+, h^- are simple too.

Consider the sequences of funcitons

$$f_n = g_n + ih_n = (g_n^+ - g_n^-) + i(h_n^+ - h_n^-)$$

where $g_n = \operatorname{Re} f_n$, $h_n = \operatorname{Im} f_n$ and g_n^+, g_n^-; h_n^+, h_n^- are their positive and negative parts.

LEMMA.

$$f_n \to f \Leftrightarrow g_n^+ \to g^+, \ g_n^- \to g^-, \ h_n^+ \to h^+, \ h_n^- \to h^-.$$

These sequences of functions converge in *terms of components*: f_n converges to f if and only if the sequences of the components of f_n converge to the respective components of f.

To define the integral, we distinguish a special class of sequences of simple functions that will be called approximating sequences.

DEFINITION. *Every linear combination of integrally bounded increasing sequences of simple functions is called an approximating sequence.*

The term *approximating sequence* is very special here, and we introduced it just to simplify the text.

From Levy's lemma it follows that approximating sequences are integrally bounded and converge almost everywhere to certain functions together with the constituting increasing sequences. We will call these increasing sequences conventionally the *monotonic components* of the approximating sequences constituted by these components. The representation of every approximating sequence as a linear combination of monotonic components is not unique.

Obviously, a linear combination of approximating sequences is also an approximating sequence. Approximating sequences form a linear hull of the set of integrally bounded increasing sequences of simple functions in the algebra of all sequences of number-valued functions on U.

Note that each integrally bounded decreasing sequence of simple functions is approximating, and each approximating sequence is equal to a linear combination of integrally bounded monotonic sequences of simple functions.

In the next section we will use the following lemma.

LEMMA. *If an increasing sequence of simple functions r_n is integrally bounded, then a sequence of $|r_n|$ is also integrally bounded.*

PROOF. If a sequence of functions r_n increases, then the sequence of their positive parts r_n^+ increases too, and the sequence of negative parts r_n^- decreases. Therefore

$$r_n^+ = r_n + r_n^- \leq r_1 + r_1^-, \quad S(r_n^+) \leq S(r_n) + S(r_1^-);$$
$$r_n^- \leq r_1^-, \quad\quad\quad\quad\quad\quad S(r_n^-) \leq S(r_1^-);$$
$$|r_n| = r_n^+ + r_n^-, \quad\quad\quad S(|r_n|) \leq S(r_n) + 2S(r_1^-).$$

The sequences of $r_n^+, r_n^- \mid r_n \mid$ are integrally bounded. ∎

The integral boundedness of a monotonic sequence of simple functions can be said to be equivalent to the *absolute integral boundedness* of this sequence.

Every approximating sequence $f_n = \Sigma a_j f_{jn}$ constituted by increasing sequences of simple functions f_{jn} with coefficients $a_j = (b_j^+ - b_j^-) + i \cdot (c_j^+ - c_j^-)$ is equal to the linear combination $f_n = (g_n^+ - g_n^-) + i \cdot (h_n^+ - h_n^-)$ of the integrally bounded increasing sequences of simple functions

$$g_n^+ = \Sigma b_j^+ f_{jn}, \quad g_n^- = \Sigma b_j^- f_{jn},$$
$$h_n^+ = \Sigma c_j^+ f_{jn}, \quad h_n^- = \Sigma c_j^- f_{jn}.$$

Here $b_j^+, b_j^-, c_j^+, c_j^- \geq 0$ are the standard components of the numbers a_j, and the functions $g_n^+, g_n^-, h_n^+, h_n^-$ need not be positive components of the functions f_n. They just constitute new monotonic components of the sequence f_n, with the standard coefficients $1, -1, i, -i$. What is meant is always clear from the context.

We call *standard* the representation of f and f_n as linear combinations with coefficients $1, -1, i, -i$. Note that under this representation

$$g = \operatorname{Re} f = g^+ - g^-, \quad h = \operatorname{Im} f = h^+ - h^-.$$

We will say that an approximating sequence of simple functions $f_n = \Sigma a_j f_{jn}$ converges to the function $f = \Sigma a_j f_j$ in *terms of components* if $f_{jn} \uparrow f_j$ almost everywhere for each j. Clearly convergence in terms of components entails the convergence $f_n \to f$ almost everywhere.

2.4.3. Integrable Functions

Convergence in terms of components enables a simple description of the class of functions for which the integral will be defined.

2.4. Integrals

DEFINITION. (i) *A real-valued function is called monotonically integrable function if an integrally bounded increasing sequence of simple functions converges to it almost everywhere.*
(ii) *A number-valued function equal to a linear combination of monotonically integrable functions is called an integrable function.*

By this definition integrable functions form a linear hull of the set of limit functions for componentwise convergent approximating sequences. A function f is integrable if a certain approximating sequence of simple functions f_n converges to it in terms of components that is, if

$$f = \sum a_j f_j, \quad f_n = \sum a_j f_{jn}, \quad f_{jn} \uparrow f_j$$

for certain numbers a_j and simple functions f_{jn}. In place of arbitrary combinations one may consider the standard linear combinations with coefficients $1, -1, i, -i$.

From the definition, it follows that each function equivalent to an integrable one is integrable too. Therefore in the representations of considered linear combinations, equalities may be deemed equivalencies. The monotoneity of sequences and simplicity of constituent functions may also be required accurate to equivalence, that is, sequences of the functions $\tilde{f}_n$ equivalent to the simple functions from an approximating sequence may also be deemed approximating sequences.

Let $\mathcal{H}$ and $\mathcal{L}$ be sets of all monotonically integrable functions and integrable functions, respectively. Whenever an explicit indication of the set U, on which considered functions are defined, the chosen algebra $\mathcal{R}$ of simple sets, and the measure μ is required, we will write $\mathcal{L}(\mu)$, $\mathcal{L}(U, \mu)$ or $\mathcal{L}(U, \mathcal{R}, \mu)$ instead of $\mathcal{L}$. Obviously, $\mathcal{S} \subseteq \mathcal{L} \subseteq \mathcal{M}$ because each simple function is integrable and each integrable function is measurable.

Another implication of this definition is that every linear combination of integrable functions is also integrable. Hence, $\mathcal{L}$ is a vector space in the algebra $\mathcal{M}$. Examples indicate that the product of integrable functions can be nonintegrable.

Now we prove some useful propositions.

LEMMA 1. *A function $f = g + ih$ is integrable if and only if its real and imaginary parts, $g = \mathrm{Re} f$ and $h = \mathrm{Im} f$, are integrable.*

PROOF. We need to prove only that the integrability of f implies the integrability of g and h. If f is integrable, then $f = (g^+ - g^-) + i(h^+ - h^-)$ for some monotonically integrable functions g^+, g^-, h^+, h^-, since $g = g^+ - g^-$, $h = h^+ - h^-$, both g and h are integrable. ∎

LEMMA 2. *The maximum $p \vee q$ of monotonically integrable functions p and q is a monotonically integrable function.*

PROOF. Let $p_n \uparrow p$, $q_n \uparrow q$ for some simple functions p_n, q_n.

Then
$$p_n \leq p_{n+1} \leq p_{n+1} \vee q_{n+1},$$
$$q_n \leq q_{n+1} \leq p_{n+1} \vee q_{n+1},$$
$$p_n \vee q_n \leq p_{n+1} \vee q_{n+1}.$$

Since the function $\vee$ is continuous in all its variables and $p_n \to p$, $q_n \to q$, it follows $p_n \vee q_n \uparrow p \vee q$.

Clearly the functions $p_n \vee q_n$ are simple. By the lemma of Section 2.4.2, sequences $|p_n|$, $|q_n|$ of functions are integrally bounded together with p_n, q_n. Since $|p_n \vee q_n| \leq |p_n| + |q_n|$, then $|S(p_n \vee q_n)| \leq S(|p_n \vee q_n|) \leq S(|p_n|) + S(|q_n|)$, and the sequence of functions $p_n \vee q_n$ is integrally bounded. Hence, the limit function $p \vee q$ is monotonically integrable. ∎

PROPOSITION. *A function f is integrable if and only if its positive components g^+, g^-, h^+, h^- are integrable.*

PROOF. Since $f = (g^+ - g^-) + i(h^+ - h^-)$, the integrability of g^+, g^-, h^+, h^- implies the integrability of f. Now we need to prove the converse.

By Lemma 1 the integrability of f implies the integrability of $g = g^+ - g^-$, $h = h^+ - h^-$. Hence, $g = p - q$ and $h = r - s$ for some p, q, r, $s \in \mathcal{K}$. By Lemma 2 the functions $p \vee q$ and $r \vee s$ are integrable, and in view of $g^+ = p \vee q - q$, $g^- = p \vee q - p$, $h^+ = r \vee s - s$, $h^- = r \vee s - r$ the integrability of $p \vee q$, $r \vee s$ and p, q, r, s implies the integrability of g^+, g^-, h^+, h^-. ∎

COROLLARY. *The maximum and minimum of a finite family of integrable real-valued functions are integrable.*

PROOF. For a pair of functions f and g this assertion follows from the proposition proved above and the equalities
$$f \vee g = f + (g - f)^+, \quad f \wedge g = f - (g - f)^-,$$
where the righthand sides stand for the positive and negative parts of the difference $g - f$.

The general case is proved by induction. ∎

Note that from the proposition proved above it follows that the absolute value $|f| = f^+ + f^-$ of the real-valued integrable function $f = f^+ - f^-$ is also integrable.

2.4.4. Definition of Integral

The integral of a function is defined as the limit of a sequence of integral sums of simple functions which approximate the given function

2.4. Integrals

in terms of components. We begin by formulating a definition, then prove that it is correct.

DEFINITION. *The integral of a function $f \in \mathcal{L}$ is the limit of a sequence of integral sums $S(f_n)$*

$$\int f = \lim S(f_n)$$

for every approximating sequence of simple functions $f_n \in S$, which converges to f almost everywhere.

The following theorem on the existence and uniqueness of the integral ensures the correctness of the given definition.

THEOREM. *For every integrable function there exists a unique integral.*

PROOF. **Existence.** Consider an integrable function $f = (g^+ - g^-) + i(h^+ - h^-)$ and a sequence of simple functions $f_n = (g_n^+ - g_n^-) + i(h_n^+ - h_n^-)$, which converges to it in terms of components. Since the sequences $S(g_n^+)$, $S(g_n^-)$, $S(h_n^+)$, $S(h_n^-)$ are monotonic and bounded, the limits $b^+ = \lim S(g_n^+)$, $b^- = \lim S(g_n^-)$, $c^+ = \lim S(h_n^+)$, $c^- = \lim S(h_n^-)$ exist. Hence so does

$$a = \lim S(f_n) = (b^+ + b^-) + i(c^+ - c^-).$$

By definition this limit is equal to $\int f$.

Uniqueness. Consider another sequence of simple functions $p_n = (q_n^+ - q_n^-) + i(r_n^+ - r_n^-)$ which converges to $f = (q^+ - q^-) + i(r^+ - r^-)$ in terms of components. Because the sequences $S(q_n^+)$, $S(q_n^-)$, $S(r_n^+)$, $S(r_n^-)$ are monotonic and bounded there exist the limits $u^+ = \lim S(q_n^+)$, $u^- = \lim S(q_n^-)$, $v^+ = \lim S(r_n^+)$, $v^- = \lim S(r_n^-)$. Since

$$\operatorname{Re} f = g^+ - g^- = q^+ - q^-, \quad \operatorname{Im} f = h^+ - h^- = r^+ - r^-;$$

and

$$g^+ + q^- = q^+ + g^-, \quad h^+ + r^- = r^+ + h^-,$$

we have

$$\lim(g_n^+ + q_n^-) = \lim(q_n^+ + g_n^-),$$
$$\lim(h_n^+ + r_n^-) = \lim(r_n^+ + h_n^-).$$

Applying Levy's lemma and taking into account the linearity of integral sums we find

$$b^+ + u^- = u^+ + b^-, \quad c^+ + v^- = v^+ + c^-,$$
$$b = b^+ - b^- = u^+ - u^- = u,$$

and

$$c = c^+ - c^- = v^+ - v^- = v.$$

Hence,

$$\lim S(f_n) = a = b + ic = u + iv = \lim S(p_n),$$

that is, the integral of f is unique. ∎

The proved independence of the integral of the choice of an approximating sequence of simple functions means that we are free to choose this sequence in accordance with the problem conditions.

Summarizing, the functional $\int$ on the vector space $\mathscr{L}$ of integrable functions associating the number $\int f$ with a function $f \in \mathscr{L}$ is called *the integral* of f. The measure μ under consideration is indicated when necessary by writing $\int f \mu, \int f \, d\mu, \int f(x) \, d\mu(x)$ instead of $\int f$ and saying the integral with respect to μ. If needed, the set U on which the integrand is defined is indicated under the integral symbol.

The algebraic properties of integrals are described by the following proposition.

PROPOSITION. *The integral is a linear functional on $\mathscr{L}$, which extends the integral sum S.*

PROOF. (i) $\int$ extends S. Indeed,

$$\int f = \lim S(f_n) = S(f) \quad \text{for} \quad f_n = f \in S.$$

(ii) The functional $\int$ is linear, since for numbers a and b, the approximating functions $f, g \in \mathscr{L}$ of sequences f_n, g_n, of simple functions

$$\int (af + bg) = \lim S(af_n + bg_n)$$
$$= a \cdot \lim S(f_n) + b \cdot \lim S(g_n) = a \cdot \int f + b \cdot \int g. \quad \blacksquare$$

Integral sums and measures are related by one-to-one correspondence: each integral sum extends some measure and, conversely, each measure may be extended up to some integral sum, different integral sums extending different measures. Therefore, extending integral sums the integrals extend the measures as well. Thus, the integral $\int$ is a linear functional on $\mathscr{L}$ extending the measure μ.

2.4. Integrals

2.4.5. Order Properties of Integrals

In addition to linearity, integrals possess several important order properties which result from the order properties of integral sums.

PROPOSITON 1. *An integral is a positive linear functional.*

PROOF. Let $f \in \mathscr{L}$ and $f \geq 0$. Then $f = g^+ - g^-$, $g^+ \geq g^-$ for some functions $g^+, g^- \in \mathscr{H}$, which are limit functions of almost everywhere convergent, integrally bounded, increasing sequences of simple functions g_n^+, g_n^-. By the proposition of Section 2.3.3, from the inequality $\lim \bar{g_n} = g^- \leq g^+ = \lim g_n^+$, which holds almost everywhere, it follows

$$\int g^- = \lim S(g_n^-) \leq \lim S(g_n^+) = \int g^+,$$

hence,

$$\int f = \int (g^+ - g^-) = \int g^+ - \int g^- \geq 0.$$

The positiveness of the integral is established. ∎

The positiveness of integrals implies their monotoneity (like for any linear functional).

PROPOSITION 2. *The integral is increasing, i.e., if $f, g \in \mathscr{L}$ and $f \leq g$, then $\int f \leq \int g$.*

From the monotoneity of the integral we obtain a useful inequality for the integral $\int f \mu$ of a real-valued integrable function f, two real numbers a and b, and the set of integration A. Let the integral $\int A$ be called the measure of A, denoted $\mu(A)$. Thus, by definition

$$\mu(A) = \int A \quad (A \in \mathscr{L}).$$

In particular, for simple sets $A \in \mathscr{R}$, we obtain the former value of $\mu(A)$.

PROPOSITION 3. *If $a \leq f(x) \leq b$ for $x \in A$, and $f(x) = 0$ for $x \notin A$, then*

$$a\mu(A) \leq \int f\mu \leq b\mu(A).$$

PROOF. Given $aA \leq f \leq bA$, we have

$$\int aA \leq \int f \leq \int bA.$$

It remains to notice that

$$\int aA = a\int A = a\mu(A),$$
$$\int bA = b\int A = b\mu(A). \quad \blacksquare$$

Now we establish the frequently used *triangle inequality for integrals*, which holds for each integrable function.

PROPOSITION 4. $\left| \int f \right| \leq \int |f|$.

PROOF. We shall prove this proposition for real-valued functions. Consider a real-valued integrable function $f = g^+ - g^-$ with positive g^+, g^-. By the proposition of Section 2.4.3, both g^+ and g^- are integrable along with $|f| = g^+ + g^-$. By virtue of the linearity and positiveness of integrals

$$\left| \int f \right| = \left| \int g^+ - \int g^- \right| \leq \int g^+ + \int g^- = \int |f|.$$

This proves the triangle inequality for integrals. ∎

Consider an integrable function f, a positive number c and an integrable set A. Propositions 3 and 4 lead us to

PROPOSITION 5. *If $|f(x)| \leq c$ for $x \in A$ and $f(x) = 0$ for $x \notin A$, then*

$$\left| \int f\mu \right| \leq c\mu(A).$$

PROOF. We prove this proposition for real-valued functions. From Propositions 3 and 4 it follows that

$$\left| \int f\mu \right| \leq \int |f|\mu \leq c\mu(A). \quad \blacksquare$$

2.5. Limit Theorems

Here we prove three principal theorems on passage to the limit under the integral sign.

2.5.1. Levy's Theorem

In Section 2.3.3 we proved Levy's lemma for integral sums. From this lemma we deduce an analogous statement for integrals.

Consider a sequence of integrable functions f_n. If the sequence of integrals $\int f_n$ is bounded, we will say that the sequence of functions f_n is *integrally bounded*: $\left| \int f_n \right| \leq c$ for all n and a given number $c > 0$. If the sequence f_n converges almost everywhere to an integrable function f and $\int f_n \to \int f$, we will say that this sequence *may be integrated term-by-term*. The termwise integrability of the sequence f_n is equivalent to $\lim \int f_n = \int \lim f_n$.

THEOREM (Levy). *If a monotone sequence of integrable functions f_n is integrally bounded, then it converges almost everywhere to an integrable function f and may be integrated term-by-term.*

2.5. Limit Theorems

PROOF. It will suffice to prove the theorem for the case of an integrally bounded increasing sequence of monotonically integrable functions f_n.

Every such sequence converges almost everywhere to an integrable function f. Indeed, by definition every f_n is approximated by an increasing sequence of simple functions f_{mn} ($m = 1, 2, ...$) and $\int f_n = \lim\limits_{m \to \infty} S(f_{mn})$. Note that $f_{mn} \leq f_n \leq f_p$ for any m whenever $n \leq p$. Consider a sequence of the simple functions $g_p = \max \{f_{pn} : n \leq p\} \leq f_p$. This sequence increases along with the sequences $f_{m1}, ..., f_{mp}$ ($m = 1, 2, ...$) and is integrally bounded together with the sequence f_p ($p = 1, 2, ...$).

By Levy's lemma there exists a function f to which an integrally bounded increasing sequence of simple functions g_p converges almost everywhere. Hence, f is an integrable function.

Since $f_{pn} \leq g_p, f_{pn} \to f_n$ and $g_p \to f$ as $p \to \infty$, we have $f_n \leq f$. Also, since $g_n \leq f_n, g_n \leq f_n \leq f$. This inequality and the convergence $g_n \to f$ ensure the desired convergence $f_n \to f$.

The term-by-term integrability of the sequence f_n is easily proved. The sequence g_n approximates f and so $\int f = \lim S(g_n)$. From $g_n \leq f_n \leq f$ and $\int g_n = S(g_n)$ it follows that

$$\int g_n \leq \int f_n \leq \int f, \quad S(g_n) \leq \int f_n \leq \int f.$$

These inequalities together with the equality $\int f = \lim S(g_n)$ yield the desired equality $\lim \int f_n = \int f$, and Levy's theorem is established. ∎

COROLLARY. *If a monotone sequence of integrable functions f_n is integrally bounded and converges almost everywhere to a function g, then this function is integrable and has*

$$\int g = \lim \int f_n.$$

PROOF. By the Levy's theorem there exists an integrable function f such that f_n converges to it and

$$\int f = \lim \int f_n.$$

If f_n also converges almost everywhere to g, then g is equivalent to and integrable with f, and has the same integral. ∎

2.5.2. Fatou's Theorem

Consider a sequence of real numbers b_n, which is bounded below, so the lower sequence $a_n = \inf\limits_{p \geq n} b_p$ may be defined. The sequence

a_n is increasing, and if it is bounded above, then the limit $a = \lim a_n$ exists. We call it the *lower limit* of the sequence b_n and denote it by $\liminf b_n$.

In exactly the same fashion the upper sequence $c_n = \sup_{p \geq n} b_n$ is defined for a sequence of real numbers b_n bounded above. The sequence c_n is decreasing and if it is bounded below then the limit $c = \lim c_n$ exists. We call it the *upper limit* of b_n and denote by $\limsup b_n$.

For bounded sequences of real numbers b_n and real number b we have the following theorem on upper and lower limits

$$\lim b_n = b \Leftrightarrow \liminf b_n = \limsup b_n = b.$$

Lower and upper sequences f_n and h_n and the corresponding lower and upper limits f and h for a sequence of real-valued functions g_n are defined in terms of the sequence of their values $g_n(x)$. If

$$f_n(x) = \inf_{p \geq n} g_p(x), \quad h_n(x) = \sup_{p \geq n} g_p(x),$$
$$f(x) = \lim f_n(x), \quad h(x) = \lim h_n(x).$$

We recognize that if they are defined for almost all x rather than all x, then we are allowed to take as f_n, h_n, f, h any functions with given values. In this case the upper and lower limits are said to be defined accurate to equivalence.

The theorem on the upper and lower limits of bounded sequences of real numbers suggests an analogous assertion for sequences of functions:

$$\lim g_n = g \Leftrightarrow \liminf g_n = \limsup g_n = g.$$

(Here all the equalities hold almost everywhere. The sequence of real-valued functions g_n is bounded above and below by some real-valued functions.)

LEMMA (Fatou). *For every integrally bounded sequence of functions $g_n \geq 0$ the lower limit $f = \liminf g_n$ is an integrable function and*

$$\int \liminf g_n \leq \liminf \int g_n.$$

PROOF. Consider $f_{qn} = \min_{q \geq p \geq n} g_p$ for all numbers n and $q \geq n$. The sequence f_{qn} ($q = n, n+1 \ldots$) is clearly decreasing. Since by the conditions of the lemma $g_p \geq 0$, then $f_{qn} \geq 0$. Consequently, there exists the limit $f_n = \lim_{q \to \infty} f_{qn} = \inf_{p \geq n} g_p$. The functions $f_{qn} = \min_{q \geq p \geq n} g_p$ are integrable along with g_p. Since $0 \leq f_{qn} \leq f_{nn}$ and $0 \leq \int f_{qn} \leq \int f_{nn}$, the sequence f_{qn} with $q = n, n+1, \ldots$ is integrally bounded. From the

2.5. Limit Theorems

Levy's theorem it follows that the limit function f_n is integrable and

$$\int f_n = \lim_{q \to \infty} \int f_{qn}.$$

Note that $0 \leq f_n \leq f_{qn} \leq g_n$ for $q \geq n$. Given the sequence of functions g_n is integrally bounded, $0 \leq \int f_n \leq \int f_{qn} \leq \int g_n \leq c$ for $q \geq n$ and some $c > 0$. At the same time the sequence $f_n = \inf_{p \geq n} g_p$ ($n = 1, 2, \ldots$) increases. By the Levy's theorem it converges almost everywhere to an integrable function f and $\int f = \lim \int f_n$. The theorem on upper and lower limits tells us that $f = \liminf g_n$. On the other hand, since $f_n = \inf_{p \geq n} g_p \leq g_p$ for each $p \geq n$, we have

$$\int f_n \leq \int g_p = b_p, \quad p \geq n,$$

$$\int f_n \leq \inf_{p \geq n} b_p = a_n.$$

Consequently,

$$\int \liminf g_n = \int f = \lim \int f_n \leq \lim a_n = \liminf b_n = \liminf \int g_n.$$

The proof of Fatou's lemma is completed. ∎

Fatou's theorem is obtained from Fatou's lemma by adding a dual assertion for upper limits and replacing the integral boundedness of the sequence g_n with its absolute boundedness by some integrable function $\bar{g} \geq 0$.

THEOREM (Fatou). *For every integrally bounded sequence of real-valued functions g_n whose absolute values $|g_n|$ are bounded by an integrable function $\bar{g} \geq 0$, the lower limit $f = \liminf g_n$ and the upper limit $h = \limsup g_n$ are integrable functions. Besides,*

$$\int \liminf g_n \leq \liminf \int g_n \leq \limsup \int g_n \leq \int \limsup g_n.$$

PROOF. To prove this theorem, it is sufficient to apply Fatou's lemma to the sequences

$$u_n = \bar{g} + g_n, \quad v_n = \bar{g} - g_n.$$

The real-valued functions u_n, v_n are integrable together with g_n and $\bar{g}$. From $|g_n| \leq \bar{g}$ it follows that

$$0 \leq u_n \leq 2\bar{g}, \quad 0 \leq v_n \leq 2\bar{g}, \quad 0 \leq \int u_n \leq 2\int \bar{g}, \text{ and } 0 \leq \int v_n \leq 2\int \bar{g}.$$

Thus, the sequences u_n, v_n meet the conditions of Fatou's lemma. Hence, the functions

$$\liminf u_n = \bar{g} + \liminf g_n,$$
$$\liminf v_n = \bar{g} - \limsup g_n$$

are integrable and

$$\int \liminf u_n \leq \liminf \int u_n,$$
$$\int \liminf v_n \leq \liminf \int v_n.$$

Consequently, the functions

$$\liminf g_n = \liminf u_n - \bar{g},$$
$$\limsup g_n = \bar{g} - \liminf v_n$$

are also integrable and

$$\int \liminf g_n = \int \liminf u_n - \int \bar{g} \leq \liminf \int u_n - \int \bar{g}$$
$$= \liminf \left(\int \bar{g} + \int g_n \right) - \int \bar{g} = \liminf \int g_n,$$
$$\int \limsup g_n = \int \bar{g} - \int \liminf v_n \geq \int \bar{g} - \liminf \int v_n$$
$$= \int \bar{g} - \liminf \left(\int \bar{g} - \int g_n \right) = \limsup \int g_n.$$

The inequality $\liminf \int g_n \leq \limsup \int g_n$ follows from the definitions of lower and upper limits for number-valued sequences. ∎

2.5.3. Integrability Criterion

A real-valued function $f = g^+ - g^-$ is integrable if and only if its positive components g^+, g^- are integrable. Hence, the integrability of f implies the integrability of $|f| = g^+ + g^-$.

From Fatou's theorem we have

LEMMA. *If a sequence of integrable number-valued functions f_n bounded in their absolute values by a nonnegative function $\bar{f} \geq 0$ converges almost everywhere to a function f, then f is integrable and $\int f = \lim \int f_n$.*

PROOF. It suffices to prove this lemma for real-valued functions.

Let the functions $f_n = g_n$ and $f = g$ be real-valued. Then the condition $\lim f_n = f$ is equivalent to the identities $\liminf g_n = \limsup g_n = g$. Since the sequence g_n satisfies the conditions of Fatou's theorem for $\bar{g} = \bar{f}$,

2.5. Limit Theorems

it follows that g is integrable and

$$\int g \leq \liminf \int g_n \leq \limsup \int g_n \leq \int g,$$

whence

$$\int f = \int g = \lim \int g_n = \int \lim f_n.$$

The lemma is established. ∎

Now we prove a convenient and useful

Integrability Criterion. *A number-valued function f is integrable if and only if f is measurable and there exists an integrable function $\bar{f} \geq 0$ such that $|f| \leq \bar{f}$.*

PROOF. **Sufficiency.** The integrability of $f = g + ih$ is tantamount to the integrability of its real and imaginary parts, g and h. The measurability of f is tantamount to the measurability of g and h. The inequality $|f| \leq \bar{f}$ implies $|g| \leq \bar{f}$ and $|h| \leq \bar{f}$. Thus, if the formulated conditions hold for f, they do so for g and h, and if g and f are integrable, then f is also integrable. Therefore, to prove sufficiency it suffices to consider a real-valued function f.

In this case the measurability of f means that a sequence of real-valued simple functions $\bar{f}_n$ converges to the former almost everywhere. But $|f| \leq \bar{f}$ does not imply $|\bar{f}_n| \leq \bar{f}$ for all n, and we cannot apply the lemma immediately. However the sequence $\bar{f}_n$ may be replaced by the sequence of functions $f_n = \bar{f} \wedge [\bar{f}_n \vee (-\bar{f})]$, which satisfies the conditions of the lemma.

The functions f_n are integrable along with $\bar{f}$ and $\bar{f}_n$. Since $\bar{f}_n \to f$, $\bar{f} \geq 0$, and $-\bar{f} \leq f \leq \bar{f}$, then $f_n = \bar{f} \wedge [\bar{f}_n \vee (-\bar{f})] \to \bar{f} \wedge [f \vee (-\bar{f})] = f$. From the above definitions it follows that $-\bar{f} \leq f_n \leq \bar{f}$, i.e., $|f_n| \leq \bar{f}$. Now the conditions of the lemma are satisfied, and it follows that f is integrable.

Necessity. The measurability of an integrable function immediately follows from the definitions given above.

A function $f = (g^+ - g^-) + i(h^+ - h^-)$ is integrable if and only if its positive components g^+, g^-, h^+, h^- are integrable, along with $\bar{f} = g^+ + g^- + h^+ + h^-$. The triangle inequality yields $|f| \leq \bar{f}$. Thus, an integrable function $\bar{f}$ meets the conditions of the criterion. ∎

COROLLARY. *A measurable function f is integrable if and only if $|f|$ is integrable.*

PROOF. If f is measurable, then so is $|f|$. By the integrability criterion, if f is integrable there exists an integrable function $\bar{f}$ such that $|f| \leq \bar{f}$. Consequently, by the same criterion, $|f|$ is integrable, as asserted.

If f is measurable and $|f|$ is integrable, then the integrability of f immediately follows from the criterion at $\bar{f} = |f|$. ∎

2.5.4. Lebesgue's Theorem

This frequently used theorem supplements the results of Levy and Fatou. It is formulated for any number-valued functions, not necessarily real-valued.

THEOREM (Lebesgue). *If a sequence of measurable number-valued functions f_n bounded in absolute values by an integrable function $\bar{f} \geq 0$ converges almost everywhere to a function f, then the functions f_n and f are integrable and $\lim \int f_n = \int f$.*

PROOF. This theorem differs from the lemma of Section 2.5.3 by the statement on the measurability of f_n and boundedness of $|f_n|$ almost everywhere. Therefore, to prove the theorem it suffices to demonstrate that the statement of the theorem implies that f_n are integrable almost everywhere and that the lemma remains true also for $|f_n|$ bounded almost everywhere.

We replace f_n and $\bar{f}$ by the equivalent functions g_n and $\bar{g}$, which are equal to zero on the set of measure zero where $|f_n|$ is not bounded by $\bar{f}$. Clearly, $|g_n| \leq \bar{g}$ everywhere. The measurability of f_n implies the measurability of g_n, and the integrability of $\bar{f}$ implies the integrability of $\bar{g}$. Also, since $f_n \to f$, then $g_n \to f$ almost everywhere.

According to the criterion, the measurability of g_n together with the inequality $|g_n| \leq \bar{g}$ ensure the integrability of g_n, and from the lemma of Section 2.5.3 it follows that the limit function f is integrable and $\lim \int g_n = \int f$. Since f_n and g_n are equivalent, then f_n is integrable together with g_n and so $\int f_n = \int g_n$. Consequently,

$$\lim \int f_n = \int f.$$

The proof of the theorem is established. ∎

COROLLARY. *Given a sequence of integrable functions f_n such that $\Sigma |f_n| < \infty$, there exists an integrable function f such that $\Sigma f_n = f$ almost everywhere and $\int f = \Sigma f_n$.*

2.5. Limit Theorems

PROOF. The sequence of functions $g_n = \sum_{m \leq n} |f_m|$ satisfies the conditions of Levy's theorem. Therefore, there exists an integrable function $g = \Sigma |f_n|$ almost everywhere. Since

$$\left| \sum_{m \leq n \leq p} f_n \right| \leq \sum_{m \leq n \leq p} |f_n|,$$

and by the Cauchy criterion we find that there exists a function $f = \Sigma f_n$ almost everywhere.

The sequence of functions $h_n = \sum_{m \leq n} f_m$ satisfies the conditions of Lebesgue's theorem: h_n are integrable, $h \to f$ almost everywhere, $|h_n| \leq g$, and g is integrable. Hence, f is integrable and

$$\int f = \lim \int h_n = \Sigma \int f_n. \blacksquare$$

2.5.5. Fourier Transform

A useful application of the integrability criterion is described below.

Consider an algebra $\mathcal{R}$ of simple sets constituted by limited intervals on the real axes $\mathbb{R} = \,]-\infty, \infty[$, and a Lebesgue measure dx on $\mathcal{R}$ equal to the total length of all the mutually disjoint intervals constituting a simple set

$$dx(\Sigma A_j) = \Sigma dx(A_j), \qquad x(|a, b|) = b - a.$$

(Here, $|a, b|$ means an arbitrary interval with end points a and b, i.e., $|a, b|$ may be $]a, b[$, $]a, b]$, $[a, b[$ or $[a, b]$ with $-\infty < a \leq b < \infty$.

For every real number u a function $g : \mathbb{R} \to \mathbb{C}$ with values $g(x) = e^{iux}$ ($-\infty < x < \infty$) is continuous and therefore measurable with respect to measure dx.

Consider also a function $f : \mathbb{R} \to \mathbb{C}$ integrable by a Lebesgue measure dx. The product $h = g \cdot f$ of measurable functions g and f is also a measurable function. Besides, $|h| = |g| \cdot |f| = 1 \cdot |f| = |f|$, and $|f|$ is integrable along with f. Consequently, h is integrable according to the criterion.

A function $\varphi : \mathbb{R} \to \mathbb{C}$ defined as

$$\varphi(u) = \int e^{iux} f(x) \, dx$$

is called the Fourier transform of f.

Consider now a Stieltjes measure on $\mathbb{R}$ given by the probability distribution function $F : \mathbb{R} \to \mathbb{R}$ (which is increasing, continuous on the left, and has the limits $F(-\infty) = 0$, $F(\infty) = 1$). The constant 1 is integrable with respect to such measure dF:

$$\int 1 \cdot dF = 1.$$

Since $|g(x)| \leq 1$, then by the criterion the function g is integrable by measure dF.

A function $\varphi : \mathbb{R} \to \mathbb{C}$ given by

$$\varphi(u) = \int e^{iux} dF(x)$$

is called the Fourier transform of the measure dF.

2.6. Measurable Functions

As a rule mathematical analysis has to do with functions measurable with respect to a given measure.

2.6.1. Measurability and Integrability

By definition, a measurable function is the limit of an almost everywhere convergent sequence of simple functions. Different algebras of simple sets and different measures define different classes of measurable functions. Measurable functions were shown to form an algebra under which a sum, a product of measurable functions, and a product of a measurable function by a number is also a measurable function with respect to the given measure. By definition, an integrable function is the limit of a special sequence of simple functions. Consequently, every integrable function is measurable, but not conversely: there exist nonintegrable measurable functions.

The converse relation between measurability and integrability is given by the following

PROPOSITION. *Every measurable function is equal to the limit of an almost everywhere convergent sequence of integrable functions.*

Indeed, every measurable function is the limit of an almost everywhere convergent sequence of simple functions, and every simple function is integrable.

2.6.2. Sequences of Measurable Functions

In what follows we will assume that there exists an increasing sequence of simple sets U_n with union equal to U. Quite appropriately relevant measures will be referred to as *locally bounded*. This assumption is equivalent to letting that there exists a sequence of mutually disjoint simple sets E_n whose sum is equal to U. Accordingly, the measures under consideration are also called σ-*finite*.

This assumption ensures the existence of a strictly positive integrable function.

2.6. Measurable Functions

LEMMA. *If a sequence of measurable functions f_n converges to f almost everywhere, then f is measurable.*

PROOF. Consider an integrable function $h > 0$ and the sequence of functions $g_n = hf_n (h + |f_n|)^{-1}$ measurable along with h and f_n, having $|g_n| \leq h$, and converging

$$g_n \to g = hf(h + |f|)^{-1}.$$

From Lebesgue's theorem it follows that g is integrable and, necessarily, measurable. Consequently, the function $f = hg(h - |g|)^{-1}$ is measurable. ∎

We call a set of functions closed if in addition to each almost everywhere convergent sequence of functions it includes all the limit functions. The proposition of Section 2.4.1 and the lemma just proved assert the following statement.

THEOREM. *Measurable functions form a closed algebra.*

The local boundedness of a measure guarantees the existence of identity $1 = \lim U_n \in \mathcal{M}$ for this algebra. Therefore, every constant is a measurable function.

The following proposition turns out to be useful in many a situation.

PROPOSITION. *A measurable function f is equal to zero almost everywhere if and only if it is integrable and $\int |f| = 0$.*

PROOF. If $f = 0$ almost everywhere, then f is integrable and $\int |f| = 0$. Conversely, if f integrable and $\int |f| = 0$, then the sequence $g_n = n|f|$ meets the conditions of Levy's theorem: it is increasing, constituted by integrable functions, and integrally bounded (by zero). Consequently, g_n converges almost everywhere to an integrable function g, i.e., $n|f(x)| \to g(x)$ for almost all x. Hence, $f(x) = 0$ for almost all x. ∎

In particular, if the integral of a positive integrable function is zero, then the function is zero almost everywhere.

2.6.3. Measurable and Integrable Sets

Identifying sets with their indicators leads us straightway to defining measurable and integrable sets as the ones possessing measurable and integrable indicators, respectively.

The theorem on algebra of measurable functions yields several corollaries for sets.

COROLLARY 1. *Measurable sets form a closed algebra of sets.*

We will call a class of integrable sets boundedly closed if along with each sequence convergent almost everywhere and bounded almost everywhere by some integrable set it contains all its limit sets.

COROLLARY 2. *Integrable sets form a boundedly closed algebra of sets.*

We denote this algebra by $\overline{\mathcal{R}}$ and formulate an important

DEFINITION. *The number*

$$\overline{\mu}(A) = \int A\mu$$

is called the measure of the integrable set $A \in \overline{\mathcal{R}}$.

THEOREM. $\overline{\mu} : \overline{\mathcal{R}} \to \mathbb{R}$ *is a countable additive measure extending the measure* $\mu : \mathcal{R} \to \mathbb{R}$.

PROOF. By the definitions

$$\overline{\mu}(A) = \int A\mu = \mu(A)$$

for every simple set A. Hence, $\overline{\mu}$ extends μ.

The additivity of $\overline{\mu}$ follows from the linearity of integral, and the positiveness of integral implies the positiveness of $\overline{\mu}$. The countable additivity of $\overline{\mu}$ follows from the corollary of Lebesgue's theorem. ∎

We will call $\overline{\mu}$ *an integral extension* of μ henceforth and omit the overbar whenever there is no risk of misunderstanding.

Real-valued measurable functions are described by the following

MEASURABILITY CRITERION. *A real-valued function f is measurable if and only if the preimage $f^{-1}[a, \infty[$ of each interval $[a, \infty[$ having $(-\infty < a < \infty)$ is a measurable set.*

REMARK. In the criterion the interval $[a, \infty[$ may be replaced by anyone of $]a, \infty[$, $]-\infty, a]$, $]-\infty, a[$ having $(-\infty < a < \infty)$.

COROLLARY. *Continuous real-valued functions are measurable with respect to Lebesgue measure.*

The structure of real-valued functions measurable with respect to Lebesgue measure λ on the interval $[a, b]$ with $(-\infty < a < b < \infty)$ is described by

THEOREM (Luzin). *If $f : [a, b] \to \mathbb{R}$ is measurable with respect to Lebesgue measure, then for each $\varepsilon > 0$ there exists a continuous function $g : [a, b] \to \mathbb{R}$ such that $\lambda(\{x : f(x) \neq g(x)\}) < \varepsilon$.*

2.7. Fubini and Tonelli's Theorems

These theorems establish a relationship between double and repeated integrals. They are frequently used in theoretical analysis and in evaluation of integrals.

2.7. Fubini and Tonelli's Theorems

2.7.1. Product of Measures

Consider two sets X, Y, albegras $\mathcal{A}, \mathcal{B}$ of their simple subsets, and measures λ, μ on these algebras, respectively.

From the Cartesian product $Z = X \times Y$ of the sets X and Y we distinguish a class of *simple rectangles* $C = A \times B$ with sides $A \in \mathcal{A}$ and $B \in \mathcal{B}$. The unions of finite families of simple rectangles will be called simple sets in $X \times Y$. The sets on $X \times Y$ will be called *plane*.

It is not hard to prove two following assertions.

LEMMA 1. *Every simple set on $X \times Y$ is the sum of a finite family of mutually disjoint simple rectangles.*

PROPOSITION 1. *Simple sets in $X \times Y$ form an algebra.*

The algebra of simple sets in $X \times Y$ is said to be the product of $\mathcal{A}$ and $\mathcal{B}$, denoted by $\mathcal{A} \times \mathcal{B}$.

The measure $\lambda \times \mu = \nu$ on the algebra $\mathcal{C} = \mathcal{A} \times \mathcal{B}$ of simple sets in $Z = X \times Y$ is defined quite naturally, but its definition is associated with rather lengthy arguments. First, the measure of a simple rectangle is defined, then the measure of a simple set is constructed by the measures of constituting rectangles.

DEFINITION 1. *The measure $\lambda \times \mu (A \times B) = \nu(C)$ of a simple rectangle $A \times B = C$ is the product of the measures $\lambda(A)$ and $\mu(B)$ of its sides:*

$$\lambda \times \mu (A \times B) = \lambda(A) \times \mu(B).$$

LEMMA 2. *On the class of simple rectangles the function ν is additive.*

LEMMA 3. $\Sigma \nu(C_i) = \Sigma \nu(Z_j)$ $(\Sigma C_i = \Sigma Z_j)$.

DEFINITION 2. *The measure $\nu(C)$ of the simple set $C = \Sigma C_i$ is the sum of measures of constituting simple rectangles C_i:*

$$\nu(C) = \Sigma \nu(C_i).$$

By Lemma 3, this definition specifies a function ν on the algebra $\mathcal{A} \times \mathcal{B} = \mathcal{C}$ of simple sets.

PROPOSITION 2. *On the algebra $\mathcal{C}$ of simple sets the function ν given by Definitions 1 and 2 is a measure.*

Thus, Definitions 1–2 and Lemmas 1–3 define the measure $\lambda \times \mu : \mathcal{A} \times \mathcal{B} \to \mathbb{R}$ which is called *the product of measures* $\lambda : \mathcal{A} \to \mathbb{R}$ and $\mu : \mathcal{B} \to \mathbb{R}$. The measure $\lambda \times \mu$ is uniquely defined by its values on the rectangles

$$\lambda \times \mu (A \times B) = \lambda(A) \cdot \mu(B) \quad (A \in \mathcal{A}, B \in \mathcal{B}).$$

PROPOSITION 3. *The product $\lambda \times \mu = \nu$ of countably additive measures λ and μ is a countably additive measure.*

It is easy to verify that the product of locally bounded measures is a locally bounded measure. Thus, the product of measures possesses all the properties desired except for possibly completeness. It can be completed when necessary.

2.7.2. Double and Iterated Integrals

Here, like in Section 2.7.1, we consider a set X, the algebra $\mathcal{A}$ of its simple parts, a measure $\lambda = dx$ on this algebra, a set Y, the algebra $\mathcal{B}$ of its simple parts, and a measure $\mu = dy$ on this algebra. Both measures $\lambda = dx$ and $\mu = dy$ are assumed to be *countably additive complete and locally bounded.* The designations dx and dy have been adopted to achieve habitual integral representations.

The product $\lambda \times \mu = dx \cdot dy$ is a countably additive and locally bounded measure on the algebra $\mathcal{A} \times \mathcal{B}$ of simple sets in $X \times Y$, completed when necessary.

Let us consider a function $h : X \times Y \to \mathbb{C}$ of two variables $x \in X$ and $y \in Y$, which assumes complex values $h(x, y) \in \mathbb{C}$. We shall denote it also by $h(\cdot, \cdot)$. For every $a \in X$ and $b \in Y$ we define *the partial functions* of one variable, $h(a, \cdot) : Y \to \mathbb{C}$ and $h(\cdot, b) : X \to \mathbb{C}$ with values $h(a, y)$ and $h(x, b)$ for $y \in Y$ and $x \in X$. The functions $h(a, \cdot)$ and $h(\cdot, b)$ are also called *the sections* of h at points $a \in X$ and $b \in Y$.

Before we formulate the Fubini and Tonelli theorems for integrals, we define the double and iterated integrals and introduce suitable designations.

(i) If $h(\cdot, \cdot)$ is integrable by measure $dxdy$, then we say that *there exists the double integral*

$$c = \iint h(x, y)\, dx\, dy.$$

(ii) Suppose that for every $x \in X$ not in a set $A \subseteq X$ of measure $dx(A) = 0$, the function $h(x, \cdot)$ is integrable with respect to measure dy. Now, let f be a function on X with values

$$\int f(x) = \int h(x, y)\, dy$$

for $x \notin A$ and arbitrary numerical values $f(x)$ for $x \in A$. If f is integrable with respect to dx, then we say that *there exists the iterated integral*

$$a = \int f(x)\, dx = \int \left(\int h(x, y)\, dy \right) dx.$$

Existence and value of this integral do not depend on the choice of set A, measure $dx(A) = 0$, and values of $f(x)$ for $x \in A$.

(iii) Suppose that for every $y \in Y$ not in a set $B \subseteq Y$ of measure $dy(B) = 0$, the function $h(\cdot, y)$ is integrable with respect to dx. Consider a function g on Y having the values

$$g(y) = \int h(x, y)\, dx$$

2.7. Fubini and Tonelli's Theorems

for $y \notin B$ and arbitrary numerical values $g(y)$ for $y \in B$. If g is integrable with respect to measure dy then we say that *there exists the iterated integral*

$$b = \int g(y)\,dy = \int (\int h(x, y)\,dx)\,dy.$$

Existence and value of this integral do not depend on the choice of B of measure $dy(B) = 0$ and values $g(y)$ for $y \in B$.

The functions f and g are sometimes called *simple integrals*. They are defined accurate to equivalence with respect to the measures dx and dy, respectively.

Since the integrability of a numerical function is tantamount to the integrability of its positive components and the integral is linear, it is sufficient to study the relationship between double and iterative integrals for positive functions.

In the forthcoming proof of Fubini's theorem we will need the following auxiliary proposition.

LEMMA. *Almost all cuts of a plane set of measure zero also have measures zero.*

This lemma abounds in useful corollaries. We formulate one of them using the notation introduced above.

COROLLARY. *If $h_n(\cdot, \cdot) \to h(\cdot, \cdot)$ almost everywhere by $dxdy$ then $h_n(a, \cdot) \to h(a, \cdot)$ almost everywhere by dy and $h_n(\cdot, b) \to h(\cdot, b)$ almost everywhere by dx for almost every $a \in X$ and $b \in Y$.*

PROOF. Let $h_n(x, y) \to h(x, y)$ for $(x, y) \notin C$, $dxdy\,C = 0$, then $h_n(a, y) \to h(a, y)$ for $y \in C(a, \cdot)$ and $h_n(x, b) \to h(x, b)$ for $x \in C(\cdot, b)$ and by the lemma $dy\,C(a, \cdot) = 0$, and $dx\,C(\cdot, b) = 0$. ∎

From this corollary follows

PROPOSITION. *Almost all cuts of a measurable function are measurable.*

PROOF. Let $h(\cdot, \cdot)$ be measurable. Then by definition the sequence of simple functions $h_n(\cdot, \cdot)$ converges to it almost everywhere by $dxdy$. It is clear that the cuts $h_n(a, \cdot)$ and $h_n(\cdot, b)$ are also simple functions. By the proven statement $h_n(a, \cdot) \to h(a, \cdot)$ almost everywhere by dy and for almost every $a \in X$, and $h_n(\cdot, b) \to h(\cdot, b)$ almost everywhere by dx for almost every $b \in Y$. Consequently, for such a and b, the function $h(a, \cdot)$ is measurable by dy and the function $h(\cdot, b)$ is measurable by dx. ∎

2.7.3. Fubini's Theorem

The notations and definitions given in Section 2.7.2 for double and iterated integrals extend to this section.

THEOREM (Fubini). *If for a number-valued function h of two variables there exists the double integral c, then for it there exist both iterated integrals a and b, and all the three integrals are equal, $a = b = c$.*

A brief formulation of Fubini's theorem reads
$$\iint h(x, y)\,dx\,dy = \int\left(\int h(x, y)\,dx\right)dy = \int\left(\int h(x, y)\,dy\right)dx.$$

We approach Fubini's theorem by two auxiliary lemmas.

LEMMA 1. *Fubini's theorem holds for simple functions.*

PROOF. Let
$$h = \sum_i c_i C_i = \sum_i \sum_j c_{ij} A_{ij} \times B_{ij},$$

where $c_i = c_{ij}$ are numbers and $C_i = \sum_j A_{ij} \times B_{ij}$ are the sums of simple rectangles. Then

$$c = \iint h(x, y)\,dx\,dy = \sum_i \sum_j c_{ij}\,dx(A_{ij})\,dy(B_{ij}),$$

$$h(x, \cdot) = \sum_i \sum_j c_{ij} A_{ij}(x) B_{ij},$$

$$f(x) = \int h(x, y)\,dy = \sum_i \sum_j c_{ij} A_{ij}(x)\,dy(B_{ij}),$$

$$a = \int f(x)\,dx = \int\left(\int h(x, y)\,dy\right)dx = \sum_i \sum_j c_{ij}\,dx(A_{ij})\,dy(B_{ij}) = c,$$

$$h(\cdot, y) = \sum_i \sum_j c_{ij}\, A_{ij} B_{ij}(y),$$

$$g(y) = \int h(x, y)\,dx = \sum_i \sum_j c_{ij}\,dx(A_{ij}) B_{ij}(y),$$

$$b = \int g(y)\,dy = \int\left(\int h(x, y)\,dx\right)dy = \sum_i \sum_j c_{ij}\,dx(A_{ij})\,dy(B_{ij}) = c.$$

Thus $a = b = c$. (All these integrals exist because the functions under consideration are simple.) The lemma is established. ∎

LEMMA 2. *Fubini's theorem holds for monotonically integrable functions.*

PROOF. Let $h \in \mathcal{H}$. Then there exists a sequence of simple functions h_n which approximates h:

$$h_n(z) \uparrow h(z) \quad (z = (x, y) \notin C),$$

$$c = \iint h(x, y)\,dx\,dy = \lim \iint h_n(x, y)\,dx\,dy. \tag{1}$$

Here C is a set of measure zero in $Z = X \times Y$.

$$dx\,dy(C) = 0.$$

2.7. Fubini and Tonelli's Theorems

Let $\{f_n\}$ be a sequence of positive simple functions

$$f_n(x) = \int h_n(x, y)\,dx \geq 0.$$

This sequence is increasing and integrally bounded along with h_n:

$$\int f_n(x)\,dx = \int \left(\int h_n(x, y)\,dy\right) dx = \iint h_n(x, y)\,dx\,dy \leq c. \quad (2)$$

Consequently $\{f_n\}$ approximates certain integrable function f and

$$\lim \int f_n(x)\,dx = \int f(x)\,dx.$$

Pick a set

$$A = \{x : dy(C(x, \cdot)) \neq 0\} \cup \{x : f_n(x) \nrightarrow f(x)\}.$$

Since the set C is of measure zero, $dy\,(C(x, \cdot)) = 0$ for almost all x. At the same time $f_n(x) \to f(x)$ also for almost every x. Hence, $dx\,(A) = 0$.

If $x \notin A$ then $dy\,(C(x, \cdot)) = 0$. By the definition of C, for $y \notin C(x, \cdot)$ the pair $(x, y) \notin C$, and the sequence $h_n(x, y) \uparrow h(x, y)$. Hence, the increasing sequence of simple functions $h_n(x, \cdot)$ converges to $h(x, \cdot)$ in the measure dy almost everywhere. At the same time $0 \leq f_n(x) \uparrow f(x)$ for $x \notin A$, and so the sequence $\{h_n(x, \cdot)\}$ is integrally bounded:

$$\int h_n(x, y)\,dy = f_n(x) \leq f(x) < \infty.$$

Thus, for $x \notin A$, the function $h(x, \cdot)$ is integrable in dy and

$$f(x) = \lim f_n(x) = \lim \int h_n(x, y)\,dy = \int h(x, y)\,dy. \quad (3)$$

The function f is integrable by measure dx. Thus, the iterated integral a exists

$$a = \int f(x)\,dx = \int \left(\int h(x, y)\,dy\right) dx$$

and equals the double integral. Indeed, from equalities (1) – (3) it follows

$$a = \int f(x)\,dx = \lim \int f_n(x)\,dx = \lim \iint h_n(x, y)\,dx\,dy$$
$$= \iint h(x, y)\,dx\,dy = c.$$

One can prove that the iterated integral b exists and $b = c$ in exactly the same fashion, but by symmetry this follows immediately from the proved after the change $x \leftrightarrow y$.

Fubini's theorem for monotonically integrable functions is established. ■

Now we proceed to the general case.

PROOF (Fubini's theorem). Let h be a numerical-valued function on $X \times Y$ integrable by measure $dx\,dy$ and having components h_1, h_2, h_3, h_4 monotonically integrable:

$$h = h_1 - h_2 + ih_3 - ih_4.$$

By Lemma 2, Fubini's theorem is valid for these components: the existence of the double integrals ($j = 1, 2, 3, 4$)

$$c_j = \iint h_j(x, y)\,dx\,dy$$

implies the existence of the iterated integrals

$$a_j = \int \left(\int h_j(x, y)\,dy\right) dx,$$
$$b_j = \int \left(\int h_j(x, y)\,dx\right) dy,$$

and the equalities $a_j = b_j = c_j$, ($j = 1, 2, 3, 4$).

Observe that for each $x \in X$ and $y \in Y$ the functions $h_j(x, \cdot)$ and $h_j(\cdot, y)$ are the monotonically integrable components of $h(x, \cdot)$ and $h(\cdot, y)$:

$$h(x, \cdot) = h_1(x, \cdot) - h_2(x, \cdot) + ih_3(x, \cdot) - ih_4(x, \cdot),$$
$$h(\cdot, y) = h_1(\cdot, y) - h_2(\cdot, y) + ih_3(\cdot, y) - ih_4(\cdot, y).$$

The integrability of $h_j(x, \cdot)$ and $h_j(\cdot, y)$ by measure dy and by measure dx, respectively, imply the integrability of $h(x, \cdot)$ and $h(\cdot, y)$ with respect to dy and dx, respectively. Thus, the simple integrals

$$f(x) = \int h(x, y)\,dy, \quad g(y) = \int h(x, y)\,dx$$

exist for almost every $x \in X$ and $y \in Y$.

The functions f and g are linear combinations of the functions f_j and g_j ($j = 1, 2, 3, 4$) given by

$$f_j(x) = \int h_j(x, y)\,dy, \quad g_j(y) = \int h_j(x, y)\,dx$$

for almost every $x \in X$ and $y \in Y$. The existence of the iterated integrals a_j and b_j by definition means the integrability of f_j and g_j with respect

2.7. Fubini and Tonelli's Theorems

to dx and dy; it entails the integrability of f and g by these measures. Consequently, the iterated integrals a, b exist and

$$a = a_1 - a_2 + ia_3 - ia_4, \quad b = b_1 - b_2 + ib_3 - ib_4.$$

From these equalities combined with

$$c = c_1 - c_2 + ic_3 - ic_4$$

and $a_j = b_j = c_j$ ($j = 1, 2, 3, 4$) it follows that

$$a = b = c,$$

which completes the proof of Fubini's theorem. ∎

2.7.4. Tonelli's Theorem

In Fubini's theorem the existence of iterated integrals is deduced from the existence of the double integral. In Tonelli's theorem, conversely, given the existence of one of the iterated integrals we prove the existence of the double. However, this concerns only positive measurable functions.

THEOREM (Tonelli). *If for a positive measurable function h of two variables there exists one of the iterated integrals a or b, then there exist the second iterated integral, the double integral c, and all the three integrals are equal: $a = b = c$.*

PROOF. Suppose that

$$a = \int f(x)\,dx = \int \left(\int h(x, y)\,dy\right) dx$$

exists. For the positive measurable h these exists an increasing sequence of positive integrable functions h_n which converges to h in the measure $dxdy$ almost everywhere:

$$0 \leq h_n(z) \uparrow h(z) \quad (z = (x, y) \notin C)$$

where $dxdy\,(C) = 0$. We must show also that $\{h_n\}$ is integrally bounded. (In Fubini's theorem, this follows directly from the integrability of h. Here one assumes only that h is measurable.)

Observe that the sequence of integrable functions f_n given by

$$f_n(x) = \int h_n(x, y)\,dy$$

increases along with $\{h_n\}$, and for almost every $x \in X$ it is bounded by the function

$$f(x) = \int h(x, y) \, dy$$

integrable with respect to the measure dx. Applying Fubini's theorem to the functions h_n, taking into account that the repeated limit a exists, and integrating the inequalities $f_n \leq f$, which hold almost everywhere with respect to dx, we obtain

$$\iint h_n(x, y) \, dx \, dy = \int \left(\int h_n(x, y) \, dy \right) dx = \int f_n(x) \, dx$$
$$\leq \int f(x) \, dx = \int \left(\int h(x, y) \, dy \right) dx = a.$$

Thus, the sequence of functions h_n is integrally bounded. By Levy's theorem, the function h is integrable with respect to the measure $dxdy$. The double integral

$$c = \iint h(x, y) \, dx \, dy = \lim \iint h_n(x, y) \, dx \, dy$$

exists. From Fubini's theorem it follows that the second related integral

$$b = \int \left(\int h(x, y) \, dx \right) dy$$

exists as well, and $a = b = c$.

In exactly the same fashion we could prove that if a positive measurable function h has iterated integral b, then it also has the double integral c and the iterated integral a, with $a = b = c$. This result immediately follows from the fact proved by symmetry arguments under the change $x \leftrightarrow y$.

Tonell's theorem is established. ∎

Taken together the Fubini and Tonelli theorems lead us to a generalizing proposition regarding number-valued measurable functions h.

Denote as usual by $|h|$ a function with values $|h(x, y)|$. The measurability of h entails the measurability of $|h|$, and the integrability of h is equivalent to that of $|h|$. We denote the two iterated and the double integrals of $|h|$ by α, β and γ, respectively:

$$\alpha = \int \left(\int |h(x, y)| \, dy \right) dx,$$
$$\beta = \int \left(\int |h(x, y)| \, dx \right) dy,$$
$$\gamma = \iint |h(x, y)| \, dx \, dy.$$

2.7. Fubini and Tonelli's Theorems

The existence of a (finite) γ is tantamount to the existence of the double integral

$$c = \iint h(x, y) \, dx \, dy,$$

but (finite) α, β need not exist when a, b exist (and are even equal). The Fubini and Tonelli theorems are united by

THEOREM (Fubini-Tonelli). *Let h be a measurable number-valued function of two variables. If one of the iterated integrals α, β, or the double integral γ for $|h|$ exists, then the iterated and double integrals a, b and c of h exist, and $a = b = c$.*

PROOF. If h is measurable, so is $|h|$. By Tonelli's theorem, the measurability of $|h|$ and the existence of one of the iterated integrals α or β for $|h|$ ensures the existence of the double integral γ for $|h|$ and, hence, the double integral c for h. The double integral c for the measurable h exists by far if one supposes from the very beginning that the double integral γ of $|h|$ exists.

By Fubini's theorem, the existence of c for h implies the existence of both iterated integrals a, b for h, and the equality $a = b = c$, as claimed. ∎

The Fubini-Tonelli theorem is equivalent to the theorems of Fubini and Tonelli.

2.7.5. Rules for Operations Under the Integral Sign

The Fubini and Tonelli theorems establish the conditions enabling one to interchange the integrals. Here we give conditions governing the interchange of the integral and the limit, and the differentiation and the improper integral.

Suppose that $X \subseteq \mathbb{R}$ and call the variable $x \in X$ a *parameter*. A function $f : X \to \mathbb{C}$ given by

$$f(x) = \int h(x, y) \, dy$$

is said to be *the integral depending on the parameter x* if for each value of the parameter x the function $h(x, \cdot) : Y \to \mathbb{C}$ is integrable with respect to the measure dy.

All other notations and terms retain their previous meaning.

Let a be an arbitrary point of the closure $\overline{X}$ of X in $\overline{\mathbb{R}} = [-\infty, \infty]$. (In particular we admit the values $\pm \infty$ for a.) We formulate in detail the conditions which must be obeyed to interchange the integration with respect to dx and the limit as $x \to a$:

(1a) *for every $x \in X$, $h(x, \cdot)$ is measurable with respect to dy;*

(1b) $h(\cdot, y)$ *converges to the value* $g(y)$ *of some integrable* $g: Y \to \mathbb{C}$ *for almost every* $y \in Y$ *as* $x \to a$;
(1c) $|h(\cdot, y)|$ *is bounded by the value* $\bar{g}(y)$ *of some integrable function* $\bar{g}: Y \to \mathbb{R}$, *for almost every* $y \in Y$.

The assertion "$\lim_{x \to a} \int h(x, y) dy$ makes sense" implies that $h(x, \cdot)$ is integrable with respect to dy and the function f, given by $f(x) = \int h(x, y) dy$, has a limit as $x \to a$. The assertion "$\int \lim_{x \to a} h(x, y) dy$ makes sense" means that there exists a limit function $g: Y \to \mathbb{C}$ given by $g(y) = \lim_{x \to a} h(x, y)$ for almost every $y \in Y$ and it is integrable with respect to dy.

THEOREM 1. *If the conditions* (1a)–(1c) *are satisfied, the both sides of the equality*

$$\lim_{x \to a} \int h(x, y) dy = \int \lim_{x \to a} h(x, y) dy$$

make sense and the equality is true.

PROOF. We reduce the setting to sequences and apply the Lebesgue theorem. Let $x(n) \in X$ be an arbitrary sequence converging to $a \in \bar{X}$. If $h(x, y) \to g(y)$ as $x \to a$, then $h(x(n), y) \to g(y)$. The conditions (1a)–(1c) satisfy Lebesgue's theorem for the functions $g_n = h(x(n), \cdot)$, g and $\bar{g}$:

(1) g_n are measurable;
(2) $|g_n| = |h(x(n), \cdot)|$ bounded almost everywhere by an integrable $\bar{g}$;
(3) g_n converges to g almost everywhere.

By Lebesgue's theorem g_n and g are integrable and $\lim \int g_n = g$, that is,

$$\lim f(x(n)) = \lim \int h(x(n), y) dy = \int g(y) dy$$

for every sequence $x(n) \to a$. Hence, f has a limit as $x \to a$ and

$$\lim_{x \to a} \int h(x, y) dy = \lim_{x \to a} f(x) = \int g(y) dy = \int \lim_{x \to a} h(x, y) dy,$$

as desired. ∎

Now we turn to derivatives. Let a be a point of an interval X in $\mathbb{R} =]-\infty, \infty[$. (The point a is not necessarily interior, the interval X is not necessarily open.)

Let $\mathcal{D}_a h(\cdot, y)$ be the derivative of $h(\cdot, y)$ at a, and $\Delta_a h(\cdot, y)$ be the

2.7. Fubini and Tonelli's Theorems

relative increment of $h(\cdot, y)$ at a, i.e., the function with the values

$$\Delta_a h(x, y) = \frac{h(x, y) - h(a, y)}{x - a}$$

at $x \neq a$. If $h(\cdot, y)$ is differentiable at a then at $x = a$ the value of $\Delta_a h(\cdot, y)$ is chosen to be $\mathcal{D}_a h(\cdot, y)$.

By definition

$$\mathcal{D}_a h(\cdot, y) = \lim_{x \to a} \Delta_a h(x, y).$$

Let us formulate in detail the conditions that enable one to interchange the integral in dy and the derivative at a:

(2a) $h(x, \cdot)$ is measurable with respect to dy for each $x \in X$ and $h(a, \cdot)$ is integrable with respect to dy;

(2b) $h(\cdot, y)$ is differentiable at a for almost every $y \in Y$;

(2c) $|\Delta_a h(\cdot, y)|$ is bounded by the value $\bar{g}(y)$ of some integrable function $\bar{g} : Y \to \mathbb{R}$ for almost every $y \in Y$.

The statement "$\mathcal{D}_a \int h(x, y) dy$ makes sense" means that the functions $h(x, \cdot)$ are integrable with respect to dy and $f : X \to \mathbb{C}$ with values $f(x) = \int h(x, y) dy$ has a derivative at a. The statemeut "$\int \mathcal{D}_a h(x, y) dy$ makes sense" means that $h(\cdot, y)$ is differentiable at a for almost every $y \in Y$, and a function with values $\mathcal{D}_a h(\cdot, y)$ for such $y \in Y$ is integrable with respect to dy.

THEOREM 2. *If the conditions (2a)–(2c) are satisfied, the both sides of the equality*

$$\mathcal{D}_a \int h(x, y) dy = \int \mathcal{D}_a h(x, y) dy$$

make sense and the equality is true.

PROOF. Theorem 2 actually follows from Theorem 1. We apply it to the function $\Delta_a h(\cdot, \cdot)$.

The conditions (2a)–(2c) ensure the fulfilment of (1a)–(1c) for the function $\bar{h} = \Delta_a h(\cdot, \cdot)$ given by

$$\bar{h}(x, y) = \frac{h(x, y) - h(a, y)}{x - a} \quad (x \neq a),$$

$$\bar{h}(a, y) = \mathcal{D}_a h(\cdot, y).$$

Indeed, $\bar{h}(x, \cdot)$ is measurable by dy along with $h(x, \cdot)$ for all $x \in X$, and $\bar{h}(\cdot, y)$ converges for almost every $y \in Y$ to $\mathcal{D}_a h(\cdot, y)$ as $x \to a$, $|\bar{h}(\cdot, y)|$ is bounded by the value $\bar{g}(y)$ of some integrable $\bar{g}$ for almost every $y \in Y$.

The integrability of $h(a, \cdot)$ is assumed and that of $\bar{h}(x, \cdot)$ follows from Theorem 1. Since

$$h(x, \cdot) = h(a, \cdot) + (x - a)\bar{h}(x, \cdot),$$

the integrability of $h(a, \cdot)$ and $\bar{h}(x, \cdot)$ implies the integrability of $h(x, \cdot)$ by measure dy for each $x \in X$.

By Theorem 1, the both sides of the equality

$$\lim_{x \to a} \int \frac{h(x, y) - h(a, y)}{x - a} dy = \int \lim_{x \to a} \frac{h(x, y) - h(a, y)}{x - a} dy$$

make sense. Since $h(x, \cdot)$ is integrable for each $x \in X$, we have

$$\lim_{x \to a} \int \frac{h(x, y) - h(a, y)}{x - a} dy$$
$$= \lim_{x \to a} \frac{1}{x - a} [\int h(x, y) dy - \int h(a, y) dy] = \mathcal{D}_a \int h(x, y) dy,$$
$$\lim_{x \to a} \frac{h(x, y) - h(a, y)}{x - a} = \mathcal{D}_a h(\cdot, y).$$

Thus, the both sides of the equality make sense and it is valid as asserted. ∎

Theorem 2 gives a frequently used rule of the differentiation of integrals depending on a parameter, known as *Leibniz's rule*.

For functions differentiable on an interval, conditions allowing one to interchange the integral and the derivative are similar to the conditions of Theorem 2.

Let X be an interval of the real axis, $\mathcal{D}h(\cdot, y)$ denote the derivative of $h(\cdot, y)$.

The detailed conditions enabling one to interchange the integral with respect to dy and the derivative with respect to x read:

(2a') $h(x, \cdot)$ *is integrable by dy for every $x \in X$*;
(2b') $h(\cdot, y)$ *is differentiable for almost every $y \in Y$*;
(2c') $|\mathcal{D}h(\cdot, y)|$ *is bounded by the value $\bar{g}(y)$ of some integrable function* $\bar{g}: Y \to \mathbb{C}$ *for almost every $y \in Y$.*

The following is similar to Theorem 2.

THEOREM 2'. *If the conditions (2a')–(2c') are satisfied, then the both sides of the equality*

$$\mathcal{D}\int h(x, y) dy = \int \mathcal{D}h(x, y) dy$$

make sense and the equality is true.

2.7. Fubini and Tonelli's Theorems

PROOF. We will demonstrate that the conditions (2a')–(2c') entail (2a)–(2c) for each $a \in X$.

The conditions (2a) and (2b) follow directly from (2a') and (2b'), while (2c) follows from (2c') and the classical Lagrange formula

$$\left| \frac{h(x, y) - h(a, y)}{x - a} \right| = |\mathcal{D}_c h(\cdot, y)| \leq \bar{g}(y),$$

where $c = c(a, x)$ is a point in the interval X. ∎

For improper integrals we take $X = \mathbb{R} =]-\infty, \infty[$, and define the Lebesgue improper integrals analogously to the Riemann integrals.

Let $f: \mathbb{R} \to \mathbb{C}$ be integrable by the Lebesgue measure dx over any interval. We will call the double limit

$$\int_{-\infty}^{\infty} f(x)\,dx = \lim_{a \to -\infty, \, b \to \infty} \int_a^b f(x)\,dx$$

the *Lebesgue improper integral* of f.

If f is integrable by the Lebesgue measure dx over the whole real axis $\mathbb{R} =]-\infty, \infty[$, then the improper integral is equal to the conventional integral. This follows from Theorem 1. Indeed,

$$\int f - \int_a^b f = \int f - \int [a, b]f = \int (1 - [a, b])f \to 0$$

for $a \to -\infty$ and $b \to -\infty$, since $1 - [a, b](x) \to 0$ for each $x \in \mathbb{R}$ and $|f|$ is integrable by dx along with f and $|(1 - [a, b]) \cdot f| \leq |f|$.

It is worth noting, however, that some functions nonintegrable with respect to the Lebesgue measure can have a Lebesgue improper integral. (For example, $f(x) = x^{-1} \sin x$ with $x \neq 0$.)

Now we formulate conditions to interpose the integral with respect to dy with the improper integral with respect to dx:

(3a) $h(\cdot, \cdot)$ is measurable by $dxdy$;

(3b) $h(\cdot, y)$ has a Lebesgue improper integral for almost every $y \in Y$;

(3c) for all $a \leq b$ from $\mathbb{R}$ and almost every $y \in Y$ the integrals $\int_a^b |h(x, y)|\,dx$ are bounded by the value $\bar{g}(y)$ of some integrable $\bar{g}: Y \to \mathbb{R}$.

The assertion "$\int_{-\infty}^{\infty} (\int h(x, y)\,dy)\,dx$ makes sense" means that $h(x, \cdot)$ are integrable by dy and f, given by $f(x) = \int h(x, y)\,dy$, has an improper integral with respect to dx. The assertion "$\int_{-\infty}^{\infty} (\int h(x, y)\,dx)\,dy$ makes sense" means that $h(\,\cdot\,,y)$ has an improper integral with respect to dx for almost every $y \in Y$ and, for such y, the function $g: Y \to C$ with values

$$g(y) = \int_{-\infty}^{\infty} h(x,y)\,dx$$ is integrable with respect to dy.

THEOREM 3. *If the conditions* (3a)–(3c) *are satisfied, then the both sides of the equality*

$$\int_{-\infty}^{\infty} \left(\int h(x,y)\,dy\right) dx = \int \left(\int_{-\infty}^{\infty} h(x,y)\,dx\right) dy$$

make sense and the equality is true.

PROOF. The proof is based on the theorems of Fubini-Tonelli and Levy.

From the condition (3c) it follows that for almost every $y \in Y$, $|h(\,\cdot\,,y)|$ has the improper integral

$$\int_{-\infty}^{\infty} |h(x,y)|\,dx = \lim_{a \to -\infty,\, b \to \infty} \int_a^b |h(x,y)|\,dx \leq \bar{g}(y).$$

Hence, for such $y \in Y$, $|h(\,\cdot\,,y)|$ is integrable by the Lebesgue measure dx over $\mathbb{R} = \,]-\infty, \infty\,[$. Indeed, from the conditions (3a)–(3c) it follows that the sequence of functions $\varphi_n = [-n, n] \cdot |h(\,\cdot\,,y)|$, given by

$$\varphi_n(x) = |h(x,y)| \quad (x \in [n, n]),$$
$$\varphi_n(x) = 0 \quad (x \notin [-n, n]),$$

is an increasing sequence of integrable (by the Lebesgue measure dx on $\mathbb{R}$) functions, which is integrally bounded and converges everywhere to $\varphi = |h(\,\cdot\,,y)|$. Consequently, by the corollary of Levy's theorem, $|h(\,\cdot\,,y)|$ is integrable with respect to dx along with $h(\,\cdot\,,y)$. (This fact is a corollary of the integrability criterion. As shown in Section 2.7.2, the $dxdy$ measurability of $h(\,\cdot\,,\,\cdot\,)$ implies the measurability of $h(\,\cdot\,,y)$ by dx.) Further, the integral with respect to the measure dx equals the improper integral:

$$\int h(x,y)\,dx = \int_{-\infty}^{\infty} h(x,y)\,dx.$$

From the integrability of $|h(\,\cdot\,,y)|$ by the Lebesgue measure dx on $\mathbb{R}$ and the condition (3c) it follows that the iterated integral

$$\beta = \int \left(\int |h(x,y)|\,dx\right) dy \leq \int \bar{g}(x)\,dy < \infty$$

2.7. Fubini and Tonelli's Theorems

exists. By Fubini-Tonelli's theorem, the measurability of $h(\cdot, \cdot)$ (condition (3a)) and the existence of the iterated integral β for $|h(\cdot, \cdot)|$ guarantee the existence and equality of the iterated integrals for $h(\cdot, \cdot)$. Since the integrals by measure dx equal the corresponding improper integrals, the both sides of the theorem's identity make sense and the identity itself is valid, as claimed. ∎

We shall prove one more theorem on interchanging of the integrals. Its conditions are as follows:

(3a') $h(x, \cdot)$ *is integrable by* dy *for every* $x \in X$;

(3b') $h(\cdot, y)$ *has a Lebesgue improper integral for almost every* $y \in Y$;

(3c') *for almost every* $y \in Y$ *and all* $u \leq v$ *from* $\mathbb{R}$, $|\Delta_u h(\cdot, y)|$ *and* $\int_u^v |h(x, y)| dx$ *are bounded by the value* $\bar{g}(y)$ *of some integrable* $\bar{g}: Y \to \mathbb{R}$.

The assertions of Theorem 3' are the same as those in Theorem 3.

THEOREM 3'. *If the conditions* (3a')–(3c') *are satisfied, then the both sides of the equality*

$$\int_{-\infty}^{\infty} (\int h(x, y) dy) dx = \int (\int_{-\infty}^{\infty} h(x, y) dx) dy$$

make sense and the equality is true.

PROOF. First, we demonstrate that the both sides of this equality make sense. We start from the right-hand side.

By the condition (3b'), for almost every $y \in Y$, $h(\cdot, y)$ is integrable by dx over each interval $[u, v] \subset \mathbb{R}$ ($-\infty < u \leq v < \infty$). Consider the function g with values

$$g(u, v, y) = \int_u^v h(x, y) dx$$

for such $y \in Y$ and arbitrary values elsewhere. We will prove that $g(u, v, \cdot)$ is integrable with respect to dy.

Observe that from the condition (3c) the continuity of $h(\cdot, y)$ follows for almost every $y \in Y$. Indeed, if

$$\left| \frac{h(x, y) - h(u, y)}{x - u} \right| \leq \bar{g}(y)$$

for $x \neq u$, then

$$|h(x, y) - h(u, y)| \leq \bar{g}(u) \cdot |x - u|.$$

On $[u, v]$ the continuous function $h(\cdot, y)$ is approximated by the step functions

$$h_n(\cdot, y) = \sum_k h(x_{kn}, y) A_{kn},$$
$$A_{kn} = [u + (k-1)(v-u)/n, u + k(v-u)/n],$$
$$x_{kn} = u + k(v-u)/n, \quad k = 1, \ldots, n,$$

whence

$$g(u, v, \cdot) = \lim_{n \to \infty} \left[\sum_n h(x_{kn}, \cdot) \cdot (v-u)/n\right].$$

By the condition (3a) the functions $h(x_{kn}, \cdot)$ are integrable by dy, i.e., necessarily measurable, and so are the sums in square brackets and their limit. At the same time the condition (3c) implies that

$$|g(u, v, y)| \leq \int_u^v |h(x, y)| \, dx \leq \bar{g}(y)$$

for almost every value of the integrable function $\bar{g} : Y \to \mathbb{R}$. Consequently $g(v, u, \cdot)$ is integrable by dy.

Consider a function ψ given by

$$\psi(u, x) = \int g(u, x, y) \, dy \quad (x \in [u, v]).$$

We intend to prove that $\psi(u, \cdot)$ is differentiable and calculate its derivative. Let us verify that $g(u, \cdot, y)$ satisfies the conditions of Theorem 2'. Indeed,

(a) $g(u, x, \cdot)$ is integrable by dy for every $x \in [u, v]$;

(b) for every $y \in Y$ at which $h(\cdot, y)$ is continuous, $g(u, \cdot, y)$ is differentiable with $\mathcal{D}g(u, \cdot, y) = h(\cdot, y)$;

(c) for almost every value of the integrable function $\bar{h} = |h(u, \cdot)| + \bar{g} \cdot (v - u)$,

$$|\mathcal{D}g(u, \cdot, g)| = |h(\cdot, y)| \leq \bar{h}(y).$$

This follows from the condition (3a') and (3c') in view of the following inequalities

$$|h(x, y)| - |h(u, y)| \leq |h(x, y) - h(u, y)| \leq \bar{g}(y) \cdot |x - u|$$
$$\leq \bar{g}(y) \cdot (v - u),$$
$$|h(x, y)| \leq |h(u, y)| + \bar{g}(y) \cdot (v - u).$$

By Theorem 2' the function $\psi(u, \cdot)$ is differentiable and its derivative

2.7. Fubini and Tonelli's Theorems

at $x \in [u, v]$ is given by

$$\mathcal{D}\psi(u, x) = \mathcal{D}\int g(y, x, y)\,dy = \int \mathcal{D}g(u, x, y)\,dy = \int h(x, y)\,dy.$$

Now we prove that $\lim\limits_{u \to \infty, v \to \infty} \psi(u, v) = \int (\int_{-\infty}^{\infty} h(x, y)\,dx)\,dy$. We need to verify that g satisfies the conditions of Theorem 1. Indeed,
(a) $g(u, v, \cdot)$ is measurable by dy for all $u \leq v$;
(b) by the condition 3b' for almost every $y \in Y$, $g(\cdot, \cdot, y)$ converges to the improper integral of $h(\cdot, y)$ as $u \to -\infty$, $v \to \infty$;
(c) $|g(\cdot, \cdot, y)| \leq \bar{g}(y)$ for almost every value of the integrable $\bar{g}: Y \to \mathbb{R}$.

By Theorem 1

$$\lim_{u \to -\infty, v \to \infty} \psi(u, v) = \lim_{u \to -\infty, v \to \infty} \int (\int_u^v h(x, y)\,dx)\,dy$$

$$= \int (\lim_{u \to -\infty, v \to \infty} \int_u^v h(x, y)\,dx)\,dy = \int (\int_{-\infty}^{\infty} h(x, y)\,dx)\,dy,$$

which proves that the right-hand side of the equality of Theorem 3' makes sense.

Proceed now to the left-hand side. By the condition (3a') the function $f: \mathbb{R} \to \mathbb{C}$ with values

$$f(x) = \int h(x, y)\,dy$$

is defined. The condition (3c') implies that f is continuous:

$$|f(x) - f(u)| \leq \int |h(x, y) - h(u, y)|\,dy$$

$$\leq |x - u| \int \bar{g}(y)\,dy \leq c|x - u|,$$

where $c = \int \bar{g}(y)\,dy < \infty$.

Consider a function φ given by

$$\varphi(u, v) = \int_u^v f(x)\,dx \quad (u \leq v).$$

Since f is continuous, $\varphi(u, \cdot)$ is differentiable and its derivative at $x \in [u, v]$ is given by

$$\mathcal{D}\varphi(u, x) = f(x) = \int h(x, y)\,dy.$$

Observe that $\mathcal{D}\varphi(u, x) = \mathcal{D}\psi(u, x)$ ($x \in [u, v]$), and consequently

$\varphi(u, x) - \psi(u, x) = c$ ($x \in [u, v]$). Since $\varphi(u, u) = \psi(u, u) = 0$, $c = 0$. Thus $\varphi(u, v) = \psi(u, v)$ ($u \leq v$), therefore

$$\int_{-\infty}^{\infty} (\int h(x, y) dy) dx = \lim_{u \to -\infty, v \to \infty} \varphi(u, v)$$

$$= \lim_{u \to -\infty, v \to \infty} \psi(u, v) = \int (\int_{-\infty}^{\infty} h(x, y) dx) dy.$$

We see that the left-hand side of the statement of Theorem 3' also makes sense and coincides with the right-hand side. Theorem 3' is established. ■

The theorems 1, 2, 3, 2', and 3' present the rules of operations under the integral sign. These theorems have a variety of frequent and effective applications.

2.8. Indefinite Integrals

This name is used to denote frequently recurring measures whose values are expressed by integrals.

2.8.1. The Radon-Nikodym Theorem

Let U be a set, $\mathcal{A}$ an algebra of parts of U, μ a measure on $\mathcal{A}$, $\mathcal{B} \supseteq \mathcal{A}$ an algebra of parts of U, and ν a measure on $\mathcal{B}$. As usual we assume that these measures are positive, countably additive, complete and locally bounded.

We will say conventionally that ν *is continuous with respect to* μ and write $\nu \ll \mu$ if for any $\varepsilon > 0$ there is a $\delta > 0$ such that $\nu(A) \leq \varepsilon$ for each $A \in \mathcal{A}$ having $\mu(A) \leq \delta$.

Let $\overline{\mathcal{A}}$ and $\overline{\mathcal{B}}$ be the algebras of sets integrable by measures ν and μ, and $\overline{\mu}, \overline{\nu}$ be integral extensions of these measures. Suppose that $\mathcal{B} \supseteq \mathcal{A}$. It can be proved that the continuity of ν with respect to μ is equivalent to the continuity of $\overline{\nu}$ with respect to $\overline{\mu}$, and the continuity of $\overline{\nu}$ with respect to $\overline{\mu}$ is equivalent to the condition $\overline{\nu}(A) = 0$ for every $A \in \overline{\mathcal{A}}$ having $\overline{\mu}(A) = 0$. In what follows we will consider the integral extensions but omit overbars in the notations of algebras and measures.

Consider a function f measurable with respect to μ. If for each $A \in \mathcal{A}$ the function Af is integrable by measure μ and

$$\nu(A) = \int Af d\mu = \int_A f d\mu$$

then the measure ν is said to be the *indefinite integral* of f by measure μ, and the function f is said to be the *derivative of* ν *by measure* μ, denoted $d\nu / d\mu$ or $d\nu = f d\mu$.

2.8. Indefinite Integrals

The derivatives of measure ν with respect to μ make up a class of functions equivalent with respect to μ. Since the measures μ and ν are positive, then $d\nu/d\mu = f \geq 0$ almost everywhere with respect to μ. All these concepts can be generalized to number-valued functions of sets considering linear combinations of measures and their derivatives.

A function f such that for each $A \in \mathcal{A}$ the function Af is integrable by measure μ will be called *locally integrable by* μ. If we assume the local boundedness of μ, then the local integrability of f implies its μ measurability. Derivatives with respect to measure μ are locally integrable by μ.

Applying the local integrability of derivative and Lebesgue's theorem it is not hard to see that indefinite integrals with respect to measure μ are continuous with respect to it. The more subtle converse statement is also true: if a measure ν is continuous with respect to another measure μ then ν is an indefinite integral by μ.

THEOREM (Radon-Nikodym). *A measure ν is the indefinite integral of a locally integrable function by measure μ if and only if ν is continuous with respect to μ.*

This theorem can be deduced from the Riesz theorem on representation of linear functionals on Hilbert space. The Radon-Nikodym theorem abounds with useful corollaries.

2.8.2. Theorem on the Change of Variables in Integration

This theorem provides one of the basic tools for computing integrals.

THEOREM. *Let f be measurable with respect to μ, integrable with respect to ν, and ν be continuous with respect to μ. Then the product $f(d\nu/d\mu)$ is integrable by μ and*

$$\int f(d\nu/d\mu)\,d\mu = \int f\,d\nu. \tag{1}$$

PROOF. By the theorem the r.h.s. integral exists, and so we need to prove the existence of the l.h.s. integral. The existence of the derivative $d\nu/d\mu$ follows from the Radon-Nikodym theorem.

Among the derivatives of ν with respect to μ equivalent with respect to μ we distinguish one $d\nu/d\mu = g$ with positive values. We will prove the integrability of the product fg and the equality of the integrals under consideration first for sets, then for simple functions, positive integrable functions, and, finally, real-valued integrable functions.

(i) Let $f = X \in \mathcal{A}$. The Radon-Nikodym theorem implies the integrability of $fg = Xg$ and

$$\int f\,d\nu = \int X\,d\nu = \nu(X) = \int Xg\,d\mu = \int fg\,d\mu.$$

(ii) Let $f = \Sigma c_i X_i$ ($c_i \in \mathbb{R}$, $X_i \in \mathcal{A}$). Then by the result of (i) fg is integrable and

$$\int f d\nu = \Sigma c_i \int X_i d\nu = \Sigma c_i \int X_i g d\mu = \int fg d\mu.$$

(iii) Let f be positive, measurable by μ and integrable by ν. The first two properties guarantee that there exists an increasing sequence of simple functions

$$f_n = \Sigma c_{in} X_n \quad (c_{in} \in \mathbb{R}, \ X_{in} \in \mathcal{A}),$$

which converges to f almost everywhere by μ. Since $g \geq 0$, $f_n \uparrow f$ implies $f_n g \uparrow f_g (\mu)$. From (ii) and the ν integrability of f by ν it follows that $f_n g$ are integrable by μ and

$$\int f_n g d\mu = \int f_n d\nu \leq \int f d\nu < \infty.$$

(Since $\mathcal{A} \subseteq \mathcal{B}$, f_n are ν integrable by ν).

Applying Levy's theorem to the sequences $\{f_n\}$ and $\{f_n g\}$ we deduce from the proved that fg is integrable by μ and

$$\int f d\nu = \lim \int f_n d\mu = \lim \int f_n g d\mu = \int fg d\mu.$$

(iv) Let $f: U \to \mathbb{R}$ be measurable by μ and integrable by ν. Then so are its positive components f^+ and f^-. By (iii), the theorem holds for them. Consequently $f^+ g$, $f^- g$, $fg = (f^+ - f^-)g = f^+ g - f^- g$ are integrable by μ with

$$\int f d\nu = \int f^+ d\nu - \int f^- d\nu = \int f^+ g d\mu - \int f^- g d\mu$$
$$= \int (f^+ - f^-) g d\mu = \int fg d\mu.$$

(v) Let $f: U \to \mathbb{C}$ be measurable by μ and integrable by ν. Then so are its real and imaginary parts. By (iv), the theorem is valid for them, and so it holds for f.

The proof of the theorem is completed. ∎

NOTE. This section makes substantial use of the assumption of integral extension of the measures. In particular, without this assumption the statement of item (iii), that there exists an increasing sequence of simple functions, is no longer valid.

Let X and Y be open sets in $\mathbb{R}^m$, $\mathcal{A}$ and $\mathcal{B}$ be algebras induced by bounded open sets in X and Y, $\lambda = dx$ and $\mu = dy$ be Lebesgue measures on $\mathcal{A}$ and $\mathcal{B}$, T be a smooth homeomorphism from X to Y, T' be its derivative and $|\det T'|$ be the absolute value of its determinant, g be a number-valued function integrable by dy on Y, and $f = gT$ be a composition on X.

2.8. Indefinite Integrals

Since T is a homeomorphism from X to Y, the Lebesgue measure μ on $\mathcal{B}$ and T defines the measure $\nu = \mu T$ on $\mathcal{A}$ with values

$$\nu(A) = \mu(T(A)) \quad (A \in \mathcal{A}).$$

The measure $\nu = \mu T$ is continuous with respect to the Lebesgue measure λ and

$$d\mu T / d\lambda = |\det T'|. \tag{2}$$

In addition, the composition $f = gT$ is integrable by measure $\nu = \mu T$ and

$$\int_Y g \, d\mu = \int_X gT \, d\mu T. \tag{3}$$

For $g = B \in \mathcal{B}$ this equality is equivalent to identity $\mu(B) = \mu(B)$. For functions, it is proved step by step like equation (1). From (1)–(3) it follows that

$$\int_Y g \, d\mu = \int_X gT \, d\mu T = \int_X gT (d\mu T / d\lambda) \, d\lambda = \int_X gT |\det T'| \, d\lambda.$$

Rewriting the first and last integrals in classical notations we obtain the formula for the change of variables in multiple integrals

$$\int_Y g(y) \, dy = \int_X gT(x) \cdot |\det T'(x)| \cdot dx. \tag{4}$$

Here the derivative $T'(x)$ of the mapping T at $x \in X$ is represented by an $m \times m$ matrix of partial derivatives. The equality (4) does not require $\det T'(x) \neq 0$.

Equation (3) is a special case of the general formula for the change of variables in integration. Let X and Y be abstract sets, $\mathcal{A}$ and $\mathcal{B}$ algebras of sets in X and Y, λ a measure on $\mathcal{A}$, $T: X \to Y$ a mapping such that $T^{-1}(B) = A \in \mathcal{A}$ for every $B \in \mathcal{B}$, $\mu = \lambda T^{-1}$ a measure on $\mathcal{B}$ with values

$$\mu(B) = \lambda(T^{-1}(B)) \quad (B \in \mathcal{B}),$$

g a number-valued integrable by μ function on Y, and $f = gT$ a composition on X.

Using the line of argument of the proof of the theorem and equation (3) it is not hard to verify the equality

$$\int_B g \, d\mu = \int_A f \, d\lambda \quad (f = gT, \ A = T^{-1}(B), \ \mu = \lambda T^{-1}) \tag{5}$$

for each $B \in \mathcal{B}$. The equality (3) may be obtained from (5) with $\lambda = \mu T$ using the local boundedness of the measures λ and μ.

The formula (5) allows one to reduce the calculation of integrals over curves and surfaces to one of integrals over domains in finite-dimensional spaces, when T describes a parametrization of the curve or surface.

NOTE. As a rule this chapter was focused on countably additive, locally bounded and complete measures. An important exclusion is the product of measures which may be an incomplete measure. The Fubini's theorem holds for its completion as well.

Chapter 3. LINEAR OPERATORS

This chapter expounds on elements of the theory of linear operators on normed spaces and describes some specific spaces.

3.1. Hilbert Spaces

Normed spaces have been defined in Chapter 1, therefore we immediately proceed to Hilbert spaces which are complete normed spaces under the norm given by an inner product. Hilbert spaces are a particular case of Banach spaces.

3.1.1. Euclidean Spaces

Consider a vector space E over a scalar field $\mathbb{F}$; E is called *real* if $\mathbb{F} = \mathbb{R}$, and *complex* if $\mathbb{F} = \mathbb{C}$. In general one can add vectors and multiply them by a number, but cannot measure their length and angles between them. To enable such measurements we define an inner product over E.

In a real case we define an inner product for E as a positive symmetric bilinear function $p = E \times E \to \mathbb{F}$. In general the definition is a little more complicated. Consider a numerical valued function p of two vector variables x and y. We will call p a *form* and write its values as a product or indicate by angular brackets

$$p(x, y) = x \cdot y = xy = \langle x, y \rangle \quad (x, y \in E).$$

We shall call p a *Hermitian form* if it is linear in the second variable and yields its complex-conjugate on transposing its arguments (remains the same in the real case):

$$x(\beta y) = \beta(xy),$$
$$x(y + z) = xy + xz,$$
$$yx = \overline{xy} \quad (\beta \in \mathbb{F}; \ x, y, z \in E).$$

3.1. Hilbert Spaces

Every form p defines a function $q: E \to \mathbb{F}$ given by

$$q(x) = p(x, x) \quad (x \in E).$$

It is called a *quadratic form* induced by p. If p is a Hermitian form, then

$$q(x) = xx = \overline{xx} = \overline{q(x)} \quad (x \in E)$$

and the quadratic form q is real-valued. Among real quadratic forms we distinguish positive and nondegenerate, which satisfy the conditions

$$q(x) \geqslant 0 \quad (x \in E),$$
$$q(x) > 0 \quad (x \neq 0, \; x \in E).$$

We use these properties of q to define over E a norm separating the points of E.

DEFINITION 1. *A Hermitian form on $E \times E$ inducing a positive quadratic form on E is called an inner product for E.*

If this quadratic form is nondegenerate, then the inner product is said to be nondegenerate, too.

Let E be an inner product vector space.

DEFINITION 2. *The number*

$$\|x\| = (xx)^{1/2}$$

is said to be the length of a vector $x \in E$.

The following inequality sets up a correspondence between length and inner product.

THEOREM (Cauchy's inequality). $|xy| \leqslant \|x\| \cdot \|y\| \; (x, y \in E).$

PROOF. Observe that

$$xx = \|x\|^2, \quad yy = \|y\|^2,$$

$$(\lambda x + y)(\lambda x + y) = \|\lambda x + y\|^2$$
$$= |\lambda|^2 \cdot \|x\|^2 + \bar{\lambda} \cdot (xy) + \lambda \cdot (\overline{xy}) + \|y\|^2 \geqslant 0$$

for any $x, y \in E$ and $\lambda \in \mathbb{F}$.

If $\|x\| \neq 0$, then for $\lambda = -xy / \|x\|^2$ we have

$$\frac{|xy|^2}{\|x\|^2} - \frac{|xy|^2}{\|x\|^2} - \frac{|xy|^2}{\|x\|^2} + \|y\|^2 \geqslant 0,$$

$$-|xy|^2 + \|x\|^2 \cdot \|y\|^2 \geqslant 0.$$

By similar reasoning we can see that the Cauchy inequality holds also for $\|y\| \neq 0$. If $\|x\| = \|y\| = 0$, then, for $\lambda = -xy$, we obtain

$$-|xy|^2 - |xy|^2 \geq 0,$$

$$|xy| = 0 \leq |xy| \leq \|x\| \cdot \|y\|.$$

Thus, the Cauchy inequality holds for all $x, y \in E$. ∎

Applying Cauchy's inequality it is straightforward to prove that the function $\|\cdot\| : E \to R$ defined by $\|x\| = (xx)^{1/2}$ ($x \in E$) is in fact a norm over E. Indeed, since

$$\|x + y\|^2 = (x + y)(x + y) = \|x\|^2 + 2\operatorname{Re}(xy) + \|y\|^2$$
$$\leq \|x\|^2 + 2\|x\| \cdot \|y\| + \|y\|^2 = (\|x\| + \|y\|)^2,$$

then

$$\|x + y\| \leq \|x\| + \|y\| \quad (x, y \in E).$$

Also

$$\|\alpha x\| = (\alpha x \cdot \alpha x)^{1/2} = (|\alpha|^2 \cdot xx)^{1/2} = |\alpha| \cdot \|x\|$$

for any $\alpha \in \mathbb{F}$ and $x \in E$.

The norm $\|\cdot\|$ given by $\|x\| = (xx)^{1/2}$ ($x \in E$) is known as the *Euclidean norm*. The Euclidean norm $\|\cdot\|$ separates points of E if and only if the inner product defining it is nondegenerate.

An inner product vector space under the Euclidean norm is a Euclidean space. If the inner product is nondegenerate, then the Euclidean space is said to be *separated*.

Cauchy's inequality enables us to define a convenient measure for vectorial angles.

DEFINITION 3. *The number*

$$\varphi(x, y) = \arccos[xy / (\|x\| \cdot \|y\|)]$$

is said to be the value of the angle between vectors $x, y \in E$ *with nonzero lengths.*

The function arccos is defined on the complex plane $\mathbb{C}$ less the intervals $]-\infty, -1[$ and $]1, \infty[$. In particular, arccos $(-1) = \pi$ and arccos $(1) = 0$. When the inner product xy is real, $\varphi(x, y) \in [0, \pi]$.

Note that

$$xy = \|x\| \cdot \|y\| \cdot \cos \varphi(x, y)$$

DEFINITION 1. *Two vectors x, y are said to be orthogonal, symbolically* $x \perp y$, *if and only if* $(xy) = 0$.

The zero vector is orthogonal to any other. If a space is separable, then this property is unique for the zero vector (then $x \cdot x = 0$ implies $x = 0$). The angle between nonzero orthogonal vectors is arccos $(0) = 2^{-1}\pi$.

THEOREM (*Pythagorean theorem*). *If* $x \perp y$, *then*

$$\| x + y \|^2 = \| x \|^2 + \| y \|^2.$$

PROOF. If $xy = 0$, then from the triangle identity it follows that

$$\| x + y \|^2 = \| x \|^2 + \| y \|^2 + 2\,\mathrm{Re}\,(xy) = \| x \|^2 + \| y \|^2,$$

thus proving the Pythagorean theorem. ∎

To provide a geometrical insight we will refer to elements of a Euclidean space E as *points* or *vectors*. Consider a subspace A of a Euclidean space E and a point $z \in E$.

DEFINITION 2. *A point* $p \in A$ *such that* $(z - p) \perp (x - p)$ *for all* $x \in A$ *is called the orthogonal projection of* z *on* A.

The point p is a base of a perpendicular dropped from z on A.

DEFINITION 3. *A point* $a \in A$ *such that* $\| z - a \| = \min \{ \| z - x \| : x \in A \}$ *is said to be closest to* $z \in E$ *in* A.

Existence and uniqueness conditions for the orthogonal projection and closest point will be found below.

3.1.3. Orthogonal Projection

First we formulate the conditions for an orthogonal projection of a point on a subspace of a Hilbert space to exist. By saying a *subspace* we shall refer to closed subspaces of a Hilbert space (themselves Hilbert spaces).

THEOREM. *Let z be a point in a Hilbert space H, and A be a subspace of H. Then*:

(i) *there exists a point* $a \in A$ *closest to* z;

(ii) *every* $a \in A$ *closest to* z *is the orthogonal projection of* z *on* A;

(iii) *every orthogonal projection p of z on A is a point of A closest to z*.

PROOF. (i) Since $\| z - x \| \geq 0$ for all $z \in H$ and $x \in A$, there is $\alpha = \inf \{ \| z - x \| : x \in A \} \geq 0$. By the definition of a lower bound, for every number $n \in \mathbb{N}$ there is a point $x(n) \in A$ such that $\| z - x(n) \| \leq \alpha + 1/n$.

We will prove that the sequence $x(n)$ converges. Consider $z - x(m)$,

3.1. Hilbert Spaces

for $x, y \in E$ such that $\|x\| \neq 0$, $\|y\| \neq 0$ (if E is separable, we may write $x \neq 0$, $y \neq 0$).

Euclidean spaces, which are complete and endowed with a norm, are called *Hilbert spaces*.

3.1.2. Geometry of Euclidean Spaces

Many elementary geometry theorems hold in Euclidean spaces. They all follow from the properties of the inner product.

In the identities to follow, x, y, z will always denote arbitrary elements of a Euclidean space E.

THEOREM (*the triangle identity*).

$$\|x - y\|^2 = \|x\|^2 + \|y\|^2 - 2\operatorname{Re}(xy).$$

PROOF.

$$\|x - y\|^2 = (x - y)(x - y) = xx - xy - yx + yy$$
$$= xx - xy - \overline{xy} + yy = \|x\|^2 - 2\operatorname{Re}(xy) + \|y\|^2. \blacksquare$$

THEOREM (*the parallelogram law*).

$$\|x - y\|^2 + \|x + y\|^2 = 2(\|x\|^2 + \|y\|^2).$$

PROOF. Adding the triangle identities for x, y and $x, -y$, we obtain

$$\|x - y\|^2 + \|x + y\|^2 = \|x\|^2 + \|y\|^2 - 2\operatorname{Re}(xy)$$
$$+ \|x\|^2 + \|y\|^2 + 2\operatorname{Re}(xy) = 2(\|x\|^2 + \|y\|^2). \blacksquare$$

Applying the triangle identity and the parallelogram law it is straightforward to derive

The polarization identity

$$xy = 4^{-1}[(\|x + y\|^2 - \|x - y\|^2) - i(\|x + iy\|^2 - \|x - iy\|^2)].$$

It can be proved that a Euclidean space can be constructed from a normed space with the aid of the polarization identity if and only if the norm obeys the parallelogram law.

In the real case, the triangle identity and the polarization identity simplify to become

$$\|x - y\|^2 = \|x\|^2 + \|y\|^2 - 2xy,$$
$$xy = 2^{-1}[\|x\|^2 + \|y\|^2 - \|x - y\|^2].$$

3.1. Hilbert Spaces

$z - x(n)$ for arbitrary $m, n \in \mathbb{N}$ and observe that

$$(z - x(n)) - (z - x(m)) = x(m) - x(n),$$
$$(z - x(n)) + (z - x(m)) = 2[z - 2^{-1}(x(m) + x(n))],$$
$$x = 2^{-1}(x(m) + x(n)) \in A, \quad \|z - x\| \geq \alpha.$$

Applying the parallelogram identity to $z - x(n)$ and $z - x(m)$ we have

$$\|x(m) - x(n)\|^2 + 4\alpha^2 \leq \|x(m) - x(n)\|^2 + \|2(z - x)\|^2$$
$$\leq 2[\|z - x(n)\|^2 + \|z - x(m)\|^2]$$
$$\leq 2[(\alpha + 1/n)^2 + (\alpha + 1/m)^2],$$
$$\|x(m) - x(n)\|^2 \leq 4\alpha(1/m + 1/n) + 2(1/m^2 + 1/n^2).$$

Hence, the sequence $x(n)$ converges.

Since H is complete, there is $a \in H$ to which a sequence $x(n) \in A$ converges, moreover, $a \in A$, for A is closed. We prove that a is closest to z in A. Since $\alpha \leq \|z - x(n)\| \leq \alpha + 1/n$, we have $\|z - a\| = \lim \|z - x(n)\| = \alpha$. Hence, $\|z - a\| = \min\{\|z - x\| : x \in A\}$, which proves statement (i).

(ii) Consider the closest a to z in A, an arbitrary $x \in A$, and an arbitrary point on the line $y = a + \lambda(x - a)$, $\lambda \in \mathbb{R}$ passing through these points. Since A is a subspace, then $y \in A$. Therefore,

$$\alpha^2 \leq \|z - y\|^2 = (z - y)(z - y)$$
$$= [(z - a) - \lambda(x - a)][(z - a) - \lambda(x - a)]$$
$$= \|z - a\|^2 - 2\lambda \cdot \text{Re}[(z - a)(x - a)] + \lambda^2 \|x - a\|^2.$$

However $\|z - a\|^2 = \alpha^2$. Hence,

$$2\lambda \cdot \text{Re}[(z - a)(x - a)] \leq \lambda^2 \|x - a\|^2,$$
$$\text{Re}[(z - a)(x - a)] \leq 2^{-1}\lambda \|x - a\|^2 \quad (\lambda > 0),$$
$$-\text{Re}[(z - a)(x - a)] \leq 2^{-1}(-\lambda)\|x - a\|^2 \quad (\lambda < 0).$$

We obtain

$$0 \leq |\text{Re}[(z - a)(x - a)]| \leq 2^{-1} \cdot |\lambda| \cdot \|x - a\|^2 \quad (\lambda \neq 0),$$
$$\text{Re}[(z - a)(x - a)] = 0.$$

By a similar argument, introducing $i = \sqrt{-1}$, we find
$$\mathrm{Im}[(z - a)(x - a)] = 0.$$
Consequently, a is the orthogonal projection of z on A.

The proof of statement (ii) is completed.

(iii) Consider the orthogonal projection p of a point z on A and an arbitrary $x \in A$. By the Pythagorean theorem
$$\|z - x\|^2 = \|(z - p) + (p - x)\|^2 = \|z - p\|^2 + \|p - x\|^2$$
$$\geq \|z - p\|^2.$$
Hence, $\|z - p\| = \min\{\|z - x\| : x \in A\}$, and p is closest to z in A. This completes the proof of the theorem. ∎

Now we find the uniqueness condition of the orthogonal projection of z on a subspace A of H.

PROPOSITION. (i) *Let $a \in A$ be the point closest to z. Then $b \in H$ is closest to z in A if and only if $\|a - b\| = 0$.*

(ii) *Let p be the orthogonal projection of z on A. Then $q \in H$ is also an orthogonal projection of z on A if and only if $\|p - q\| = 0$.*

PROOF. (i) Let $\|z - a\| = \alpha = \{\min\|z - x\| : x \in A\}$, $b \in H$ and $\|a - b\| = 0$. Then $b \in A$, for A is closed, and $a_n = a \to b$, $a_n \in A$. By the triangle inequality,
$$\alpha \leq \|z - b\| \leq \|z - a\| + \|a - b\| = \|z - a\| = \alpha,$$
that is, $\|z - b\| = \alpha$ and b is closest to z in A.

Now, let a, b be closest to z in A. Then, in view of the parallelogram law,
$$\|a - b\|^2 = 2[\|z - a\|^2 + \|z - b\|^2] - \|(z - a) + (z - b)\|^2$$
$$\leq 4\alpha^2 - 4\|z - 2^{-1}(a + b)\|^2 \leq 4\alpha^2 - 4\alpha^2 = 0.$$
Hence, $\|a - b\| = 0$.

(ii) The second assertion follows from the first and statement (iii) of the orthogonal projection theorem. Thus, the proposition is established. ∎

If a Hilbert space is separated, then the closest point a and the orthogonal projection p are unique.

The orthogonal projection theorem tells us that each point $z \in H$ can be represented as the sum of its orthogonal projection $a \in A$ and $c = z - a$. In view of $0 \in A$,
$$ac = -(0 - a)(z - a) = 0,$$

3.1. Hilbert Spaces

that is, $a \perp c$. Moreover, since $x + a \in A$ for any $x \in A$,
$$xc = ((x + a) - a)(z - a) = 0,$$
that is, $x \perp c$ for all $x \in A$. This means that the vector $c = z - a$ is orthogonal to A, i.e., belongs to its *orthogonal complement* $A^\perp = \{ y \in H : xy = 0 \ (x \in A) \}$. The equality
$$x + y = z \quad (x \in A, \ y \in A^\perp)$$
is called an *orthogonal decomposition* of z on A and its orthogonal complement $A^\perp$. If $x_1 + y_1 = x_2 + y_2 = z$ $(x_1, x_2 \in A; y_1, y_2 \in A^\perp)$, then
$$r = x_1 - x_2 = y_2 - y_1 \in A \cap A^\perp,$$
$$\|r\|^2 = rr = (x_1 - x_2)(y_2 - y_1) = 0,$$
$$\|r\| = \|x_1 - x_2\| = \|y_2 - y_1\| = 0.$$

If H is separated, then this means that $x_1 = x_2$ and $y_1 = y_2$. Stated differently, for a separated space H, the orthogonal decomposition $x + y = z$ is unique.

If we call vectors, whose difference is of zero length, *equivalent*, one can say that in the general case any orthogonal decomposition is unique, accurate to equivalence. For separated spaces equivalence turns into equality.

3.1.4. Linear Functionals

Scalar-valued functions of vector variables are commonly referred to as *functionals*. Linear functionals are linear operators, therefore all we have said about linear operators in Section 1.1.3 relates to them. Linear functionals have many important special properties in addition to the general properties of linear operators.

Continuous linear functionals on a normed space E provide a convenient system of coordinates for E. They also constitute a normed space said to be *conjugate* to E and denoted by E'. The vector space of all linear functionals on E is denoted by E^*. It is also called conjugate to E. If E is finite-dimensional, then $E' = E^*$. In the general case, $E' \subseteq E^*$.

The set
$$\text{Ker } \varphi = \varphi^{-1}(0) = \{ z \in E : \varphi(z) = 0 \}$$
is called the kernel of a linear functional φ.

PROPOSITION. *A linear functional is continuous if and only if its kernel is closed.*

PROOF. (1) Let $\varphi \in E'$, $z_n \in \text{Ker } \varphi$, $z \in E$, and $z_n \to z$. Then $\varphi(z) = \lim \varphi(z_n) = 0$ and $z \in \text{Ker } \varphi$. Thus, $\text{Ker } \varphi$ is closed.

(2) Let $\varphi \in E^*$ and $\text{Ker } \varphi$ is closed, to prove that $\varphi \in E'$, it suffices to demonstrate that φ is continuous at 0. Suppose the converse is true, i.e., $\varphi(z_n) \not\to 0$ for a sequence $z_n \to 0$, $z_n \in E$. Then $|\varphi(z_{n(k)})| \geq \alpha$ for some $\alpha > 0$ and $n(1) < n(2) < ... < n(k) < ...$. Hence,

$$\|y_k/\varphi(y_k)\| = \|y_k\|/|\varphi(y_k)| \leq \|y_k\|/\alpha \to 0$$

for $y_k = z_{n(k)}$. However

$$\varphi(y_k/\varphi(y_k)) = \varphi(y_k)/\varphi(y_k) = 1.$$

Therefore

$$\varphi(x_k) = \varphi[y_1/\varphi(y_1) - y_k/\varphi(y_k)] = 1 - 1 = 0,$$

and

$$x_k = y_1/\varphi(y_1) - y_k/\varphi(y_k) \in \text{Ker } \varphi.$$

Since $\|y_k/\varphi(y_k)\| \to 0$ and $\text{Ker } \varphi$ is closed,

$$y_1/\varphi(y_1) = \lim[x_k + y_k/\varphi(y_k)] = \lim x_k + \lim[y_k/\varphi(y_k)]$$
$$= \lim x_k + 0 = \lim x_k \in \text{Ker } \varphi.$$

In other words, $\varphi[y_1/\varphi(y_1)] = 0$. This contradicts to $\varphi[y_1/\varphi(y_1)] = 1$. ∎

Kernels of linear functionals are also called *hyperplanes*. We proved above that the kernel of a continuos linear functional is a closed hyperplane. Because functionals are linear, their kernels are subspaces of a vector space on which the functionals are defined.

In the following we intend to characterize continuous linear functionals on a Hilbert space. Consider a Hilbert space H, a vector $c \in H$, a space H' conjugate to H, of linear functionals on H.

LEMMA 1. *The equality $\varphi(z) = cz$ ($z \in H$) defines a continuous linear functional φ on E.*

PROOF. Indeed, this equality assigns to each vector $z \in H$ a scalar $cz \in \mathbb{F}$, thus defining a function φ on H which yields the value $\varphi(z) = cz$ for a vector z. Since the inner product is linear in the second variable, φ is a linear functional:

$$\varphi(\alpha_1 z_1 + \alpha_2 z_2) = c(\alpha_1 z_1 + \alpha_2 z_2)$$
$$= \alpha_1(cz_1) + \alpha_2(cz_2) = \alpha_1 \varphi(z_1) + \alpha_2 \varphi(z_2)$$

for all $z_1, z_2 \in H$ and $\alpha_1, \alpha_2 \in \mathbb{F}$.

3.1. Hilbert Spaces

By the Cauchy inequality, $|\varphi(z)| = |cz| \leq \|c\| \cdot \|z\|$ $(z \in H)$. Thus, φ is continuous at zero, and hence continuous. ∎

Let $b, c \in \mathbb{F}$.

LEMMA 2. *If H is separated, then $bz = cz$ $(z \in H) \Leftrightarrow \|b - c\| = 0$.*

PROOF. If $bz = cz$ $(z \in H)$, then $(b - c)z = 0$ $(z \in H)$. In particular, $\|b - c\|^2 = (b - c)(b - c) = 0$ for $z = b - c$. Consequently, $\|b - c\| = 0$.

If $\|b - c\| = 0$, then the Cauchy inequality implies that $0 \leq \|bz - cz\| = |(b - c)z| \leq \|b - c\| \cdot \|z\| = 0$, therefore $bz = cz$, as desired. ∎

COROLLARY. *If H is separated, then*

$$bz = cz \ (z \in H) \Leftrightarrow b = c.$$

LEMMA 3. *Let $\varphi \in H'$ be given by $\varphi(z) = cz$ $(z \in H)$ for some $c \in H$. Then $\|\varphi\| = \|c\|$.*

PROOF. By definition $\|\varphi\| = \sup\{|\varphi(u)| : u \in H, \|u\| \leq 1\}$. The Cauchy inequality yields

$$|\varphi(u)| = |cu| \leq \|c\| \cdot \|u\| \leq \|c\| \quad (u \in H, \|u\| \leq 1).$$

Consequently, $\|\varphi\| \leq \|c\|$.

Now we prove the opposite inequality. Note that $\varphi(c) = cc = \|c\|^2$. If $\|c\| = 0$, then $\|\varphi\| = \|c\| = 0$. If $\|c\| \neq 0$, then

$$\varphi(u) = \varphi(\gamma c) = \gamma \varphi(c) = \gamma \cdot \|c\|^2 = \|c\|$$

for $u = \gamma c$ and $\gamma = \|c\|^{-1}$. Since $\|u\| = \gamma \cdot \|c\| = 1$, we have $|\varphi(u)| \leq \|\varphi\|$, hence $\|c\| \leq \|\varphi\|$. ∎

The vector $c \in H$ defining a continuous linear functional $\varphi \in H'$ with values $\varphi(z) = cz$ $(z \in H)$ will be called the *coefficient* of φ.

Lemmas 1–3 define a class of continuous linear functionals on a Hilbert space H. A natural question is then whether or not this class contains all continuous functionals on H. The answer is given by

THEOREM (Riesz). *For each $\varphi \in H'$ there is such $c \in H$ that $\varphi(z) = cz$ $(z \in H)$.*

PROOF. If $\varphi = 0$, then $c = 0$.

Let $\varphi \neq 0$. Then $A = \operatorname{Ker} \varphi$ is a closed hyperplane in H. By the orthogonal projection theorem, for each $z \in H$ there is $x \in A$ and $y \in B = A^\perp$ such that $x + y = z$, and there is $b \in B$ and $b \notin A$. (Otherwise $B \subseteq A$, $x + y \in A$ for all $x \in A$, $y \in B$ and $A = H$, which contradicts to $\varphi \neq 0$). As a consequence we have $H = A + \mathbb{F} b$.

Thus, for every $z \in H$ there exist $x \in H$, $t \in \mathbb{F}$ such that $z = x + tb$, where $b \notin A$ is a vector chosen from $B = A^\perp$. We note that $\|b\| \neq 0$. (Otherwise, $b = \lim a_n$ for $a_n = 0 \in A$, and since A is closed, we would have $b \in A$.)

Now we choose the coefficient c for φ. Because $H = A + \mathbb{F} b$, for every $c, z \in H$ there are $a, x \in A$ and $\gamma, t \in \mathbb{F}$ such that $c = a + \gamma b$, $z = x + tb$. Therefore,

$$\varphi(z) = \varphi(x + tb) = \varphi(x) + t\varphi(b) = t\varphi(b),$$
$$cz = (a + \gamma b)(x + tb) = ax + \bar{\gamma} t \cdot \|b\|^2.$$

(The inners $bx = ab = 0$ for $a, x \in A$ and $b \in A^\perp$, and $bb = \|b\|^2$.) If $a = 0$ and $\bar{\gamma} = \|b\|^{-2} \cdot \varphi(b)$, then

$$\varphi(z) = t\varphi(b) = t\varphi(b) \cdot \|b\|^{-2} \cdot \|b\|^2 = \bar{\gamma} t \cdot \|b\|^2 = cz.$$

Thus,

$$c = 0 + \overline{\varphi(b)} \cdot \|b\|^{-2} \cdot b = \overline{\varphi(b)} \cdot \|b\|^{-2} \cdot b.$$

The Riesz theorem is established. ∎

A brief formulation of this theorem may be as follows: *every continuous linear functional on a Hilbert space has a coefficient.*

The Riesz theorem allows us to consider points in a Hilbert space instead of the linear functionals, which often proves useful.

3.1.5. Hilbert Space $\mathscr{L}^2$

The points of this space are measurable square-integrable functions.

As before we consider a (positive, countably additive, complete, locally bounded) measure μ on an algebra A of subsets of a set U. Together with μ use is often made of its integral extension $\bar{\mu}$ to the algebra $\bar{A}$ of all the sets integrable with respect to μ that also denotes μ.

Define $\mathscr{L}^2 = \mathscr{L}^2(U, \mu)$ to be the set of functions $f \colon U \to \mathbb{F}$ measurable with respect to μ whose squares $f^2 \colon U \to \mathbb{F}$ are integrable with respect to μ. Thus, by definition, for a function f measurable with respect to μ we have

$$f \in \mathscr{L}^2 \Leftrightarrow f^2 \in \mathscr{L},$$

where $\mathscr{L}$ is the set of number-valued functions on U integrable with respect to μ.

Since

$$|fg| \leq 2^{-1}(|f|^2 + |g|^2),$$

3.1. Hilbert Spaces

we have $fg \in \mathscr{L}$ for all $f, g \in \mathscr{L}^2$. (We do not claim that $fg \in \mathscr{L}^2$. In addition f, g constituting the integrable product $f \cdot g$ need not be integrable themselves.)

It is straightforward to verify that $\mathscr{L}^2$ endowed with usual operations sets up a vector space:

$$(\alpha f + \beta g)^2 = \alpha^2 f^2 + 2\alpha\beta fg + \beta^2 g^2 \in \mathscr{L},$$

if $f^2, g^2 \in \mathscr{L}$, and so $fg \in \mathscr{L}$.

The equality

$$\langle f, g \rangle = \int \bar{f} g \quad (f, g \in \mathscr{L}^2)$$

defines an inner product for $\mathscr{L}^2$. Indeed,

$$\langle f, g \rangle = \int \bar{f} g = \overline{\int \bar{g} f} = \overline{\langle g, f \rangle},$$

$$\langle \alpha_1 f_1 + \alpha_2 f_2, g \rangle = \int (\overline{\alpha_1} \bar{f_1} + \overline{\alpha_2} \bar{f_2}) g = \overline{\alpha_1} \int \bar{f_1} g + \overline{\alpha_2} \int \bar{f_2} g$$
$$= \overline{\alpha_1} \langle f_1, g \rangle + \overline{\alpha_2} \langle f_2, g \rangle;$$

$$\langle f, \beta_1 g_1 + \beta_2 g_2 \rangle = \beta_1 \langle f, g_1 \rangle + \beta_2 \langle f, g_2 \rangle,$$

$$\langle f, f \rangle = \int \bar{f} f = \int |f|^2 \geq 0.$$

By the general rule of Euclidean spaces, the equality

$$\|f\| = \langle f, f \rangle^{1/2} = (\int |f|^2)^{1/2} \quad (f \in \mathscr{L}^2)$$

defines the norm $\|\cdot\|$ for $\mathscr{L}^2$. Under this norm $\mathscr{L}^2$ can be nonseparated.

EXAMPLES.
1. $\mathscr{L}^2 = \mathscr{L}^2(\mathbb{R}, dx)$.
2. $\mathscr{L}^2 = \mathscr{L}^2([0,1], dx)$.
3. $\mathscr{L}^2 = \mathscr{L}^2(\mathbb{N}, dn) = l^2$.

$\mathscr{L}^2$ of Examples 1–2 is nonseparated, and in Example 3, under a counting measure l^2, it is separated. The points of l^2 are numerical sequences $\{x_n\}_{n=1}^{\infty}$ which have $\sum_{n=1}^{\infty} |x_n|^2 < \infty$.

Before we prove that $\mathscr{L}^2$ is a Hilbert space, we will prove a lemma relating convergence in $\mathscr{L}^2$ to convergence almost everywhere. A sequence converging *in* $\mathscr{L}^2$ is said to converge in the (quadratic) mean.

DEFINITION. *A sequence* $\{f_n\} \in \mathscr{L}^2$ *converges in the mean to* $f \in \mathscr{L}^2$

if and only if

$$\|f_n - f\| = (\int |f_n - f|^2)^{1/2} \to 0.$$

Convergence in the mean does not imply convergence almost everywhere and conversely.

LEMMA (Riesz). *If a sequence of functions in $\mathscr{L}^2$ converges in the mean, then it has a subsequence which converges almost everywhere.*

PROOF. (i) Suppose that a sequence $\{f_n\} \in \mathscr{L}^2$ converges in the mean, i.e.,

$$\|f_q - f_p\| \to 0 \quad (q, p \to \infty).$$

We are to prove that there is a subsequence $\{g_n\} = \{f_{r(n)}\}$ which converges almost everywhere

$$|g_q(x) - g_p(x)| \to 0 \quad (q, p \to \infty)$$

for almost all $x \in U$ with respect to a measure μ.

We note that for every number n there is a number $r(n)$ for which

$$\|g_{n+1} - g_n\| \leq 2^{-n}$$

for $g_n = f_{r(n)}$. The numbers $r(n)$ can be chosen strictly increasing.

Indeed, since $\{f_n\}$ converges in the mean there is a number $r(1)$ for which $\|f_q - f_p\| \leq 2^{-1}$ ($p, q \geq r(1)$). Pick an arbitrary number n and suppose that for $m \leq n$ we have chosen strictly ascending numbers $r(m)$ such that $\|f_q - f_p\| \leq 2^{-m}$ ($p, q \geq r(m)$). Since $\{f_n\}$ converges in the mean, there is a number $r(n+1) > r(n)$ for which $\|f_q - f_p\| \leq 2^{-(n+1)}$ ($p, q \geq r(n+1)$). We conclude by induction that there is a strictly increasing sequence $\{r(n)\}$ for which $\|f_q - f_p\| \leq 2^{-n}$ ($p, q \geq r(n)$). In particular, this inequality holds for $p = r(n)$, $q = r(n)$, as stated.

(ii) For each number n consider the function

$$s_n = |g_1| + \sum_{1 \leq m < n} |g_{m+1} - g_m|.$$

(In particular, $s_1 = |g_1|$). Since, $g_1, g_{m+1} - g_m \in \mathscr{L}^2$, we have $|g_1|$,

3.1. Hilbert Spaces

$|g_{m+1} - g_m| \in \mathscr{L}^2$ and thus $s_n \in \mathscr{L}^2$. Clearly, $\{s_n\}$ is an increasing sequence.

(iii) The sequence of functions $h_n = s_n^2$ meets the conditions of Levy's theorem. Indeed, since $s_n \in \mathscr{L}^2$ we have $h_n \in \mathscr{L}$; $\{h_n\}$ increases together with $\{s_n\}$ and is integrally bounded:

$$\int h_n = \int s_n^2 = \|s_n\|^2 = \left\| |g_1| + \sum_{1 \leq m < n} |g_{m+1} - g_m| \right\|^2$$

$$\leq \left(\|g_1\| + \sum_{1 \leq m < n} \|g_{m+1} - g_m\| \right)^2$$

$$\leq \left(\|g_1\| + \sum_{1 \leq m < n} 2^{-m} \right)^2 \leq (\|g_1\| + 1)^2$$

for each n.

By the Levy theorem, $\{h_n\}$ converges almost everywhere to an $h \in \mathscr{L}$. Since $h_n = s_n^2 \geq 0$, then from $\{h_n\} \to h$ it follows that $s_n \to s = h^{1/2}$.

(iv) We conclude that the sequence of $|g_1|, |g_{m+1} - g_m|$ $(m \geq 1)$ sums up to s almost everywhere. Consequently, the series of $g_1, g_{m+1} - g_m$ $(m \geq 1)$ are also summable almost everywhere, that is, the sequence of partial sums of this series

$$g_n = g_1 + \sum_{1 \leq m < n} (g_{m+1} - g_m)$$

converges almost everywhere. Thus the subsequence $\{f_{r(n)}\}$ of $\{f_n\}$ converges almost everywhere, and so the lemma is established. ∎

Now we prove the central result of this section.

THEOREM. $\mathscr{L}^2$ *is a Hilbert space.*

PROOF. We must prove that a normed space $\mathscr{L}^2$ is complete, that is, each sequence of functions $f_n \in \mathscr{L}$ convergent in the mean converges to an $f \in \mathscr{L}^2$ ($\|f_n - f\| \to 0$).

Consider a sequence $f_n \in \mathscr{L}^2$ convergent in the mean. By the Riesz lemma there is an almost everywhere convergent subsequence $g_n = f_{r(n)}$. Consider $f: U \to \mathbb{F}$ with

$$f(x) = \begin{cases} \lim g_n(x), & \text{for } g_n(x) \text{ convergent,} \\ 0, & \text{for } g_n(x) \text{ divergent.} \end{cases}$$

We prove that: (i) $f \in \mathscr{L}^2$, and (ii) $f_n \to f$ in the mean.

(i) The sequence $g_n = f_{r(n)}$ converges in the mean together with $\{f_n\}$, that is, for every $\varepsilon > 0$ there is a number $n(0)$ such that $\|g_q - g_p\| \leq \varepsilon$ $(p, q \geq n(0))$.

Pick $p \geq n(0)$ and consider the sequence $h_n = |g_{p+n} - g_p|^2$ $(n \geq 1)$. It satisfies the conditions of Fatou's lemma. Indeed, since $h_n \geq 0$ and $g_{p+n} - g_p \in \mathcal{L}^2$, then $h_n \in \mathcal{L}$. It is integrally bounded:

$$\int h_n = \int |g_{p+n} - g_p|^2 = \|g_{p+n} - g_p\|^2 \leq \varepsilon^2$$

for each n. Finally, since $g_{p+n} \to f$ $(n \to \infty)$ almost everywhere, $h_n \to h = |f - g_p|^2$ almost everywhere.

Now, by Fatou's lemma, $h \in \mathcal{L}$, and

$$\int h = \int |f - g_p|^2 \leq \varepsilon^2 \quad (p \geq n(0)),$$

and $f - g_p \in \mathcal{L}^2$. Since $g_p \in \mathcal{L}^2$, then $f = (f - g_p) + g_p \in \mathcal{L}^2$.

(ii) The last inequality means that for every $\varepsilon > 0$ there is an $n(0)$ for which $\|f - g_p\| \leq \varepsilon$ $(p \geq n(0))$. Since $r(p) \geq p$ and $\{f_n\}$ converges in the mean, $\|g_p - f_n\| = \|f_{r(p)} - f_n\| \leq \varepsilon$, $(p, n \geq n(1))$ for some $n(1) \geq n(0)$. Consequently,

$$\|f - f_n\| \leq \|f - g_p\| + \|g_p - f_n\| \leq 2\varepsilon \quad (p, n \geq n(1)).$$

Hence, for every $\varepsilon > 0$, there is an $n(1)$ for which $\|f_n - f\| \leq 2\varepsilon$, $n \geq n(1)$, and $f_n \to f$ in the mean. ∎

COROLLARY. *Every continuous linear functional φ on $\mathcal{L}^2$ is defined by an $f \in \mathcal{L}^2$ and is given by*

$$\varphi(g) = \int \bar{f} g \quad (g \in \mathcal{L}^2).$$

PROOF. Since $H = \mathcal{L}^2$ is Hilbert, the Riesz theorem on representing continuous linear functionals holds for H as well. Therefore, every $\varphi \in H'$ has a coefficient $c = f \in H$ and

$$\varphi(g) = \langle f, g \rangle = \int \bar{f} g \quad (g \in H),$$

as stated. ∎

The $\mathcal{L}^2$ spaces are of common use in the theory of differential and integral equations.

3.2. Fourier Series

This section discusses orthogonal series in separable Hilbert spaces. Their separatedness obtainable through factorization is often implicitly supposed.

3.2.1. Orthonormal Bases

Consider a Hilbert space which contains a countable, everywhere dense set Q. This implies that for each $z \in H$ there exists a sequence $x_n \in Q$ that converges to z. Such spaces are said to be *separable*. They contain orthonormal base sequences.

Let us take a finite or infinite $m = 2, 3, ..., \infty$. Choose a maximum family of linearly independent vectors $x_n \in Q$ ($1 \leq n < m$). Its linear span contains Q and so is dense in H together with Q. It is not hard to verify that the equalities

$$y_n = x_n - \sum_{1 \leq k < n} (x_n e_k) e_k, \quad e_n = \|y_n\|^{-1} \cdot y_n$$

define an orthonormal family of vectors e_n ($1 \leq n < m$): $e_k e_n = 0$ ($k \neq n$), $\|e_n\| = 1$. The multiplied pairs here are inner products, and in the inseparated case the equalities mean equivalences. The linear independence is also defined accurate to equivalence. This ensures $\|y_n\| \neq 0$ ($\|y_n\| = 0$ implies that $x_1, ..., x_n$ are linearly dependent).

An inductive procedure to construct an orthonormal family of vectors e_n from x_n with the aid of the said equalities is known to be the *Schmidt method*. The vectors e_n are linear combinations of x_n, and conversely. Hence, the orthonormal vector families of e_n and x_n span the same vector space, and the span is dense in H. In other words, vectors e_n form an *orthonormal basis* in H: for each $z \in H$ there is a sequence of linear combinations of e_n which converges to z. Furthermore, as will be shown, each vector z is a sum of the series e_n with some scalar coefficients c_n:

$$z = \sum c_n e_n \quad (1 \leq n < m).$$

EXAMPLES. Consider an open interval of the real line, $U =]a, b[$, ($-\infty \leq a < b \leq \infty$), an algebra $\mathcal{R}$ induced by bounded intervals in U, a measure μ on $\mathcal{R}$, the space $\mathscr{L}^2 = \mathscr{L}^2(U, \mu)$ of numerical-valued functions on U, measurable and square-integrable with respect to μ, and a sequence of power functions x^n given by

$$x^n(t) = t^n \quad (t \in U).$$

It is easy to verify that the functions x^n, $n = 0, 1, 2, ...$, are linearly independent (if a polynomial is identically zero on a nondegenerate interval, then all its coefficients are zeros). Applying the Schmidt orthogonalization procedure to the sequence x^n we obtain an orthonormal sequence of polynomials p_n ($n = 0, 1, 2, ...$):

$$\int p_j(t) \cdot p_k(t) \cdot d\mu(t) = 0 \quad (j \neq k),$$

$$\int p_n^2(t) \cdot d\mu(t) = 1.$$

1. Let $a = -1$, $b = 1$, and $d\mu(t) = dt$ be a Lebesgue measure. Then p_n are *Legendre polynomials*.

2. Let $a = -1$, $b = 1$, and $d\mu(t) = (1 - t^2)^{-1/2} dt$ be a measure with derivative $f(t) = (1 - t^2)^{-1/2}$ ($-1 < t < 1$) with respect to the Lebesgue measure dt. Then p_n are *Chebyshev polynomials*.

3. Let $a = -\infty$, $b = \infty$, and $d\mu(t) = \exp(-t^2) dt$. Then p_n are *Hermit polynomials*.

4. Let $a = 0$, $b = \infty$, and $d\mu(t) = \exp(-t) dt$. Then p_n are *Laguerre polynomials*.

5. Let $u = [-\pi, \pi]$, and $d\mu(t) = dt$ be a Lebesgue measure. A family of e_k given on $[-\pi, \pi]$ by

$$e_k(t) = (2\pi)^{-1/2} e^{-ikt} \quad (k = 0, \pm 1, \pm 2, ...)$$

forms an orthonormal basis in $\mathcal{L}^2$. It is referred to as *trigonometric*, for, by the Euler formula,

$$e^{-ikt} = \cos(kt) - i \cdot \sin(kt).$$

If $U = [a, b]$ ($-\infty < a < b < \infty$), then the orthonormal basis on $[a, b]$ is formed by the functions e_k given by

$$e_k(t) = (2\pi)^{-1/2} \exp\{-i \cdot 2\pi(b - a)^{-1} k(t - 2^{-1}(a + b))\}.$$

REMARK. Strictly speaking, in the trigonometric basis, vectors e_k should be numbered to obtain a series, but the above reasoning holds for any orthonormal family, and so no numeration is necessary. On passing to sine and cosine functions a numeration evolves in a natural way.

3.2.2. The Bessel Inequality

If a Hilbert space H with an orthonormal basis of vectors e_n ($1 \leq n < m$) is finite-dimensional ($m < \infty$), then

$$x = \sum_{1 \leq n < m} \xi_n e_n, \quad \xi_n = e_n x \quad (x \in H).$$

This emerges to be true also for $m = \infty$. In this case, for scalar coefficients

3.2. Fourier Series

ξ_n, this equality means that

$$x = \lim s_n, \quad s_n = \sum_{1 \le k \le n} \xi_k e_k,$$

and

$$\|x - s_n\| \to 0.$$

The coefficients $\xi_n = e_n x$ are given by the inner products of the basis vectors e_n and the vector x. These products are said to be the *Fourier coefficients* of x with respect to the basis of e_n, or its *coordinates* into this basis. The sequence $\xi_n e_n$ ($\xi_n = e_n x$) is called the *Fourier series* of x with respect to e_n.

We shall prove that the Fourier series does have the right sum. Let $n < m$ be a natural number, and A be a subspace of H spanned by the first n basis vectors $e, ..., e_n$. Find the distance from a point $x \in H$ with coordinates $\xi_k = e_k x$ and a point $a = \sum \alpha_k \cdot e_k \in A$ ($\alpha_k \in \mathbb{F}$, $1 \le k \le n$).

LEMMA. $\|x - a\|^2 = \|x\|^2 - \sum |\xi_k|^2 + \sum |\xi_k - \alpha_k|^2$.

PROOF. Using the orthonormality of e_k we have

$$\|x - a\|^2 = (x - a)(x - a) = xx - xa - ax + aa$$

$$= \|x\|^2 - \sum \alpha_k \bar{\xi}_k - \sum \bar{\alpha}_k \xi_k + \sum |\alpha_k|^2$$

$$= \|x\|^2 - \sum |\xi_k|^2 + \sum |\xi_k - \alpha_k|^2,$$

since

$$|\xi_k - \alpha_k|^2 = (\xi_k - \alpha_k)(\bar{\xi}_k - \bar{\alpha}_k)$$

$$= |\xi_k|^2 - \alpha_k \bar{\xi}_k - \bar{\alpha}_k \xi_k + |\alpha_k|^2$$

for any numbers ξ_k, α_k. ■

Let $B = A^\perp$ be the orthogonal complement of A in H. Let P and Q be orthogonal projectors of H onto the subspaces A and B. The lemma has the following

COROLLARY. $Px = \sum \xi_k e_k$ ($1 \le k \le n$).

PROOF. By the lemma, for $\alpha_k = \xi_k$ and $a = \sum \xi_k e_k$ the distance from x to a is least, that is, a is the point of A closest to x. By the orthogonal projection theorem, this means $Px = a$, as required. ■

By the definition of orthogonal projection, for any numbers α_k, we have

$$(x - \sum \xi_k e_k)(\sum \alpha_k e_k) = 0.$$

The lemma immediately yields

$$\|x\|^2 = \sum |\xi_k|^2 + \|x - \sum \xi_k e_k\|^2.$$

(This equality also follows from the Pythagorean theorem.) Whence

$$\sum_{1 \leq k \leq n} |\xi_k|^2 \leq \|x\|^2$$

for each $n < m$. This relation establishes

BESSEL'S INEQUALITY.

$$\sum_{1 \leq n < m} |\xi_n|^2 \leq \|x\|^2.$$

This inequality allows us to solve the problem of the sum of a Fourier series.

3.2.3. Fourier Series Expansion

In a Hilbert space H having an orthonormal basis e_n, for orthogonal series $\alpha_n e_n$ ($\alpha_n \in \mathbb{F}$) ($1 \leq n < m \leq \infty$), there is a convenient summability criterion.

THEOREM (summability criterion). *A series $\alpha_n e_n$ has a sum in H if and only if $\sum |\alpha_n|^2 < \infty$.*

PROOF. Let $p > n$ and

$$s_n = \sum_{1 \leq k \leq n} \alpha_k e_k, \quad s_p = \sum_{1 \leq k \leq p} \alpha_k e_k.$$

Then, in view of the orthonormality of e_k

$$\|s_p - s_n\|^2 = \|\sum_{n < k \leq p} \alpha_k e_k\|^2 = \sum_{n < k \leq p} |\alpha_k|^2.$$

By the Cauchy criterion these equalities establish the theorem. ∎

The theorem of existence and uniqueness of a Fourier series expansion is easily proved with the aid of the summability criterion and Bessel's inequality.

THEOREM. (i) *For a vector $x \in H$, the Fourier series $\xi_n e_n$ ($\xi_n = e_n x$) has a sum equal to x.*

(ii) *If for some numbers η_n the series $\eta_n e_n$ has a sum equal to x, then $\eta_n = \xi_n$.*

PROOF. (i) By Bessel's inequality $\sum |\xi_n|^2 < \infty$. The summability criterion then implies that the series $\xi_n e_n$ has a sum s. It remains to prove that $s = x$, that is, $\|s - x\| = 0$.

3.2. Fourier Series

Since e_n makes an orthonormal basis in H, to prove $\|s - x\| = 0$ it suffices to demonstrate that $e_j(s - x) = 0$ $(1 \leq j < m)$.

Consider the partial sum

$$s_n = \sum_{1 \leq k \leq n} \xi_k e_k.$$

Notice that

$$e_j(x - s_n) = \xi_j - \sum_k \xi_k e_j e_k = \xi_j - \xi_j = 0$$

for $1 \leq j \leq n$. Whence by the continuity of the inner product,

$$e_j(x - s) = e_j(x - \lim_n s_n) = \lim_n e_j(x - s_n) = 0$$

for $1 \leq j < m$, which proves the first part of the theorem.

(ii) Let

$$\sum \eta_n e_n = x = \sum \xi_n e_n.$$

Then

$$\sum (\eta_n - \xi_n) e_n = 0.$$

Therefore,

$$\eta_j - \xi_j = e_j(\eta_j - \xi_j)e_j = e_j \sum (\eta_n - \xi_n)e_n = 0 \quad \text{for } 1 \leq j < m.$$

The theorem is proved. ∎

EXAMPLE. Let $H = \mathscr{L}^2([-\pi, \pi], dt)$, and $\{e_k\}$ be a trigonometric basis. Then the Fourier coefficients of a function $x \in H$ are given by

$$\xi_k = e_k x = (2\pi)^{-1/2} \int_{-\pi}^{\pi} e^{ikt} x(t)\, dt.$$

The Fourier series is summable in the mean

$$\left\| x - \sum_{|k| \leq n} \xi_k e_k \right\|^2 = \int_{-\pi}^{\pi} \left| x(t) - \sum_{|k| \leq n} \xi_k e_k(t) \right|^2 dt \to 0$$

as $n \to \infty$.

For the sequence of partial sums

$$s_n = \sum_{|k| \leq n} \xi_k e_k,$$

convergence to a function x at each point, and moreover the uniform convergence $s_n \to x$, all require additional conditions with reference to the Riesz lemma. We can state only that there exists a subsequence of s_n which converges to x at almost every point. L. Karleson has proved that the sequence s_n also converges at almost every point.

3.2.4. The Parseval Equation

This equality specifies the Bessel inequality.

THEOREM (Parseval's equation). *For each vector $x \in H$ and its Fourier coefficients $\xi_n = e_n x$,*

$$\|x\|^2 = \sum |\xi_n|^2.$$

PROOF. Construct the partial sums

$$s_n = \sum_{1 \leq k \leq n} \xi_k e_k,$$

$$\sigma_n^2 = \sum_{1 \leq k \leq n} |\xi_k|^2.$$

We have shown above that

$$\|x\|^2 = \sigma_n^2 + \|x - s_n\|^2, \quad \sigma_n^2 \to \sigma^2 = \sum |\xi_k|^2, \quad \|x - s_n\| \to 0.$$

Consequently, $\|x\|^2 = \sigma^2$, and the Parseval equation is established. ∎

The Parseval equation generalizes the Pythagorean theorem to orthonormal families of vectors, and represents the length of a vector x in terms of its Fourier coefficients $\xi_n = e_n x$,

$$\|x\| = (\sum |\xi_n|^2)^{1/2}.$$

3.2.5. The Riesz-Fisher Theorem

Let l^2 be a space of sequences of numbers ξ_n ($1 \leq n < m$). By what we proved above, to each vector $x \in H$ there corresponds a sequence of its Fourier coefficients $\xi_n = e_n x$ in l^2. The Parseval equation states that such a correspondence preserves the length of vectors. Obviously, it is linear, that is, sets up the linear isometric transformation from H onto l^2. If we prove that every sequence in l^2 is constituted by the Fourier coefficients of a vector in H, we establish the isomorphism of the Hilbert spaces H and l^2 accurate to equivalence of the vectors.

As before, we assume that H has an orthonormal basis of e_n ($1 \leq n < m \leq \infty$), that is, H is a finite- or infinite-dimensional, separable Hilbert space. Accordingly, l^2 is made up of finite or infinite sequences $\{\xi_n\}$ ($1 \leq n < m \leq \infty$).

THEOREM (Riesz-Fisher). *Let $\{\xi_n\}$ be a number-valued sequence such* that $\Sigma |\xi_n|^2 < \infty$. Then there exists a vector $x \in H$ such that

$$\xi_n = e_n x \quad (1 \leq n < m).$$

PROOF. By the criterion of summability of the series $\xi_n e_n$ in H, we infer that there exists a vector $x = \Sigma \xi_n e_n \in H$. In view of the uniqueness of the Fourier series expansion we have $\xi_n = e_n x$, as required. ∎

COROLLARY. *All the separable Hilbert spaces of a given dimension are isomorphic to each other (accurate to equivalence).*

EXAMPLE. The space $\mathscr{L}^2 = \mathscr{L}^2([a,b], dt)$ of functions on a closed interval $[a,b]$ ($-\infty < a < b < \infty$) is isomorphic to the space $l^2 = \mathscr{L}^2(\mathbb{N}, dn)$ of number-valued sequences.

This isomorphism is widely used in theory and applications.

3.3. Spaces of Functions

In this section we present some popular spaces of functions.

3.3.1. Metric Spaces

The mathematical definition of distance describes its general properties.

Metric. Consider an abstract set E together with its Cartesian product by itself

$$E \times E = \{(x, y) : x \in E, y \in E\}.$$

DEFINITION. *A distance in E is a positive symmetric function $d : E \times E \to \mathbb{R}$ which is zero on the diagonal and obeys the triangle inequality*, viz.,
(1) $d(x, y) \geq 0$;
(2) $d(x, y) = d(y, x)$;
(3) $d(x, x) = 0$;
(4) $d(x, y) \leq d(x, z) + d(z, y)$ $(x, y, z \in E)$.

The distance d is also called a *metric*. The pair (E, d) is said to be a *metric space*. Elements of the set E are referred to as points of the metric space (E, d), occasionally denoted simply by E.

Usually it is also required that the distance d should separate points of E:
(5) $d(x, y) > 0$ for $x \neq y$ $(x, y \in E)$.
It is convenient not to do so, calling metric spaces with such metrics *separated*. As a rule, like with normed spaces, we will suppress the adjective "separated." A distance d with properties 1–4 is sometimes called a *semimetric* (by analogy with seminorm). When $d(x, y) = 0$, points x and y are said to be *equivalent*.

The metric space $(A, d_{A \times A})$, made up of a set $A \subseteq E$ and the restriction

$d_{A \times A}$ of the distance d to the set $A \times A \subseteq E \times E$, is said to be a *subspace* of the metric space (E, d).

An example of the distance is the norm of the difference of two vectors

$$d(x, y) = \|x - y\|.$$

We will carry over the terms and notations from normed spaces to metric spaces, ignoring the fact that for points of the metric space the difference may not be defined. With such an agreement the definitions of convergence of point sequences and completeness can be carried over to metric spaces automatically. Together with the definition of convergence they receive the definitions of open and closed balls and sets, neighborhood of a point, closure of a set, and compactness of a set in a metric space. The product of metric space is one defined by analogy with the product of normed spaces.

One often has to deal with parts of normed spaces which are not subspaces. The norm generates a natural metric for them. Keeping the terms and notations in this case is especially convenient. Such metrics are invariant under translations and are absolutely uniform:

$$\|(x + z) - (y + z)\| = \|x - y\|,$$

$$\|\alpha x\| = |\alpha| \cdot \|x\|,$$

for any $\alpha \in F$ and $x, y, z \in E$ such that $x + z, y + z, \alpha x \in E$. If E is a vector space, then these conditions are fulfilled and the equality $\|x\| = \|x - 0\|$ ($x \in E$) defines a norm for E.

Completion of Metric Spaces. Consider two metric spaces (E_1, d_1), (E_2, d_2) and a map $\varphi : E_1 \to E_2$. If

$$d_1(x_1, y_1) = d_2(\varphi(x_1), \varphi(y_1)) \quad (x_1, y_1 \in E_1)$$

we will say that φ *preserves distance*, or that φ is an *isometry*. If (E_1, d_1) is separated, the isometry φ is one-one. Covers of the isometry are called the *isomorphisms* of the metric spaces (up to equivalence).

We will say that a metric space $(\overline{E}, \overline{d})$ *completes the metric space* (E, d) if

(1) $(\overline{E}, \overline{d})$ is complete,

(2) (E, d) is isomorphic to an everywhere dense subspace of $(\overline{E}, \overline{d})$.

Such an isomorphism allows us to identify the set E with its image in $\overline{E}$ and treat E as a part of $\overline{E}$, which simplifies the terminology. The following important theorem allows us to consider complete metric spaces instead of arbitrary ones, and to use all the advantages of completeness.

3.3. Spaces of Functions

THEOREM (completion). *For any metric space there exists a completion. All completions of a given metric space are isomorphic to one another.*

EXAMPLES.

1. If $E = Q$ is the space of rational numbers, then its completion $\overline{E} = \mathbb{R}$ is the space of real numbers (under usual metrics).

2. If $E = \mathcal{P}[a, b]$ is the space of polynomials on $[a, b]$, then its completion $\overline{E} = C[a, b]$ is the space of continuous functions on $[a, b]$ (under uniform metrics). This fact follows from the Weierstrass approximation theorem.

3. If $E = S[a, b]$ is the space of step functions on $[a, b]$, then its completion $\overline{E} = \mathcal{D}[a, b]$ is the space of functions without nonjump discontinuities on $[a, b]$, that is functions having left-hand and right-hand limits at every point on $[a, b]$ (uniform metrics).

In the general case, elements of the completion $\overline{E}$ are convergent point sequences in E. If a sequence converges to a point in E, then it is identified with this point. If a sequence converges to itself (Cauchy) but does not converge to a point in E, then it provides a new element to complete E. The metric $\overline{d}$ for the completion $\overline{E}$ is defined by the metric d of E with the passage to the limit. The completion preserves the invariance and absolute homogeneity of the metric. Therefore, normed spaces are completed to Banach spaces.

Continuous Extension. In completing metric and normed spaces, it is often required to extend functions by continuity. Let E and F be metric spaces, a set $A \subseteq E$, its closure $\overline{A}$ and a function $f: A \to F$. We will say that $b \in F$ is the limit of f at $a \in \overline{A}$ and write $\lim_{x \to a} f(x) = b$, if $f(x(n)) \to b$ for every sequence of $x(n) \in A$ such that $x(n) \to a$. Equivalently, for any $\varepsilon > 0$ there is a $\delta > 0$ having $\|f(x) - b\| \le \varepsilon (\|x - a\| \le \delta, x \in A)$.

The limit b is unique if and only if F is separated. In the general case, it would be more correct to write $b \in \lim_{x \to a} f(x)$. By the definition, given $a \in A$, then the existence of b implies $\|b - f(a)\| = 0$. If $a \in A$ and the limit $b = f(a)$ exists, then the function f is said to be continuous at a. Functions continuous at each point of the domain are defined as continuous.

THEOREM (extension). *Suppose that the metric space F is separated and $f: A \to F$ has a limit at each point $\overline{x} \in \overline{A}$. Then*

$$\overline{f}(\overline{x}) = \lim_{x \to \overline{x}} f(x)$$

defines the unique continuous function $\overline{f} = \overline{A} \to F$ that extends f.

PROOF. Since F is separated the limit $\overline{f}(\overline{x})$ is unique at each $\overline{x} \in \overline{A}$.

If we denote the appropriate proximity of points by the symbol $\approx$, the proof of continuity of $\bar{f}: \bar{A} \to F$ with values $\bar{f}(\bar{x})$ can be written as follows. If $\bar{x} \approx \bar{a}$ ($\bar{a} \in \bar{A}$), there exists $x \in A$ such that $x \approx \bar{x}$, $x \approx \bar{a}$. By the condition, $f(x) \approx \bar{f}(\bar{x})$ and $f(x) \approx \bar{f}(\bar{a})$, hence $\bar{f}(\bar{x}) \approx \bar{f}(\bar{a})$ and $\bar{f}$ is continuous at every point $\bar{a} \in \bar{A}$. From the existence of the limit of f at $x \in A$ follows $\bar{f}(x) = f(x)$. Therefore, $\bar{f}$ extends f. Finally we note that each continuous $\bar{g}: \bar{A} \to F$ extending f is given by

$$\bar{g}(\bar{x}) = \lim_{x \to \bar{x}} f(x) = \bar{f}(\bar{x}) \quad (\bar{x} \in \bar{A})$$

and so is equal to $\bar{f}$. ∎

In the general case, in the continuous extension theorem instead of the equality relating the functions $\bar{f}$ and f one should write

$$\bar{f}(\bar{x}) \in \lim_{x \to \bar{x}} f(x)$$

and should not state that the continuous extension is unique. Alternatively, one should identify points of zero separation, i.e., replace the considered space by the separated space which is induced by such point identification, or else consider equivalences.

The extension theorem has a corollary for uniformly continuous functions. A function $f = A \to F$ is said to be *uniformly continuous* if for any $\varepsilon > 0$ there is a $\delta > 0$ such that

$$\|f(x) - f(y)\| \leq \varepsilon \quad (\|x - y\| \leq \delta; \; x, y \in A).$$

Obviously, from the uniform continuity follows the usual one. The converse need not be true.

COROLLARY. *Let F be a complete separated metric space, and $f: A \to F$ be a uniformly continuous function. Then there exists a unique uniformly continuous function $\bar{f}: \bar{A} \to F$ which extends f.*

The completeness of F is needed in order to apply Cauchy's limit criterion.

The uniformly-continuous-extension theorem can also be formulated as follows. If F is separated and complete and f is uniformly continuous, then a uniform extension of f exists and is uniformly continuous. In particular, this theorem allows bounded linear operators to be extended on Banach spaces from vector spaces to their closures. For example, this is a way to obtain generalized solutions of some differential equations.

Metric Topology. For qualitative analysis it is often convenient to

3.3. Spaces of Functions

describe the proximity to a given point of a metric space in terms of its neighborhoods. To this end we introduce a metric topology.

Let E be a metric space. For each point $a \in E$ we define a family of balls $B(a, r)$ of radius $r > 0$ centered at a. Every set $V \subseteq A$ that contains a ball $B(a, r)$ is called a *neighborhood* of a. A set $U \subseteq E$ is called *open* if it is a neighborhood of each of its point. (In particular, the empty set 0 and the set E are open.) The class $\mathcal{T} = \mathcal{T}(E)$ of all open sets is defined as a *metric topology* for E. Complements of open sets are called *closed sets*. The intersection of all the closed sets that contain $A \subseteq E$ is called the *closure* of A, denoted by $\overline{A}$.

The topological language is convenient owing to its transparency. For example, the convergence of a sequence $x_n \in E$ to $a \in E$ means that for any neighborhood V of a there exists a number n_0 such that $x_n \in V$ for all $n \geq n_0$. The continuity of a mapping $f: E \to F$ the metric space E to the metric space F at $a \in E$ means that each neighborhood $V(b)$ of $b = f(a)$ contains the image $f(U(a))$ of a neighborhood $V(a)$ of a. In other words, the preimage $f^{-1}(V(b))$ of the neighborhood $V(b)$ of $b = f(a)$ is a neighborhood of a. The continuity of f implies that the preimage $U = f^{-1}(V)$ of any open set V in F is an open set in E.

For metric spaces, compactness can be defined in three languages: metric, topological, and that of sequences. Compact sets possess many properties of finite sets and can be considered close to finite. Such a representation agrees best with the metric definition of compactness.

We will refer to any finite family of balls of radius $\varepsilon > 0$ which cover the set A in the metric space E as an ε-*net* for A. If for any $\varepsilon > 0$ there exists an ε-net, then the set A is called *relatively compact* (or completely bounded). Closed relatively compact sets are said to be compact. In separated complete spaces, closedness may be replaced with completeness. In finite-dimensional normed spaces, the only compact sets are those closed and bounded.

The topological definition of compactness is given in the following

THEOREM (Borel). *A closed set A in a metric space is compact if and only if for any family of open sets which covers A there is a finite subfamily which covers A, too.*

The statement may be briefly put as follows: every open cover has a finite subcover. If a metric space is separated, there is no need to require closedness – if follows from the existence of a finite subcover.

The following theorem defines compactness in terms of sequences.

THEOREM (Weierstrass). *A closed set A in a complete metric space is compact if and only if for every sequence of points in A there exists a subsequence which converges to a point in A.*

In brief, every sequence has a convergent subsequence. If a metric space is separated, the requirement of closedness is again unnecessary: it follows from the existence of the convergent subsequence.

For metric spaces, Borel and Weierstrass' theorems are proved practically like for intervals of the real axis.

Compactness and Continuity. Let E and F be metric spaces. In order to avoid repeated specifying of the closedness of the sets, we will assume that these spaces are separated. Using the topological definitions of compactness and continuity it is easy to verify that the continuous maps $f: E \to F$ preserve compactness.

THEOREM. *Under a continuous map $f: E \to F$ the image $B = f(A)$ of a compact set $A \subseteq E$ is a compact set in F.*

In short this theorem is spelled as: "a continuous image of a compact set is a compact set." In the general case when F may not be separated, the closure $\overline{B} = \overline{f(A)}$ of the image $B = f(A)$ of a compact set A will be compact.

Since on the real axis a compact set is bounded and closed, it requires the maximum and minimum elements. Hence, the theorem on the continuous image of a compact set leads to

COROLLARY 1. *On a compact set a continuous real-valued function takes on its maximum and minimum values.*

Using the topological definition of compactness, it is easy to deduce

COROLLARY 2. *Let $f: A \to B$ be a continuous bijection of a compact set $A \subseteq E$ on a set $B \subseteq F$. Then the inverse map $f^{-1}: B \to A$ is also continuous.*

In other words, a continuous bijection of compact sets is a homeomorphism.

Finally, for functions on compact sets ordinary and uniform continuities are equivalent. This result is established by the following

THEOREM (Cantor). *Let $f: A \to F$ be a continuous mapping of a compact set $A \subseteq E$ into the space F. Then f is uniformly continuous.*

Usual continuity follows from uniform continuity for mappings of any sets which need not be compact. The fact that for compact sets the converse is also true follows from their having finite subcovers of open covers.

3.3.2. Smooth Functions

We will first consider continuous functions since they are the largest class of smooth functions, then proceed with some other classes.

Space of Continuous Functions. Let E be a metric space, F a Banach space, $A \subseteq E$ a compact set, and $\mathscr{C} = \mathscr{C}(A, F)$ the vector space of continuous mappings $A \to F$. As usual to avoid unnecessary complications, we suppose that F is separated.

For every $f \in \mathscr{C}$ the set $f(A) \subseteq F$ is compact, hence, bounded. Therefore,

3.3. Spaces of Functions

a norm for $\mathscr{C}$ can be defined by

$$\|f\| = \sup\{\|f(x)\| : x \in A\}.$$

The vector space $\mathscr{C}$ endowed with such a norm forms a Banach space.

From the above definitions it follows that the convergence of a sequence of functions $f_n \in \mathscr{C}$ is equivalent to uniform convergence.

If F has a countable, everywhere dense set of points, then $\mathscr{C}$ has to possess such a set, too. That is, $\mathscr{C}$ is separated along with F. In particular, the space of number-valued continuous functions is separated as well as the space of finite-dimensional vector functions. This fact is not difficult to verify using the uniform convergence and compactness of the domain of functions in $\mathscr{C}$.

Compact sets in $\mathscr{C}$ can be described with the aid of the equicontinuity concept.

A set $\mathscr{A} \subseteq \mathscr{C}$ is said to be *equicontinuous* if for any $\varepsilon > 0$ there is a $\delta > 0$ such that

$$\|f(x) - f(y)\| \leq \varepsilon \quad (\|x - y\| \leq \delta; \ x, y \in A)$$

for all $f \in \mathscr{A}$. In other words, sufficient proximity of argument values ensures the required proximity of the corresponding values of the function at once for all the functions of the equicontinuous set. We should emphasize that $\|x - y\|$ means the distance between points x, y of the metric space E.

Let $\mathscr{A}(x)$ be the set of values at $x \in A$ of functions f from $\mathscr{A} \subseteq \mathscr{C}$:

$$\mathscr{A}(x) = \{f(x) : f \in \mathscr{A}\}.$$

Like any set in F, $\mathscr{A}(x)$ can be relatively compact, then we will say that the set of functions $\mathscr{A}$ is relatively compact at the point x.

THEOREM (Ascoli). *A closed set $\mathscr{A} \subseteq \mathscr{C}$ is compact if and only if $\mathscr{A}$ is compact at each point and equicontinuous.*

PROOF. Necessity is verified quite easily. To prove sufficiency we must demonstrate that the conditions of the theorem ensure that for $\mathscr{A}$ there exists an ε-net for all $\varepsilon > 0$. Indeed, it enables the product $A \times F$ to be partitioned into a finite set of rectangles with sides of lengths ε and δ. These rectangles may be combined into step stripes on A which cover $\mathscr{A}$. Having set a function on each stripe, we obtain an ε-net for $\mathscr{A}$, which proves sufficiency. ∎

If $F = \mathbb{R}^m$, then the relative compactness of a set $\mathscr{A} \subseteq \mathscr{C}$ at every point is tantamount to the uniform boundedness of functions from $\mathcal{A}$:

$$\|f(x)\| \leq \alpha \quad (x \in A;\ f \in \mathcal{A})$$

for some $\alpha > 0$. Therefore, Ascoli's theorem infers

THEOREM (Arzelà). *If $F = \mathbb{R}^m$, then a closed set $\mathcal{A} \subseteq \mathcal{C}$ is compact if and only if $\mathcal{A}$ is uniformly bounded and equicontinuous.*

Observe that the equicontinuity of $\mathcal{A}$ means its boundedness in a normed space $\mathcal{C}$:

$$\|f\| \leq \alpha \quad \text{for all } f \in \mathcal{A} \text{ and some } \alpha > 0.$$

EXAMPLES.
Let $E = F = \mathbb{R}$ and $A = [0, 1]$. Then the set $\mathcal{A}$ of linear functions on $[0, 1]$ with a coefficient from $[-1, 1]$ is closed, equicontinuous and therefore compact. On the other hand, the set $\mathcal{B}$ of all linear functions on $[0, 1]$ is not compact.

A conjugate of space $\mathcal{C}$ is the space $\mathcal{V}$ of number-valued measures on the algebra generated by closed parts of A. The norm of this measure is its variation equal to the sum of positive components. Every number-valued measure is equal to a linear combination of some positive measures. A correspondence between linear functionals $\varphi \in \mathcal{C}'$ on $\mathcal{C}$ and number-valued measures $\mu \in \mathcal{V}$ is established by integration.

THEOREM (Riesz). *The equality*

$$\varphi(f) = \int f d\mu \quad (f \in \mathcal{C})$$

defines the isometric correspondence between the spaces $\mathcal{C}'$ and $\mathcal{V}$.

EXAMPLE.
Let $\mu = \mu_a$ be a measure which describes the distribution of unit mass at a point $a \in A$:

$$\mu_a(X) = 1 \ (a \in X), \quad \mu_a(X) = 0 \ (a \notin X)$$

for each $X \subseteq A$. Then the corresponding functional $\varphi = \delta_a$ has the values

$$\delta_a(f) = \int f d\mu_a = f(a) \quad (f \in \mathcal{C}).$$

Such functionals are called *delta functions*.

Spaces of Smooth Functions. Let U be an open set in the space $E = \mathbb{R}^q$ ($q \geq 1$), and $\mathcal{C}^p = \mathcal{C}^p(U)$ be the vector space of p-times continuously differentiable number-valued functions on U ($0 \leq p \leq \infty$). By definition, continuous functions are deemed 0-times differentiable.

3.3. Spaces of Functions

Points of $\mathscr{C}^\infty$ are infinitely differentiable functions. Functions from $\mathscr{C}^p$ are said to be *p*-smooth, or, shortly, *smooth functions*.

To write out derivatives, it is convenient to use multiindices $\alpha = (\alpha(1), ..., \alpha(q))$ and their absolute values $|\alpha| = \alpha(1) + ... + \alpha(q)$. Positive integers $\alpha(1), ..., \alpha(q)$ indicate the number of times the function is differentiated with respect to the 1st, ..., *q*th variable. The corresponding derivative is denoted by

$$\partial^\alpha = \partial_1^{\alpha(1)} ... \partial_q^{\alpha(q)}.$$

The number $|\alpha| = \alpha(1) + ... + \alpha(q)$ is the order of derivative ∂^α. It is assumed that for $|\alpha| \leq p$, the function differentiated is from $\mathscr{C}^p$. The derivatives $\partial^\alpha f$ are continuous number-valued functions on U. Their values at $u \in U$ are denoted by $\partial^\alpha f(u)$ or $\partial^\alpha fu$. In particular, $\partial^0 f = f$ and $\partial^0 f(u) = f(u)$.

One often writes $\dfrac{\partial}{\partial x_1}, ..., \dfrac{\partial}{\partial x_q}$ instead of $\partial_1, ..., \partial_q$. Other notations are used as well.

We will also write $\alpha! = \alpha(1)! ... \alpha(q)!$ and $v^\alpha = v_1^{\alpha(1)} ... v_q^{\alpha(q)}$ for $v = (v_1, ..., v_q) \in \mathbb{R}^q$. Then for $u \in U$, $v \in U - u$ and $f \in \mathscr{C}^p$, the Taylor formula is written as ($p < \infty$):

$$f(u + v) = \sum_{|\alpha| < p} (\alpha!)^{-1} \cdot \partial^\alpha fu \cdot v^\alpha + r_p fu(v),$$

$$r_p fu(v) = p \sum_{|\alpha| = p} (\alpha!)^{-1} \int_0^1 (1 - t)^{p-1} \partial^\alpha f(u + tv) \, dt \cdot v^\alpha.$$

Consider a family of smooth number-valued functions c_α on U and a polynomial $P(u, \cdot)$ of degree m given by

$$P(u, \xi) = \sum_{|\alpha| \leq m} c_\alpha(u) \cdot \xi^\alpha$$

for $\xi = (\xi_1, ..., \xi_q)$. By a formal substitution of ∂ for ξ, we define the linear differential operator

$$P(u, \partial) = \sum_{|\alpha| \leq m} c_\alpha(u) \cdot \partial^\alpha$$

on $\mathscr{C}^p$. By definition

$$P(u, \partial) f = \sum_{|\alpha| \leq m} c_\alpha(u) \cdot \partial^\alpha f$$

for $f \in \mathscr{C}^p$.

EXAMPLE.
If $m = 2$, $c_{jj} = 1$ and $c_{jk} = 0$ for $j \neq k$ ($1 \leq j, k \leq q$), then

$$P(u, \partial) = \partial_1^2 + \ldots + \partial_q^2 = \Delta$$

is the *Laplace operator* (Laplacian).

We define a metric for $\mathscr{C}^p = \mathscr{C}^p(U)$ in three steps. First, for each $f \in \mathscr{C}^p$, a positive integer $m \leq p$ and a compact $K \subseteq U$ we consider the norm

$$\|f\|_{mK} = \sup\{|\partial^\alpha f(u)| : |\alpha| \leq m, u \in K\}.$$

Pick an increasing sequence of compact sets $K(n) \subseteq U$ which covers U. For each number n and functions $f, g \in \mathscr{C}^p$ we consider the number

$$\|f - g\|_n = \sum_{0 \leq m \leq p} \frac{1}{2^m} \cdot \frac{\|f - g\|_{mK(n)}}{1 + \|f - g\|_{mK(n)}}.$$

Finally, we define the distance between f and $g \in \mathscr{C}^p$ as

$$\|f - g\| = \sum_{0 \leq n < \infty} \frac{1}{2^n} \cdot \frac{\|f - g\|_n}{1 + \|f - g\|_n}.$$

The triangle inequality for this distance is verified easily. If $p = \infty$, then the equality $m = p$ is excluded from $m \leq p$. The distance is independent of the choice of a sequence of $K(n)$.

Observe that this distance for $\mathscr{C}^p$ is invariant under translations, but it is not absolutely homogeneous. Therefore, the equality $\|f\| = \|f - 0\|$ does not define a norm for $\mathscr{C}^p$. (Some authors sometimes consider norms that do not obey the property of absolute homogeneity. See, e.g., [2, Vol. 1, Sect 2.1].)

Consider, in particular, the space $\mathscr{C}^\infty$. By the definition of metric for $\mathscr{C}^\infty$, the convergence $f_n \to f$ in this space means a uniform convergence on the compacts $\partial^\alpha f_n \to \partial^\alpha f$ of derivatives of all orders. Whence it follows that the metric space $\mathscr{C}^\infty$ is complete.

Complete metric vector spaces are called *Frechet spaces*. Banach spaces are a particular case of Frechet spaces. $\mathscr{C}^p$ may be also viewed as an example of *multinormed* (locally convex) metrizable spaces.

Among the functions of $\mathscr{C}^p$ we distinguish the functions of compact support (which vanish outside some compact set). These functions are sometimes called *finite*. They constitute a subspace $\mathscr{C}^p_0$ of $\mathscr{C}^p$.

3.3. Spaces of Functions

EXAMPLE.
A function f defined on $\mathbb{R}^m$ as

$$f(x) = \exp\{-(1-\|x\|^2)^{-1}\} \quad (\|x\| < 1),$$
$$f(x) = 0 \quad (\|x\| \geq 1)$$

belongs to $\mathscr{C}_0^\infty = \mathscr{C}_0^\infty(\mathbb{R}^m)$.

Inner Product. Consider the vector space $\mathscr{H}^p = \mathscr{H}^p(U)$ of functions $f \in \mathscr{C}^p = \mathscr{C}^p(U)$ such that

$$\sum_{|\alpha| \leq p} \int_U |\partial^\alpha f(x)|^2 \, dx < \infty.$$

The equality

$$\langle f, g \rangle = \sum_{|\alpha| \leq p} \int_U \partial^\alpha \bar{f}(x) \cdot \partial^\alpha g(x) \cdot dx$$

defines an inner product for $\mathscr{H}^p$.

In exactly the same way one may define the vector space $\mathscr{H}_0^p$ of functions from $\mathscr{C}_0^p$, and an inner product for this space.

Bounded Functions. In addition to the space $\mathscr{C}^p = \mathscr{C}^p(U)$ $(0 \leq p < \infty)$ of smooth functions on the open part U of $\mathbb{R}^m$, we consider spaces $\mathscr{C}^p = \mathscr{C}^p(\bar{U})$ of smooth functions which may be extended together with all their derivatives up to bounded continuous number-valued functions on $\bar{U}$, the closure of U. Specifically, these may be the restrictions on $\bar{U}$ of smooth functions defined on an open neighborhood of the compact set $\bar{U}$, if U is bounded.

A norm for $\mathscr{C}^p = \mathscr{C}^p(\bar{U})$ is defined as

$$\|f\|_p = \sup\{|\partial^\alpha f(x)|: |\alpha| \leq p, x \in \bar{U}\}.$$

Since the extended functions $\partial^\alpha f$ are bounded on $\bar{U}$ the norm is finite.

3.3.3. Lebesgue Spaces

Points of these spaces are measurable functions whose powers are integrable with respect to Lebesgue's measure.

Let X be an abstract set, $\mathscr{A}$ an algebra of its parts, and μ a measure on $\mathcal{A}$. As usual we suppose that μ is positive, countably additive, complete, and locally bounded (σ-finite). Pick a number $p \in [1, \infty[$. Denote $\mathcal{L}^p = \mathcal{L}^p(X, \mu)$ the vector space of number-valued functions f on X, measurable with respect to μ and with the integrable p-th power:

$$\int |f|^p d\mu < \infty.$$

The equality

$$\|f\|_p = (\int |f|^p d\mu)^{1/p}$$

defines a norm $\|\cdot\|_p$ for $\mathcal{L}^p$. The triangle inequality

$$\|f + g\|_p \leq \|f\|_p + \|g\|_p \quad (f, g \in \mathcal{L}^p)$$

is known as *Minkowski's inequality*. It is deduced from the Hölder inequality

$$|\int fg\, d\mu| \leq \|f\|_p \cdot \|g\|_q \quad (f \in \mathcal{L}^p, g \in \mathcal{L}^q)$$

that holds for $p > 1$, $q > 1$ and $1/p + 1/q = 1$. For $p = 1$, the Minkowski inequality is obtained by integrating the triangle inequality for absolute values.

The normed space $\mathcal{L}^p$ is complete.

Another space worth considering is the space $\mathcal{L}^\infty$, made up of almost everywhere bounded number-valued functions on X, measurable with respect to μ. The norm for $\mathcal{L}^\infty$ is defined by

$$\|f\|_\infty = \inf \{ \sup_{x \notin Z} |f(x)| : Z \subseteq X, \mu(Z) = 0 \}.$$

This space is also complete.

Observe that the spaces $\mathcal{L}^p$ can be inseparated: a zero integral does not imply that the function is zero. Therefore, $\mathcal{L}^p$ are often factorized, and the classes of functions equivalent with respect to a given measure are considered as points of the space. Then the space becomes separated, but at the expense of complicated operations with its elements. It would be more natural to consider functions as points of $\mathcal{L}^p$, and allow for their inseparatedness when necessary.

The spaces $\mathcal{L}^p$, factorized by the equivalence relation, will be denoted by L^p or L_p.

We will say that numbers p and q are conjugate if

$$1 \leq p < \infty, \quad 1 \leq q \leq \infty, \quad 1/p + 1/q = 1.$$

3.3. Spaces of Functions

In particular, the pairs $p = 1$ and $q = \infty$, and $p = 2$ and $q = 2$ are conjugate. It can be proved that $\mathscr{L}^p$ and $\mathscr{L}^q$ with conjugate p and q are also conjugate. Alternatively, the space of continuous linear functionals on $\mathscr{L}^p$ is isomorphic to the space $\mathscr{L}^q$ for $q = p(p-1)^{-1}$.

Among the spaces $\mathscr{L}^p$ ($1 \leq p \leq \infty$) we distinguish the space $\mathscr{L}^2$, which is a Hilbert space endowed with the inner product

$$\langle f, g \rangle = \int \bar{f} g \, d\mu \quad (f, g \in \mathscr{L}^2).$$

This space has been described in detail in Section 3.1.5. For any other value of p, it is straightforward to verify that the norm $\|\cdot\|_p$ does not meet the parallelogram law, therefore for $p \neq 2$ $\mathscr{L}^p$ are not Hilbert. A norm on these spaces cannot be defined as a square root of an inner product of an element by itself.

If X is a compact metric space, then $\mathscr{L}^p = \mathscr{L}^p(X, \mu)$ is separable: it contains a countable everywhere dense set of functions (specifically, simple and continuous).

Convergence in $\mathscr{L}^p$ for $1 \leq p < \infty$ is called *convergence in the mean* (of order p). By definition, $f_n \to f$ for $f_n, f \in \mathscr{L}^p$ means

$$\|f_n - f\|_p = \left(\int |f_n - f|^p \, d\mu\right)^{1/p} \to 0.$$

Clearly, fractional power may be dropped and we can write

$$\|f_n - f\|_p^p = \int |f_n - f|^p \, d\mu \to 0.$$

Let us examine several examples of Lebesgue's space.

EXAMPLES.
(1) Let $X = \{1, ..., m\} \subseteq \mathbb{N}$, and $\mu = dn$ be a counting measure for X. The functions on $X = \{1, ..., m\}$ are points $f = (f(1), ..., f(m))$ of $\mathbb{R}^m$. Therefore, $\mathscr{L}^p = \mathbb{R}^m$ for all $p \in [1, \infty]$. The norm is

$$\|f\|_p = \left(\sum_{1 \leq k \leq m} |f(k)|^p\right)^{1/p} \quad (1 \leq p < \infty).$$

In particular, for $p = 1, 2, \infty$

$$\|f\|_1 = \sum |f(k)|,$$

$$\|f\|_2 = \left(\sum |f(k)|^2\right)^{1/2},$$

$$\|f\|_\infty = \sup |f(k)| \quad (1 \leq k \leq m).$$

The space $\mathscr{L}^p = \mathscr{L}^p(\{1, ..., m\}, dn)$ is separated.

(2) Let $X = \mathbb{N}$ and $\mu = dn$ be a counting measure for $\mathbb{N}$. Then the functions on $X = \mathbb{N}$ are number-valued sequences $f = \{f(1), f(2), ..., f(n), ...\}$. Therefore, for $1 \leq p < \infty$, the points of $l^p = \mathscr{L}^p(\mathbb{N}, dn)$ are the number sequences $f = \{f(n)\}$, for which

$$\|f\|_p = (\Sigma |f(n)|^p)^{1/p} < \infty.$$

The points of $l^\infty = \mathscr{L}^\infty(\mathbb{N}, dn)$ are bounded number-valued sequences. The norm is

$$\|f\|_\infty = \sup \{|f(n)|: n = 1, 2, ...\}.$$

The space $\mathscr{L}^p = \mathscr{L}^p(\mathbb{N}, dn)$ is separated.

(3) Let $X = \mathbb{R}^m$ and $\mu = dx$ be the Lebesgue measure for $\mathbb{R}^m$. The points of $\mathscr{L}^p = \mathscr{L}^p(\mathbb{R}^m, dx)$ for $1 \leq p < \infty$ are number-valued functions f on $\mathbb{R}^m$, measurable with respect to the Lebesgue measure, for which

$$\|f\|_p = (\int |f(x)|^p \, dx)^{1/p} < \infty.$$

The space $\mathscr{L}^p = \mathscr{L}^p(\mathbb{R}^m, dx)$ is inseparated.

For compact sets in $\mathscr{L}^p = \mathscr{L}^p(\mathbb{R}^m, dx)$ with $1 \leq p < \infty$, we formulate a theorem that is deduced by averaging from the Arzela theorem and is similar to the latter. We will say that a set $\mathscr{A} \subseteq \mathscr{L}^p$ is *uniformly bounded in the mean*, if

$$\|f\|_p = (\int |f(x)|^p \, dx)^{1/p} \leq \alpha \quad (f \in \mathscr{A})$$

for some $\alpha > 0$. Observe that this means that $\mathscr{A}$ is bounded in the normed space $\mathscr{L}^p$.

Consider a point $z \in \mathbb{R}^m$, function $f \in \mathscr{L}^p$, and a function $\Delta_z f \in \mathscr{L}^p$ given by

$$\Delta_z f(x) = f(x + z) - f(x) \quad (x \in \mathbb{R}^m).$$

It will be said that a set $\mathscr{A} \subseteq \mathscr{L}^p$ is *equicontinuous in the mean* if

$$\|\Delta_z f\|_p = (\int |f(x + z) - f(x)|^p \, dx)^{1/p} \to 0$$

for $z \to 0$ uniformly on $\mathscr{A}$.

3.3. Spaces of Functions

Let $B^c(0, n) = \{ x \in \mathbb{R}^m : \|x\| \geq n \}$ be the complement of a ball $B(0, n) \subseteq \mathbb{R}^m$, and $h_n = B^c(0, n) \cdot f$ be the product of the indicator of $B^c(0, n)$ and a function $f \in \mathcal{L}^p$. Obviously, $h_n \in \mathcal{L}^p$ along with f; it equals f outside the ball $B(0, n) = \{ x \in \mathbb{R}^m : \|x\| < n \}$. We will say that the set $\mathcal{A} \subseteq \mathcal{L}^p$ is e*quipotentially small in the mean at infinity*, if

$$\| B^c(0,n) \cdot f \|_p = (B^c(0,n) \int |f(x)|^p dx)^{1/p} \to 0$$

for $n \to \infty$ uniformly on $\mathcal{A}$.

THEOREM (Riesz criterion). *A closed $\mathcal{A} \subseteq \mathcal{L}^p(\mathbb{R}^m, dx)$ is compact if and only if $\mathcal{A}$ is uniformly bounded, equicontinuous and equipotentially small in the mean at infinity.*

These compactness conditions differ from the Arzela theorem by the requirement of equipotential smallness and the words "in the mean." If we consider the space $\mathcal{L}^p(U, dx)$ for a bounded part U of $\mathbb{R}^m$ then the requirement of the set to be equipotentially small at infinity is no longer needed, and Riesz criterion simplifies in this case.

3.3.4. Distributions

Distributions is the name of continuous linear functionals on some functional spaces. An alternative name of these functionals is *generalized functions*. These names stem from the fact that some of such functionals are quite naturally identified with measures and locally integrable functions.

These distributions are infinitely differentiable and the Fourier transformation is defined for these functionals. These properties made the distributions a convenient tool of analysis, widely used in the theory of differential equations.

Consider three functional spaces on which distributions will be defined. These spaces are referred to as basic, and functions on them are called *basic* or *test functions*.

As before, we will consider number-valued functions on open set $U \subseteq \mathbb{R}^m$.

The Space $\mathcal{E}$. Of all functional spaces used for distributions, the largest one is the space $\mathcal{E}(U) = \mathcal{C}^\infty(U)$ of all the smooth functions on U; it has been described is Section 3.3.2.

We draw attention to the space $\mathcal{E} = \mathcal{E}(\mathbb{R}^m) = \mathcal{C}^\infty(\mathbb{R}^m)$ of smooth functions defined on the entire space $\mathbb{R}^m$.

The Space S. A frequently used space is $S = S(\mathbb{R}^m)$ constituted by smooth functions on $\mathbb{R}^m$ which along with all their derivatives rapidly vanish at infinity. The points of S are functions $\varphi \in \mathcal{C}^\infty(\mathbb{R}^m)$ such that

$$\|\varphi\|_{\alpha\beta} = \sup\{|x^\alpha \partial^\beta \varphi(x)|: x \in \mathbb{R}^m\} < \infty$$

for all $\alpha = (\alpha(1), \ldots, \alpha(m))$ and $\beta = (\beta(1), \ldots, \beta(m))$. As before $x^a = x_1^{a(1)} \ldots x_m^{a(m)}, \partial^\beta = \partial_1^{\beta(1)} \ldots \partial_m^{\beta(m)}$ and $x = (x_1, \ldots, x_m) \in \mathbb{R}^m$. The functions in S are said to be of *rapid decrease*. These functions along with their derivatives vanish at infinity faster than the inverse of any polynominal.

A metric for S may be introduced with the aid of any family of numbers $c(\alpha, \beta) > 0$ with $\Sigma c(\alpha, \beta) = 1$. The distance between two functions $\varphi, \psi \in S$ is defined by

$$\|\varphi - \psi\| = \Sigma c(\alpha, \beta) \cdot \frac{\|\varphi - \psi\|_{\alpha,\beta}}{1 + \|\varphi - \psi\|_{\alpha\beta}}$$

where the summation is over all α and β. Usually it is reduced to series and the distance is defined as follows. For any integer $q \geq 0$,

$$\|\varphi\|_q = \sup_{|\beta| \leq q} (\sup\{(1 + \|x\|^2)^{q/2} |\partial^\beta \varphi(x)|: x \in \mathbb{R}^m\}),$$

is considered and the distance between $\varphi, \psi \in S$ is defined by

$$\|\varphi - \psi\| = \frac{1}{2} \sum_{q \geq 0} \frac{1}{2^q} \cdot \frac{\|\varphi - \psi\|_q}{1 + \|\varphi - \psi\|_q}.$$

Since

$$|x^\alpha \partial^\beta \varphi(x)| \leq (1 + \|x\|^2)^{|\alpha|/2}) \partial^\beta \varphi(x) \leq c(\alpha, \beta) \cdot |x^\alpha \cdot \partial^\beta \varphi(x)|$$

for some $c(\alpha, \beta) > 0$, the above metrics are equivalent, and functions $\varphi \in S$ may be defined as $\varphi \in \mathscr{C}^\infty(\mathbb{R}^m)$ for which $\|\varphi\|_q < \infty$ for all integers $q \geq 0$.

The set S of rapidly decreasing functions forms an algebra under ordinary operations. Indeed, linear combinations and the products of those functions rapidly decrease, too.

By definition a sequence of functions $\varphi_n \in S$ converges in S to $\varphi \in S$ if and only if $\|\varphi_n - \varphi\| \to 0$, that is, if $\|\varphi_n - \varphi\|_{\alpha\beta} \to 0$ for all multiindices α, β. Alternatively,

$$x^\alpha \partial^\beta (\varphi_n - \varphi)(x) \to 0$$

uniformly in $x \in \mathbb{R}^m$ for all α, β. The condition $\|\varphi_n - \varphi\|_{\alpha\beta} \to 0$ for all α, β can be replaced by $\|\varphi_n - \varphi\|_q \to 0$ for any integer $q \geq 0$.

3.3. Spaces of Functions

The metric space S is complete, and so it is a Frechet space. A metric on S does not meet the absolute uniformity condition. It is more convenient to consider S as a multinormed space under the indicated countable families of norms which define the convergence in S.

The Space $\mathcal{D}$. The points of $\mathcal{D} = \mathcal{D}(U)$ are functions from $\mathcal{C}_0^\infty(U)$, that is, smooth finite functions on U. A family of norms which establish convergence in $\mathcal{D}(U)$ can be defined as follows.

Let

$$K(n) = \{ x \in U : d(x, \overline{U} \setminus U) \geq 1/n, \|x\| \leq n \}$$

be a sequence of compact sets covering U. Observe that the sequence of pairwise disjoint sets

$$A(n) = K(n) \setminus K(n-1) \quad (K(0) = O)$$

also covers U, indeed, each point $x \in U$ belongs to exactly one of sets $A(n)$. The compact support supp φ of a finite function $\varphi \in \mathcal{D}(U)$ is separated from the boundary of U, $\partial U = \overline{U} \setminus U$, by a strictly positive distance, if ∂U is nonempty ($\partial U \neq O$). Therefore, $\partial^\beta \varphi(x) = 0$ at $x \in A(n)$ for all numbers n greater than some $n(\varphi)$, and all indices $\beta = (\beta(1), ..., \beta(m))$.

For each sequence $\alpha = (\alpha(1), ..., \alpha(n), ...)$ of integers $\alpha(n) \geq 0$ and $\varphi \in \mathcal{D}(U)$ we define the norm

$$\|\varphi\|_\alpha = \sum_{n \geq 1} \alpha(n) \cdot \sup \{ |\partial^\beta \varphi(x)| : x \in A(n), |\beta| \leq \alpha(n) \}.$$

(For any $\varphi \in \mathcal{D}(U)$ the number of nonzero terms in this sum is finite.) The family of norms under consideration is noncountable.

By definition, $\varphi_n \in \mathcal{D}(U)$ converges to $\varphi \in \mathcal{D}(U)$ in $\mathcal{D}(U)$ if $\|\varphi_n - \varphi\|_\alpha \to 0$ for all α. It can be demonstrated that such a convergence takes place if and only if (1) *the supports of φ_n and φ are contained in a compact set $K \subseteq U$, (2) the sequence $\{\partial^\beta \varphi(n)\}$ converges to $\partial^\beta \varphi$ uniformly on K for all multiindices β*. Such a convergence cannot be defined by a metric, i.e., the multinormed space $\mathcal{D}(U)$ is nonmetrizable.

Compare the spaces $\mathcal{D} = \mathcal{D}(\mathbb{R}^m)$, $S = S(\mathbb{R}^m)$, $\mathcal{E} = \mathcal{E}(\mathbb{R}^m)$. By the definitions of these spaces, $\mathcal{D} \subset S \subset \mathcal{E}$. Examples demonstrate that $\mathcal{D} \neq S$ and $S \neq \mathcal{E}$. In particular, $\varphi(x) = \exp(-\|x\|^2)$ on $\mathbb{R}^m$ belongs to S but does not belong to $\mathcal{D}$. Any constant belongs to $\mathcal{E}$, but does not belong to S. Identical embeddings of $\mathcal{D}$ in S and S in $\mathcal{E}$ are continuous under the convergences defined on these spaces: if $\varphi_n \to \varphi$ in $\mathcal{D}$, then $\varphi_n \to \varphi$ in S, and if $\varphi_n \to \varphi$ in S then $\varphi_n \to \varphi$ in $\mathcal{E}$.

It can be proved that $\mathcal{D}$ is dense in S and $\mathcal{E}$, and S is dense in $\mathcal{E}$. (A simple proof for the case of $m = 1$ may be found in the book by Kirillov and Gvishiani [3].) The sets $\mathcal{D}$ and S are also dense in the spaces $\mathcal{L}^p = \mathcal{L}^p(\mathbb{R}^m, dx)$ for $1 \leq p \leq \infty$. (For a $\varphi \in S$ the integrability of φ^p with respect to the Lebesque measure dx follows from the inequalities

$$(1 + \|x\|^2)^{1/2} |\partial^0 \varphi(x)| \leq c,$$

$$|\varphi^p(x)| \leq c^p (1 + \|x\|^2)^{-2p}$$

valid for all $x \in \mathbb{R}^m$ and some $c > 0$.)

Now we turn to the respective spaces of distributions.

The Space $\mathcal{D}'$. Of all the spaces considered $\mathcal{D}$ was the narrowest one, therefore, the dual space $\mathcal{D}'$ will be the widest. The elements of $\mathcal{D}'(U)$ are continuous linear functionals on $\mathcal{D}(U)$. They are called *distributions*, or generalized functions if in the Russian literature. In particular, $\mathcal{D} = \mathcal{D}(\mathbb{R}^m)$ and $\mathcal{D}' = \mathcal{D}'(\mathbb{R}^m)$ for $U = \mathbb{R}^m$.

From the above definitions it follows that a linear functional F on $\mathcal{C}_0^\infty(U)$ is a distribution on $\mathcal{D}(U)$ if and only if for each compact set $K \subseteq U$ there is a $c > 0$ and an integer $q \geq 0$ such that

$$|F(\varphi)| \leq c \cdot \sup \{ |\partial^\beta \varphi(x)| : |\beta| \leq q, x \in K \}$$

for all $\varphi \in \mathcal{C}_0^\infty(U)$ with supports in K. This criterion is convenient for applications.

In the following examples, U means an open set in $\mathbb{R}^m$.

EXAMPLE 1. We will say that a numerical-valued function f measurable with respect to a Lebesgue measure dx on U is *locally integrable* if a restriction of f to any compact set $K \subseteq U$ is integrable with respect to dx. The set of all locally integrable number-valued functions on U will be denoted by $\mathcal{L}_{loc}(U)$. In particular, $\mathcal{L}_{loc} = \mathcal{L}_{loc}(\mathbb{R}^m)$ for $U = \mathbb{R}^m$. All constants on U belong to $\mathcal{L}_{loc}(U)$.

Let $f \in \mathcal{L}_{loc}(U)$, then the equality

$$F_f(\varphi) = \int_U \varphi(x) \cdot f(x) \cdot dx \quad (\varphi \in \mathcal{D}(U))$$

defines a functional $F_f \in \mathcal{D}'(U)$. Indeed, the linearity of F_f follows from the linearity of the integral. The continuity of F_f follows from the inequality

$$|F_f(\varphi)| \leq c(K, f) \cdot \sup \{ |\varphi(x)| : x \in K \}$$

3.3. Spaces of Functions

in view of the local integrability of f, which is true for

$$c(K, f) = \int_K |f(x)|\, dx$$

for all $\varphi \in \mathscr{C}_0^\infty(U)$ having supports in the compact $K \subseteq U$.

It can be demonstrated (see, e.g. Section 1.8 in Ref. [4]) that the equality $F_f = F_g$ for $f, g \in \mathscr{L}_{\text{loc}}(U)$ is true if and only if $f = g$ almost everywhere with respect to measure dx. This enables us to identify a functional F_f with the function f and to write f instead of F_f where there is no risk of confusion. (More precisely, F_f is identified with a class of functions which are equivalent to f with respect to the Lebesgue measure dx.) The possibility to identify some functionals in $\mathscr{D}'(U)$ with functions explains the name "generalized functions." These functionals make up a larger class than that of locally integrable functions, $\mathscr{L}_{\text{loc}}(U) \subseteq \mathscr{D}'(U)$. Hence $\mathscr{D}(U) \subseteq \mathscr{D}'(U)$.

Generalized functions defined by locally integrable functions are said to be *regular*.

EXAMPLE 2. Call *numerical-valued measures* the linear combinations of locally bounded, countably additive, positive, complete measures on an algebra $\mathscr{B}(U)$ of parts of U, which is induced by a class of compact sets in U. We will call sets from $\mathscr{B}(U)$ and measures on $\mathscr{B}(U)$ *Baire*, along with sets from the closure $\overline{\mathscr{B}}(U)$ of the algebra $\mathscr{B}(U)$ under simple convergence, and numerical-valued measures on $\overline{\mathscr{B}}(U)$.

Integrals with respect to numerical-valued measures will be defined as linear combinations of integrals with respect to positive components of these measures.

Let μ be a Baire numerical-valued measure on $\mathscr{B}(U)$. Then

$$F_\mu(\varphi) = \int \varphi\, d\mu \quad (\varphi \in \mathscr{D}(U))$$

defines a functional $F_\mu \in \mathscr{D}'(U)$. Indeed, the linearity of F_μ follows from the linearity of the integral. The continuity follows from the relation

$$|F_\mu(\varphi)| \leq c(K, \mu) \cdot \sup\{|\varphi(x)| : x \in K\},$$

where $c(K, \mu)$ stands for the sum of positive components of μ on K. This inequality holds for all $\varphi \in \mathscr{C}_0^\infty(U)$ with supports in a compact set $K \subseteq U$.

Since Baire measures are defined by their values on compact sets, and the indicators of compact sets are approximated by smooth functions, the equality $F_\mu = F_\nu$ for Baire measures μ and ν on $\mathscr{B}(U)$ holds if and only if $\mu = \nu$. This fact enables us to identify a functional F_μ with the measure μ and to write μ in place of F_μ where this would not lead to confusion.

In this case the functional F_μ, as well as the Baire measure μ, is said to be a *distribution*. All the functionals from $\mathcal{D}'(U)$ are also called distributions.

EXAMPLE 3. Outstanding among all distributions are those concentrated at separate points. They are called *delta functions*. We have already encountered them in Section 3.3.2.

Let $\mu = \delta_a$ be a measure describing a distribution of unit mass at a point $a \in U$:

$$\delta_a(B) = 1 \; (a \in B), \quad \delta_a(B) = 0 \; (a \notin B)$$

for each $B \in \mathcal{B}(U)$. Then

$$F_\mu(\varphi) = \int_U \varphi \, d\mu = \varphi(a) \quad (\varphi \in \mathcal{D}(U)).$$

The functional F_μ is called a *delta function* (δ function) at a, denoted δ_a, like the defining measure. If $a = 0$, the symbol is δ rather than δ_0.

By analogy with the regular case, generalized functions are frequently written as integrals although these integrals may have no other sense. In particular, in the expression

$$F_\delta(\varphi) = \int \varphi(x) \cdot \delta(x) \cdot dx = \varphi(0)$$

the integral gives the value of the generalized function F_δ for $\varphi \in \mathcal{D}(U)$. The product $\varphi \cdot \delta$ is not integrable with respect to the Lebesgue measure dx. In this integral form, δ can be viewed as a generalized derivative of the pure point measure $\mu = d^0 x$ with respect to dx, written $d^0 x = \delta \, dx$, without having to give a formal definition to the concept of generalized derivative of measure with respect to measure. The equality

$$\int \varphi(x) \cdot \delta(x) \cdot dx = \int \varphi(x) \cdot d^0 x$$

is then a formula for change of variables in the integral.

Thus, functionals from $\mathcal{D}'(U)$ describe, in particular, numerical-valued functions on U locally integrable with respect to a Lebesgue measure and Baire measures on $\mathcal{B}(U)$. It often proves useful to join measures and functions in one model.

The Space S'. By definition, the elements of S' (which is the dual of the space S of functions of rapid decrease on $\mathbb{R}^m$) are continuous linear functionals on S. They are said to be *tempered distributions*.

Since $\mathcal{D} \subseteq S$, and an identity embedding of the space $\mathcal{D} = \mathcal{D}(\mathbb{R}^m)$ into $S = S(\mathbb{R}^m)$ is continuous, the restriction of each functional $F \in S'$ to $\mathcal{D}$ belongs to $\mathcal{D}'$. That is, the restrictions of tempered distributions to $\mathcal{D}$ are

3.3. Spaces of Functions

then themselves distributions. Since $\mathcal{D}$ is dense in S, different functionals from S' have different restrictions to $\mathcal{D}$. Hence, S' is embeddable in $\mathcal{D}'$, that is, $S' \subseteq \mathcal{D}'$.

EXAMPLE 4. We will say that a numerical-valued function f on $\mathbb{R}^m$ measurable with respect to a Lebesgue measure dx is of *slow growth* whenever for some $q \geq 0$ the function g_q on $\mathbb{R}^m$,

$$g_q(x) = (1 + \|x\|^2)^{-q} f(x),$$

is integrable with respect to dx. We denote by $\mathcal{M}$ the set of all functions of slow growth on $\mathbb{R}^m$. All the polynomials on $\mathbb{R}^m$ belong to $\mathcal{M}$. A function $h(x) = \exp\|x\|$ on $\mathbb{R}^m$ does not belong to $\mathcal{M}$.

Let $f \in \mathcal{M}$. Then the equality

$$F_f(\varphi) = \int \varphi(x) \cdot f(x) \cdot dx \quad (\varphi \in S)$$

defines a functional $F_f \in S'$, that is, a tempered generalized function; F_f is identified with f.

Applying the Hölder inequality, it is easy to verify that $\mathcal{L}^p \subseteq \mathcal{M}$ for $1 \leq p \leq \infty$ (see Section 7.12 in Rudin [5]). Thus, $\mathcal{L}^p \subseteq \mathcal{M} \subseteq S'$.

EXAMPLE 5. A positive Baire measure μ on an algebra $\mathcal{B} = \mathcal{B}(\mathbb{R}^m)$ induced by compact sets in $\mathbb{R}^m$ is said to be of slow growth if, for some $q \geq 0$, $h_q \in \mathbb{R}^m$ given by

$$h_q(x) = (1 + \|x\|^2)^{-q}$$

is integrable with respect to μ.

Let μ be a numerical-valued Baire measure of slow growth on $\mathcal{B}$. Then the equality

$$F_\mu(\varphi) = \int \varphi \, d\mu \quad (\varphi \in S)$$

defines a functional $F_\mu \in S'$. This immediately follows from the definitions. The functional is identified with the measure μ which is also called a *tempered distribution*.

Obviously pure point measures $\mu = \delta_a$ are of slow growth.

EXAMPLE 6. Let $f \in \mathcal{L}^p$ ($1 \leq p < \infty$). Then there is a positive Baire measure $d\nu = |f| \, dx$ under which $|f|$ is a derivative with respect to a Lebesgue measure dx. This fact follows from the Radon-Nikodym theorem. Applying Hölder's inequality, it is not hard to see that $d\nu$ is of slow growth.

Let $d\mu = fdx$ be a Baire measure, where f is a derivative with respect to a Lebesgue measure dx. Each function $\varphi \in S$ is integrable with respect to $d\mu$. Changing the variables, we obtain

$$F_f(\varphi) = \int \varphi f dx = \int \varphi d\mu \quad (\varphi \in S).$$

These equalities establish relations between functions of slow growth and measures.

The Space $\mathscr{E}'$. By definition, the elements of the space $\mathscr{E}'(U)$, which is dual to the space $\mathscr{E}(U)$ of smooth functions on U, are continuous linear functionals on $\mathscr{E}(U)$.

Since $\mathscr{D}(U) \subseteq \mathscr{E}(U)$ and an identity embedding of $\mathscr{D}(U)$ in $\mathscr{E}(U)$ is continuous, the restriction of each functional $F \in \mathscr{E}'(U)$ to $\mathscr{D}(U)$ belongs to $\mathscr{D}'(U)$. Because $\mathscr{D}(U)$ is dense in $\mathscr{E}(U)$, different functionals from $\mathscr{E}'(U)$ have different restrictions to $\mathscr{D}(U)$. Thus, $\mathscr{E}'(U)$ is embedded in $\mathscr{D}'(U)$, and we may assume that $\mathscr{E}'(U) \subseteq \mathscr{D}'(U)$.

Similarly, for $U = \mathbb{R}^m$, $\mathscr{E}' = \mathscr{E}'(\mathbb{R}^m)$ is embeddable in $S' = S'(\mathbb{R}^m)$ and so

$$\mathscr{E}' \subseteq S' \subseteq \mathscr{D}'.$$

Let f be a numerical-valued function on U, locally integrable with respect to a Lebesgue measure dx and having a compact support. Then the equality

$$F_f(\varphi) = \int \varphi(x) \cdot f(x) \cdot dx \quad (\varphi \in \mathscr{E}(U))$$

defines a functional $F_f \in \mathscr{E}'(U)$.

In general, $\mathscr{E}'(U)$ contains those and only those functionals from $\mathscr{D}'(U)$ which have compact supports. The support of a distribution $F \in \mathscr{D}'(U)$ is defined as follows. A point $x \in U$ does not belong to supp F if there exists an open neighborhood V of x such that $F(\varphi) = 0$ for each $\varphi \in \mathscr{D}(U)$ with supp $\varphi \subseteq V$. Points devoid of this property constitute the support of F.

It would be natural to refer to the distributions from $\mathscr{E}'(U)$ as *finite* (by analogy with finite functions). For example, delta functions are finite. They have compact supports. In general, if a distribution has a point support $\{0\}$ then it is a linear combination of derivatives of the delta function.

For $U = \mathbb{R}^m$, finite distributions make up the narrow class $\mathscr{E}'$, tempered distributions constitute a wider class S', and arbitrary distributions belong to the widest class $\mathscr{D}'$.

3.3. Spaces of Functions

3.3.5. Sobolev Spaces

Before we move on to describe these spaces, we define operations with distributions.

Operations with Distributions. Linear combinations of distributions are defined like for any other linear functionals.

The equality

$$f \cdot F(\varphi) = F(f\varphi) \quad (\varphi \in \mathcal{D}(U))$$

defines the product $fF \in \mathcal{D}'(U)$ of a distribution $F \in \mathcal{D}'(U)$ and a function $f \in \mathcal{E}(U)$. Observe that $f\varphi \in \mathcal{D}(U)$. The product is written as fF in order not to confuse it with the number $Ff = F(f)$ for $f \in \mathcal{D}(U)$.

EXAMPLE. Since $f\delta(\varphi) = \delta(f\varphi) = f(0) \cdot \varphi(0)$, we have $f\delta = f(0) \cdot \delta$.

The equality

$$\partial^\alpha F(\varphi) = (-1)^{|\alpha|} F(\partial^\alpha \varphi) \quad (\varphi \in \mathcal{D}(U))$$

defines the derivative $\partial^\alpha F \in \mathcal{D}'(U)$ of a distribution $F \in \mathcal{D}'(U)$ with a multiindex α.

EXAMPLES.

(1) Let $F = F_h$ be a distribution identified with a function $H = \mathbb{R} \to \mathbb{R}$ with values $h(x) = 0$ $(x < 0)$ and $h(x) = 1$ $(x \geq 0)$. Then $\partial F_h = \delta$. Indeed,

$$-F_h(\varphi') = -\int_{-\infty}^{\infty} \varphi'(x) \cdot h(x) \cdot dx = -\int_0^{\infty} \varphi'(x) \cdot dx$$

$$= \varphi(0) - \varphi(\infty) = \varphi(0) - 0 = \varphi(0) = \delta(\varphi)$$

for each $\varphi \in \mathcal{D}(\mathbb{R})$.

(2) Let $F = F_f$ be a distribution identified with a smooth function $f : \mathbb{R} \to \mathbb{R}$. Then $\partial F_f = F_{\partial f}$. Indeed, integrating by parts and observing the compactness of the support $\varphi \in \mathcal{D}(\mathbb{R})$ we obtain

$$-F_f(\varphi') = -\int \varphi'(x) \cdot f(x) \cdot dx = \int \varphi(x) \cdot f'(x) \cdot dx = F_{f'}(\varphi).$$

This example demonstrates that the differentiation of distributions agrees with the differentiation of smooth functions

(3) $\partial^\alpha \delta(\varphi) = (-1)^{|\alpha|} \partial^\alpha \varphi(0) \quad (\varphi \in \mathcal{D}(\mathbb{R}))$.

Observe that the derivatives of tempered distributions are themselves tempered distributions: if $F \in S'$, then $\partial^\alpha F \in S'$ for any multiindex α. That is, S' is closed under the operation of differentiation.

The multiplication by a function and the differentiation are related by Leibnitz's formula.

THEOREM (Leibnitz's formula).

$$\partial_j(fF) = \partial_j f \cdot F + f \cdot \partial_j F.$$

This formula follows from the definition and the differentiation rule for products of smooth functions.

It can be demonstrated that each distribution $F \in \mathscr{E}'(U)$ is equal to the derivative $\partial^\alpha f$ of a function f on U locally integrable with respect to dx. Each distribution $F \in S'(\mathbb{R}^m)$ is equal to the derivative $\partial^\alpha f$ of a function f of slow growth on $\mathbb{R}^m$.

Generalized Derivatives. We note that $\mathscr{L}^p(U) \subseteq \mathscr{L}_{loc}(U)$ for $1 \leq p < \infty$. This is not hard to verify with the Hölder inequality. Therefore, functions $g \in \mathscr{L}^p(U)$ can be identified with the distributions $F_g \in \mathscr{D}'(U)$.

We take a function $f \in \mathscr{L}_{loc}(U)$, the distribution $F_f \in \mathscr{D}'(U)$ identified with the function, and its derivative $\partial^\alpha F_f \in \mathscr{D}'(U)$ with multiindex α. We will write f instead of F_f, and $\partial^\alpha f$ instead of $\partial^\alpha F_f$ (although f need not be differentiable). If $\partial^\alpha f = F_g \in \mathscr{D}'(U)$ for some $g \in \mathscr{L}_{loc}(U)$, then we will call g a generalized derivative of f and write $\partial^\alpha f = g$. Observe that any $h \in \mathscr{L}_{loc}(U)$ which is equivalent to g with respect to a Lebesgue measure, is also a generalized derivative of f.

If f is smooth, its ordinary derivative is also a generalized derivative. However, there exist functions differentiable almost everywhere and locally integrable with respect to a Lebesgue measure whose common derivatives cannot be treated as generalized. An example may be functions increasing on an interval, which are continuous on the left but not absolutely continuous (see Section 6.14 in Ref. [5]).

The Spaces $W_q^p(U)$. Let p, q be integers such that $p \geq 1$ and $q \geq 0$. The points of a Sobolev space $W_q^p(U)$ are locally integrable functions $f \in \mathscr{L}_{loc}(U)$ whose generalized derivatives $\partial^\alpha f \in \mathscr{L}^p(U)$ for $|\alpha| \leq q$. A norm for $W_q^p(U)$ is defined by

$$\|f\|_{p,q} = \left(\sum_{|\alpha| \leq q} \int_U |\partial^\alpha f(x)|^p \, dx\right)^{1/p}.$$

For $p = 2$ this norm is obtained from the inner product

$$\langle f, g \rangle = \sum_{|\alpha| \leq q} \int_U \overline{\partial^\alpha f(x)} \cdot \partial^\alpha g(x) \cdot dx.$$

The normed space $W_q^p(U)$ is Banach. The normed space $W_q^2(U)$ is Hilbert.

The Hilbert space $W_q^2(U)$ is related to the Euclidean space $\mathscr{H}_0^q(U)$ of

3.4. Fourier Transforms

smooth functions described in Section 3.3.2. It can be proved that the completion $\mathcal{H}_0^q(U)$ is a subspace of the Sobolev space $W_q^2(U)$. When $U = \mathbb{R}^m$ the completion $\mathcal{H}_0^q = \mathcal{H}_0^q(\mathbb{R}^m)$ coincides with $W_q^2 = W_q^2(\mathbb{R}^m)$. Therefore, some authors frequently write $\mathcal{H}_0^q$ instead of W_q^2. Functions from $\mathcal{H}_0^q$ are approximated by sequences of q-smooth functions under the respective norm.

Sobolev's Lemma. This lemma sets up a relation between spaces $W_q^2(U)$ and $\mathcal{C}^p(U)$ for $U \subseteq \mathbb{R}^m$ ($q \geq 0, p \geq 0, m \geq 1$).

LEMMA (Sobolev). *If $q > p + m/2$, then for every $f \in W_q^2(U)$ there is $g \in \mathcal{C}^p(U)$ equal to f almost everywhere.*

We emphasize that the order of smoothness p of g, which is equivalent to f with respect to a Lebesgue measure, is strictly below the order q of the derivatives of f having integrable squares. In this sense, replacing the functions from Sobolev's spaces by smooth functions is associated with some loss in the order of smoothness.

A proof of Sobolev's lemma presented for a more general formulation can be found in Section 7.25 of Ref. [5].

3.4. Fourier Transforms

The theory of Fourier transforms for distributions extends classical harmonic analysis. Fourier transforms are effectively used for solving differential equations.

3.4.1. Fourier Transforms of Functions of Rapid Decrease

It is convenient to define the Fourier transformation first for functions of rapid decrease, then for tempered distributions.

The Fourier transform $g = \Phi(f)$ of a function $f \in S = S(\mathbb{R}^m)$ is defined as

$$g(y) = (2\pi)^{-m/2} \int e^{-iyx} f(x)\, dx.$$

Here $yx = y_1 x_1 + \ldots + y_m x_m$ is the inner product of two vectors $y = (y_1, \ldots, y_m)$ and $x = (x_1, \ldots, x_m)$ in $\mathbb{R}^m$, dx is the Lebesgue measure for $\mathbb{R}^m$. Since functions on S are integrable with respect to dx and $|e^{-iyx} f(x)| \leq |f(x)|$, the above integral exists.

EXAMPLE. Let $f(x) = \exp(-\|x^2\|/2)$. Then $\Phi(f) = f$.

The Fourier transform is linear, continuous and one-one transformation of S. The inverse transformation $f = \Phi^{-1}(g)$ for $g \in S$ is given by

$$f(x) = (2\pi)^{-m/2} \int e^{ixy} g(y) \, dy.$$

This equality is the so-called "inversion formula." Here $xy = yx$, and dy is the Lebesgue measure for $\mathbb{R}^m$. The transformation Φ^{-1} is called the inverse Fourier transformation. It is also a linear, continuous and one-one transformation of S. Stated differently, Φ and Φ^{-1} are linear homeomorphisms $S \to S$. This can be shown by applying the rules of integration and differentiation under the integral sign (see Chapter 6 in Ref. [4]).

Some authors transpose the definitions of Φ and Φ^{-1}, and ascribe the factor 1 to one and $1/(2\pi)^m$ to the other.

Notice that under these definitions

$$\Phi(\bar{f}) = \overline{\Phi^{-1}(f)}$$

for each $f \in S$.

We define the symmetry operator S on S by

$$S\varphi(x) = \varphi(-x) \quad (\varphi \in S, \ x \in \mathbb{R}^m).$$

From this definition it follows that $S^2 = I$, where I stands for the identity transformation on S. It is straightforward to verify using the inversion formula that

$$\Phi = \Phi^{-1} S, \quad \Phi^2 = S, \quad \Phi^3 = \Phi^{-1}, \quad \Phi^4 = I.$$

For functions in S another product is defined, called a convolution and denoted by $*$. By definition, the convolution $h = g * f$ of $f, g \in S$ is a function h on $\mathbb{R}^m$ given by

$$h(z) = \int g(z - x) \cdot f(x) \cdot dx.$$

It is not hard to see that $h \in S$. The convolution is commutative, associative and distributive with respect to sum. It is analogous to the rule of computing coefficients in a product of polynomials.

Fourier transformation relates convolution to ordinary product. Using the definitions and integration rules we obtain

$$(2\pi)^{-m/2} \Phi(g * f) = \Phi(g) \cdot \Phi(f),$$
$$(2\pi)^{-m/2} \Phi(g) * \Phi(f) = \Phi(g \cdot f).$$

3.4. Fourier Transforms

3.4.2. Transformations of Tempered Distributions

The Fourier transformation ΦF of a distribution $F \in S'$ is defined to be a composition $F \circ \Phi$ of the Fourier transformation $\Phi : S \to S$ and the distribution $F : S \to \mathbb{C}$. By definition

$$\Phi F(\varphi) = F(\Phi(\varphi)) \quad (\varphi \in S).$$

Since both Φ and F are linear and continuous, we have $\Phi F = F \circ \Phi \in S'$.

Fourier transform of distributions possesses the same properties as that of functions. We choose the weak convergence mode for S': by definition, given $F_n \to F$ for $F_n, F \in S$, we have $F_n(\varphi) \to F(\varphi)$ ($\varphi \in S$). This convergence allows us to say that the Fourier transformation $\Phi : S' \to S'$ is continuous. For $S \subseteq S'$, it extends $\Phi : S \to S$, and therefore has the same notation. The Fourier transformation Φ for S' is a linear, continuous and one-one map $S' \to S'$. The inverse Fourier transformation Φ^{-1} for S' is defined by

$$\Phi^{-1} F(\varphi) = F(\Phi^{-1}(\varphi)) \quad (\varphi \in S).$$

The equality

$$SF(\varphi) = F(S\varphi) \quad (\varphi \in S, \ F \in S)$$

extends the symmetry operator S to S. The equalities $S^2 = I$ and $\Phi = \Phi^{-1}S$, $\Phi^2 = S$, $\Phi^3 = \Phi^{-1}$, $\Phi^4 = I$ remain true for it (I is the identity transformation on S'). Indeed,

$$S^2 F(\varphi) = F(S^2 \varphi) = F(\varphi),$$

$$\Phi^{-1} S F(\varphi) = F(\Phi^{-1} S \varphi) = F(\Phi(\varphi)) = \Phi F(\varphi).$$

The Fourier transform sets up a relation between the operations of differentiation and multiplication by a function. Consider the operator $D^\alpha = i^{-|\alpha|} \partial^\alpha$, and the operator M^α of multiplication by the function $f(x) = x^\alpha$. These operators are related to the Fourier transform by

$$D^\alpha \Phi = \Phi M^\alpha, \quad \Phi D^\alpha = (-1)^{|\alpha|} M^\alpha \Phi.$$

These equalities are easily verified for functions from S. Since S is dense in S', they are extended by continuity to distributions from S'.

EXAMPLE. The Fourier transform of the delta function is

$$\Phi(\delta) = (2\pi)^{-m/2} \cdot 1$$

where 1 stands for the distribution represented by the identical unity. Indeed,

$$\Phi\delta(\varphi) = \delta(\Phi(\varphi)) = \psi(0) = (2\pi)^{-m/2} \cdot 1(\varphi),$$

where $\psi = \Phi(\varphi)$ is given by

$$\psi(y) = (2\pi)^{-m/2} \int e^{-iyx} \varphi(x) \, dx,$$

and

$$1(\varphi) = \int \varphi(x) \, dx.$$

From the equality for $\Phi(\delta)$ we obtain the equality

$$\Phi(1) = (2\pi)^{m/2} \delta$$

for the Fourier transform of identical unity, viz.,

$$\delta = S\delta = \Phi^2(\delta) = \Phi((2\pi)^{-m/2} \cdot 1) = (2\pi)^{-m/2} \cdot \Phi(1).$$

The first equality results from the following chain of equalities

$$S\delta(\varphi) = \delta(S\varphi) = S\varphi(0) = \varphi(-0) = \varphi(0) = \delta(\varphi),$$

which hold for every $\varphi \in S$.

Taking Fourier transforms often simplifies the solution of differential equations. Effective tools are regularization multipliers, specifically exponential. The Fourier transformation of the product of a tempered distribution by a rapidly decreasing exponential function is the Laplace transformation widely used in applications (see Section 3.10.3).

3.4.3. The Fourier-Plancherel Transform

The set $\mathcal{L}^2 = \mathcal{L}^2(\mathbb{R}^m, dx)$ is embedded in $S' = S'(\mathbb{R}^m)$ on identifying the functions from $\mathcal{L}^2$ with the induced distributions from S'. The Fourier transformation maps functions from $\mathcal{L}^2$ to functions from $\mathcal{L}^2$. Therefore the restriction of the Fourier transformation for S' to the set $\mathcal{L}^2$ is a transformation of this set. It is known as the *Fourier-Plancherel transform*.

For functions $f, g \in \mathcal{L}^2$, the inner product and norm are defined by

$$\langle f, g \rangle = \int \bar{f} g \, dx,$$
$$\|f\|^2 = \langle f, f \rangle.$$

3.4. Fourier Transforms

It is not difficult to check that

$$\langle \Phi(f), g \rangle = \langle f, \Phi^{-1}(g) \rangle,$$
$$\langle \Phi(f), \Phi(g) \rangle = \langle f, g \rangle$$
$$\|\Phi(f)\| = \|f\|.$$

This means that the Fourier-Plancherel transformation is unitary. The following theorem holds (see Chapter 6 in Ref. [4]).

THEOREM (Plancherel). *The Fourier transformation maps $\mathcal{L}^2$ onto the whole $\mathcal{L}^2$, preserving the inner product and norm.*

Let $f \in \mathcal{L}^2$ and $f_n = B_n \cdot f$, where B_n is a characteristic function of a ball $\overline{B}(0, n) \subseteq \mathbb{R}^m$. Since $B_n \in \mathcal{L}^2$, then all $f_n = B_n f$ are integrable with respect to a Lebesgue measure dx, and their Fourier transforms $g_n = \Phi(f_n)$ obey the equalities

$$g_n(y) = (2\pi)^{-m/2} \int_{\overline{B}(0,n)} e^{-iyx} f(x) \, dx.$$

From the Plancherel theorem it follows that $g_n \to g = \Phi(f)$ in $\mathcal{L}^2$, that is, in the mean.

3.4.4. The Fourier-Stieltjes Transform

This name is given to the restriction of the Fourier transformation to the set of measures μ of slow growth on an algebra $\mathcal{B} = \mathcal{B}(\mathbb{R}^m)$ induced by compact sets in $\mathbb{R}^m$. Particular cases of such measures are probability distributions, the Lebesgue measure and counting measure. The transformations of counting and Lebesgue measures are related to Fourier series and integrals. For probability distributions, characteristic functions are introduced.

For a probability distribution μ, the characteristic function is defined by

$$\hat{\mu}(y) = \int e^{iyx} \, d\mu(x).$$

Characteristic functions are described by the following theorem.

THEOREM (Bochner). *Characteristic functions are those normed, continuous, and positive definite on $\mathbb{R}^m$, no other.*

Normalization means that $\hat{\mu}(0) = 1$. Positive definiteness is expressed by

$$\Sigma \hat{\mu}(y_j - y_k) \bar{z}_j z_k \geq 0 \quad (1 \leq j, k \leq n)$$

for all finite families of vectors $y_1, ..., y_n \in \mathbb{R}^m$ and complex numbers $z_1, ..., z_n$.

3.4.5. The Radon Transform

Radon has proved that a smooth function on $\mathbb{R}^3$ is defined by the values of its integrals over planes in $\mathbb{R}^3$. The transformations relating functions to their integrals over sets of a given class, are named Radon. They play an important role in integral geometry.

Let $a \neq 0$ be a point in $\mathbb{R}^m$, c a number, $H = \{ x : ax = c \}$ a hyperplane in $\mathbb{R}^m$, $d_H x$ the $(m - 1)$-dimensional Lebesgue measure for H. The equality

$$R_{a,c}(f) = \|a\|^{-1} \int f(x) d_H x$$

defines the Radon transform $R(f)$ of a function f on $\mathbb{R}^m$, which is integrable over every hyperplane.

This integral is often written out with the delta function δ_H, which is viewed as the generalized derivative of the Lebesgue measure $d_H x$ for a hyperplane with respect to the Lebesgue measure dx for $\mathbb{R}^m$, without giving a formal definition of generalized derivative of measure with respect to measure. The equality

$$\int f(x) \cdot \delta_H(x) \cdot dx = \int f(x) \cdot d_H x$$

is then a formula for the change of variables $d_H x = \delta_H(x) \cdot dx$ in the integral. If $H = \mathbb{R}^{m-1}$, $d_H x = d^{m-1} x$, $\delta_H = \delta^{m-1}$ then $d^{m-1} x = \delta^{m-1}(x) \cdot d^m x$.

In particular, one may consider functions of rapid decrease from $S = S(\mathbb{R}^m)$. The Radon transformation R is a linear bijection S onto a vector space $\mathcal{R} = \mathcal{R}(\mathbb{R}^m)$ defined by a pair a, c of points in $\mathbb{R}^m$ and numbers (see Section 1.2 in Ref. [6]). The inverse Radon transform R^{-1} allows one to determine a function f by its integrals over hyperplanes.

Radon and Fourier transforms are related by a straightforward expression

$$R_{a,c}(f) = (2\pi)^{m/2 - 1} \int_{\mathbb{R}} \Phi^{-1} f(ta) e^{-itc} dt.$$

This equality may be used for defining Radon transforms.

3.5. Bounded Linear Operators

The general definitions were formulated in Chapter 1. Here we present the main principles of linear analysis in the form of theorems on extension, boundedness and inversion. These theorems are given here only for normed spaces.

3.5.1. Extension of Functionals

Every bounded linear functional can be extended from a subspace of a normed space to the whole space, with preservation of the norm.

THEOREM (Hahn-Banach). *For any bounded linear functional x' defined on a subspace X of a normed space Y there is a linear functional y' on Y which extends x' and has the same norm.*

PROOF. If Y is spanned by X and a $b \notin X$, then y' is defined by

$$y'(x + tb) = x'(x) + t\beta$$

for an arbitrary number t and some $\beta = y'(b)$ chosen so that the inequality

$$|x'(x) + t\beta| \le \|x'\| \cdot \|x + tb\|$$

is satisfied for all t. Clearly, this linear functional is bounded, extends x', and $\|y'\| \le \|x'\|$. It is easy to show that $\|x'\| \le \|y'\|$ as well.

The general case is proved by induction from the particular case using the arguing of Zorn's lemma. ∎

We emphasize that the extension under consideration is not necessarily unique.

The principle of global extension formulated by the Hahn-Banach theorem has a variety of applications. It provides a basis for proof of important theorems on separation of sets by hyperplanes.

Two simple examples of extension follow.

COROLLARY 1. *For each $x \ne 0$ in the space Y, there exists a bounded linear functional y' on Y such that $\|y'\| = 1$ and $y'(x) = \|x\|$.*

PROOF. Consider a subspace X constituted by vectors tx, with t being an arbitrary number. We define a bounded linear functional x' on X by $x'(tx) = t \cdot \|x\|$. Clearly, $\|x'\| = 1$ and $x'(x) = \|x\|$. By the Hahn-Banach theorem, there is a functional y' on Y extending x' and having the desired properties. ∎

COROLLARY 2. *For every subspace Z of Y and $b \notin Z$ there is a bounded linear functional y' on Y such that, for all $z \in Z$, $y'(z) = 0$ and $y'(b) = 1$.*

PROOF. Let X be a subspace of Y constituted by vectors $X = z + tb$, where $z \in Z$ and t is an arbitrary number. By $x'(z + tb) = t$ we defined a bounded linear functional x' on X. Clearly, for all $z \in Z$, $x'(z) = 0$, and $x'(b) = 1$. By the Hahn-Banach theorem, there exists a bounded linear functional y' on Y extending x' and possessing the desired properties. ∎

Both corollaries have an intuitive geometrical meaning.

3.5.2. Uniform Boundedness of Operators

Let E be a Banach space, F be a normed space, and $T_i : E \to F$ be a family of continuous linear operators with an arbitrary index set. We will say that *the family T_i is bounded at $x \in E$* if the set of values $T_i(x) \in F$ is bounded in F. We will also say that *the family T_i is uniformly bounded* if the union of sets $\{ T_i(x) : \|x\| \leq 1 \}$ is bounded in F (i.e., if the set of the numbers $\|T_i\|$ is bounded). We can speak about this as of the uniform boundedness of the family of T_i on a unit ball.

THEOREM (Banach-Steinhaus). *If a family of continuous linear operators in a Banach space is bounded at each point, then it is uniformly bounded.*

PROOF. The Banach-Steinhaus theorem is proved with the aid of the classical Baire theorem which asserts that in a complete metric space the union of a sequence of closed sets with an empty interior has an empty interior as well.

Consider a sequence of sets $E(n) = \{ x : \|T_i(x)\| \leq n \text{ for all } i \}$. It is not hard to verify that these sets are closed and their union is the whole of E. Since E has a nonempty interior, in some $A = E(n(0))$ there is an interior point a. Consequently, A contains a closed ball $\overline{B}(a, r)$ with radius $r > 0$ centered on a. Therefore,

$$\|T_i x\| = \|T_i a + T_i z\| \leq n(0),$$
$$\|T_i z\| \leq \|T_i a\| + \|T_i x\| \leq c(a) + n(0)$$

for $x = a + z$, $\|z\| \leq r$, $\|T_i(a)\| \leq c(a)$. Whence $r \|x\|^{-1} \cdot \|T_i x\| = \|T_i (r \cdot \|x\|^{-1} \cdot x)\| \leq c(a) + n(0)$, for $\|x\| \neq 0$ and $\|T_i x\| \leq c \cdot \|x\|$, where $c = r^{-1}(c(a) + n(0))$. Thus, $\|T_i x\| \leq c$ for $\|x\| \leq 1$ for all i. ∎

The principle of uniform boundedness formulated by the Banach-Steinhaus theorem is applied to advantage. Consider a simple example.

COROLLARY. *The limit T of a sequence of continuous linear operators T_n, which converges at each x of a Banach space E, is a continuous linear operator on E.*

PROOF. The linearity of T follows from the linearity of T_n and the limit. The continuity of T, equivalent to its boundedness on the unit ball,

3.5. Bounded Linear Operators

follows from the Banach-Steinhaus theorem. Indeed, since $T_n x \to T x$ for each $x \in E$, then

$$\|T_n x\| \leq \|Tx\| + \|T_n x - Tx\| \leq c(x)$$

for some $c(x) > 0$. Therefore, by the Banach-Steinhaus theorem, there is some $c > 0$ such that $\|T_n x\| \leq c$ for all n and $x \in \overline{B}(0, 1) \subseteq E$. Thus, the limit function obeys the inequality $\|Tx\| \leq c$ for $\|x\| \leq 1$. Hence, the operator T is bounded, as stated. ∎

The reader is cautioned that for a sequence of nonlinear operators the pointwise convergence does not guarantee continuity of the limit function.

The Baire Theorem. The Baire theorem on the interior point is used in proving many important theorems and provides insight into the properties of completeness. Therefore it is worth discussing in greater detail.

Let E be a complete metric space endowed with a metric d. A sequence of closed balls

$$\overline{B}_n = \overline{B}(c_n, r_n) = \{x \mid d(x, c_n) \leq r_n\} \subseteq E$$

is said to be *contracting* if they are embedded into each other and their radii tend to zero, $\overline{B}_{n+1} \subseteq \overline{B}_n$, $r_n \to 0$. For example, for $c_n = c \in E$, $r_n = 1/n$ we obtain a sequence of concentric balls about c. We will say that a sequence of balls *contracts* to $c \in E$ if c belongs to their intersection.

As usual, we suppose that the space (E, d) is nonempty and has a nonempty interior.

LEMMA. *In a complete metric space each contracting sequence of closed balls contracts to a point.*

PROOF. If a sequence of balls contracts, then the sequence of their centers c_n converges in itself: $d(c_{n+m}, c_n) \leq 2r_n$ for any m and n. Since E is complete, c_n converges to a $c \in E$, and since $c_{n+m} \in \overline{B}_n$ for any m, n and $\overline{B}_n$ is closed, then $c \in \overline{B}_n$ for any n. The lemma is established. ∎

This lemma paves the way for

THEOREM (Baire). *In a complete metric space, the union of every sequence of closed sets with an empty interior has an empty interior as well.*

PROOF. Let $F_n \subseteq E$ be a sequence of closed sets having an empty interior and the union $F = \cup F_n$. Suppose that the theorem is not true and there exists a closed ball $\overline{B} = \overline{B}(c, r) \subseteq F$ ($c \in E$, $r > 0$). We note that the intersections $E_n = \overline{B} \cap F_n$ are closed, have no interior points, and $\overline{B} = \cup E_n$. The closed ball $\overline{B}$ endowed with metric d makes up a complete metric space. Therefore, it will suffice to prove the theorem for $E = F$. This will simplify the procedure. We shall assume that $E = \cup F_n$.

We determine a contracting sequence of closed balls in E, which will lead to contradiction with the lemma. Let $U_n = F_n' = E - F_n$. Since $E = \cup F_n$, $\cap U_n = O$. By the conditions of the theorem F_n is closed and has an empty interior whereas E has interior points. Thus $F_n \neq E$ and, for each n, U_n is an open nonempty set. All its points are interior. Hence, there exists a closed ball $\overline{B}_1 = \overline{B}(c_1, r_1) \subseteq U_1$ ($c_1 \in E$, $0 < r_1 < 1$). Pick an $n \geqslant 1$ and suppose that we have already chosen a closed ball $\overline{B}_n = \overline{B}(c_n, r_n) \subseteq U_n$ with a radius $0 < r_n < n^{-1}$ centered on $c_n \in E$. Since F_{n+1} has an empty interior, the open ball $B_n = B_n(c_n, r_n)$ cannot be contained as a whole in F_{n+1}, and so intersects with its complement U_{n+1}. The intersection $B_n \cap U_{n+1}$ is open along with B_n and U_{n+1}, that is, all its points are interior. Thus, there is a closed ball $\overline{B}_{n+1} = \overline{B}(c_{n+1}, r_{n+1})$ $\subseteq B_n \cap U_{n+1}$ about $c_{n+1} \in E$ with the radius $0 < r_{n+1} < (n+1)^{-1}$. Observe that $\overline{B}_{n+1} \subseteq B_n \subseteq \overline{B}_n$ and $\overline{B}_{n+1} \subseteq U_{n+1}$.

By induction this argument implies that there exists a contracting sequence of the closed balls $\overline{B}_n \subseteq U_n$. By the lemma, $\cap \overline{B}_n \neq O$. Since $\cap \overline{B}_n \subseteq \cap U_n$, this assertion contradicts $\cap U_n = O$. ∎

Here is an equivalent formulation of Baire's theorem: *if in a complete metric space the union of a sequence of closed sets has an interior point, then at least one of these sets has an interior point.*

3.5.3. Inverting Operators

An operator inverse to the bijective linear operator is also one-one and linear. The inverse operator may be discontinuous even when the direct operator is continuous.

COUNTER-EXAMPLE. Let $E = \mathscr{C}[0, 1]$ be the Banach space of continuous functions on $[0, 1]$ under the norm

$$\|f\| = \sup\{|f(x)| : x \in [0, 1]\}.$$

The indefinite integral $A = \mathscr{I}$ defined by

$$Af = g, \quad g(y) = \int_0^y f(x)\,dx \quad (y \in [0, 1])$$

is a continuous injective linear operator of E into E. The inverse of A is the differential operator $A^{-1} = D$ defined on the subspace F constituted by smooth functions vanishing at zero.

3.5. Bounded Linear Operators

The operator D is discontinuous. Indeed, let $g_n(y) = n^{-1} y^n$ ($0 \leq y \leq 1$). Then $\|g_n\| = n^{-1}$ and $g_n \to 0$. However, for $f_n = Dg_n$ we have $f_n(x) = x^{n-1}$ ($0 \leq x \leq 1$), $\|f_n\| = 1$. Therefore, $Dg_n = f_n \not\to 0 = D0$.

Thus, the direct operator $\mathcal{I} : E \to F$ is continuous while the inverse $D : F \to E$ is discontinuous. Observe that if $A = \mathcal{I}$ is treated as a mapping of E into E then $A(E) = F \neq E$. If A is considered as a mapping of E onto F, the space F is incomplete. ∎

If we remove this incompleteness, the inverse operator becomes continuous as well.

THEOREM (Banach). *Let T be a continuous bijective linear operator of the Banach space E onto the Banach F. Then the inverse operator T^{-1} is a continuous linear operator bijectively mapping F onto E.*

PROOF. The bijectiveness of the inverse operator T^{-1} follows from its definition. The linearity is easily verified. We need to prove that T^{-1} is continuous. To do so, it suffices to demonstrate that it is continuous at 0, that is, for any $\varepsilon > 0$, there is a $\delta > 0$ such that $T^{-1} B(0, \delta) \subseteq B(0, \varepsilon)$. Since T^{-1} is one-one, this is tantamount to the inclusion $B(0, \delta) \subseteq TB(0, \varepsilon)$.

Let $F_n = \overline{TB(0, n\varepsilon/8)}$ be a sequence of closed sets, closures of the images of balls about 0 with radius $n\varepsilon/8$. Clearly $E = \cup B(0, n\varepsilon/8)$. Under the conditions of the theorem, T maps E onto all F, therefore $F = \cup F_n$. Since F is Banach, by Baire's theorem for some $n = n(0)$, the set $F_n = F_{n(0)}$ has an interior point, that is, $B(c, r) \subseteq F_{n(0)}$ $= \overline{TB(0, n(0)\varepsilon/8)}$ for some $c \in F$ and $r > 0$. It is easy to verify that $\overline{TB(0, n(0)\varepsilon/8)} = n(0) \cdot \overline{TB(0, \varepsilon/8)} = \{z = n(0) \cdot y : y \in \overline{TB(0, \varepsilon/8)}\}$.

Since $V = (c + V) - c$, then $B(0, r) \subseteq B(c, r) - B(c, r) = \{z_1 - z_2 : z_1, z_2 \in B(c, r)\}$. Since $|x_1 - x_2| \leq |x_1| + |x_2|$, then $B(0, \varepsilon/8) - B(0, \varepsilon/8) \subseteq \overline{B}(0, \varepsilon/4) \subseteq B(0, \varepsilon/2)$. Thus,

$$B(0, r/n(0)) = (1/n(0)) \cdot B(0, r) \subseteq \overline{TB(0, \varepsilon/8)} - \overline{TB(0, \varepsilon/8)}$$

$$\subseteq \overline{TB(0, \varepsilon/8) - TB(0, \varepsilon/8)} = \overline{T(B(0, \varepsilon/8) - B(0, \varepsilon/8))} \subseteq \overline{TB(0, \varepsilon/2)},$$

that is $B(0, \delta) \subseteq \overline{TB(0, \varepsilon/2)}$ for $\delta = r/n(0)$.

Considering sequences of balls $B_n = B(0, \varepsilon/2^n)$ and their images TB_n, and using the properties of T one can prove that $\overline{TB(0, \varepsilon/2)} \subseteq TB(0, \varepsilon)$. Consequently, $B(0, \delta) \subseteq TB(0, \varepsilon)$, which completes the proof of the theorem. ∎

Two corollaries to the Banach theorem are worth mentioning. Consider two norms $\|\cdot\|_1, \|\cdot\|_2$ for a vector space E. Equipped with these norms, E forms the Banach spaces $E_1 = (E, \|\cdot\|_1)$, $E_2 = (E, \|\cdot\|_2)$. We will say that $\|\cdot\|_1$ is *subordinate* to $\|\cdot\|_2$ if, for some $c_2 > 0$, $\|x\|_1 \leq c_2 \cdot \|x\|_2$ ($x \in E$), and that the norms $\|\cdot\|_1$ and $\|\cdot\|_2$ are equivalent if they are mutually subordinate.

COROLLARY 1. *If one of the norms is subordinate to the other, then they are equivalent.*

PROOF. Clearly, if $\|\cdot\|_1$ is subordinate to $\|\cdot\|_2$ then the identity operator $T : E_2 \to E_1$ is continuous. By the Banach theorem the inverse operator $T^{-1} : E_1 \to E_2$ is also continuous. Hence, $\|x\|_2 \leq c_1 \cdot \|x\|_1$ ($x \in E$) for $c_1 = \|T^{-1}\|$, as required. ∎

Consider a linear equation $Tx = y$, where $x \in E$, $y \in F$ and T is a continuous linear mapping of the Banach space E onto F.

COROLLARY 2. *If for each $y \in F$ there is a unique solution $x = x(y)$ of the equation $Tx = y$, then the solution x depends continuously on the right-hand side y.*

PROOF. Under the condition of Corollary 2, T maps E one-one onto the entire F, while its assertion means that T^{-1} is continuous. This follows from the Banach theorem. ∎

Corollary 2 formulates the condition which guarantees the correctness of the solution of the equation $Tx = y$.

3.5.4. Closedness of Operator Graphs

The closed graph theorem is equivalent to the Banach inverse operator theorem.

Let E, F be Banach spaces with product $E \times F$, A be a vector subspace dense in E ($\overline{A} = E$), $T: A \to F$ be a linear operator. The set

$$G(T) = \{(x, Tx) : x \in A\} \subseteq E \times F$$

is said to be the *graph* of T. Formally, the operator is identified with its graph.

The product $E \times F$ equipped with the norm $\|(x, y)\| = \max(\|x\|, \|y\|)$ is a Banach space and the graph $G = G(T)$ is a vector subspace in it. If it is closed it is a Banach space too. An operator T having a closed graph is said to be a *closed operator*.

EXAMPLE. Let $E = F = \mathscr{C}[0, 1]$, A – a subspace of E constituted by smooth functions, and $T = D$ – the differential operator. According to the termwise differentiation theorem, D is closed. As has been shown it is unbounded.

We emphasize that D is not everywhere defined. It turns out that a closed operator defined everywhere is bounded.

3.5. Bounded Linear Operators

THEOREM. *A closed operator $T: E \to F$ defined on the whole Banach space E and mapping it into the Banach F is bounded.*

PROOF. Let $P: G \to E$ and $Q: G \to F$ be the projectors defined by $P(x, Tx) = x$, $Q(x, Tx) = Tx$ ($x \in E$). They are continuous linear operators and P is a bijection of G onto all E. By the inverse mapping theorem, P^{-1} is continuous. Consequently, T, equal to the composition QP^{-1}, is also continuous and, hence, bounded, as stated. ∎

The identity $\|x\|_G = \|x\|_E + \|Tx\|_F$ ($x \in A$) defines another norm on A. It is called the *norm of the graph*. If T is closed, then A equipped with this norm is a Banach space. The operator T is bounded on this space.

3.5.5. Weak Compactness

Continuous linear functionals on a normed space play the role of coordinates. A proper choice of the system of coordinates simplifies solution of many problems.

Consider a normed space E, its dual space E' of continuous linear functionals on E, and its dual space E'' of continuous linear functionals on E'. Among the elements of E'' we distinguish functionals δ_x defined by points $x \in E$ with the values

$$\delta_x(x') = x'(x) \quad (x' \in E').$$

It is easy to verify that $\delta_x \in E''$. The correspondence $x \to \delta_x$ is the norm-preserving linear embedding of E into E'', therefore the functionals δ_x are often identified with points x.

The functionals δ_x make up a convenient system of coordinates for E'. If $E'' = \{\delta_x : x \in E\}$, then the space E is said to be *reflexive*. It is then necessarily Banach.

If we choose as a system of coordinates for the initial space E the dual space E' of all continuous linear functionals x' on E, then by the Hahn-Banach theorem, these coordinates separate points in E: if, for all $x' \in E'$, $x'(x) = 0$, then $x = 0$.

Different properties of sets and sequences related to coordinates are referred to as *weak* (e.g., weak boundedness, weak convergence, weak compactness), while those related to norms are referred to as *strong*.

Using the embedding into the second dual space and the principle of uniform boundedness it is easy to prove the following

THEOREM (Mackey). *A set $A \subseteq E$ is strongly bounded if and only if it is weakly bounded.*

From this theorem it is easy to deduce

LEMMA (weak convergence criterion). *A sequence of points $x(n) \in E$ converges weakly to $a \in E$ if and only if $\{x(n)\}$ is bounded and*

$z'(x(n)) \to z'(a)$ *for each* z' *from a set* $Z' \subseteq E'$ *with a dense linear span in* E'.

The following theorem often proves useful.

THEOREM (Mazur). *If a sequence* $x(n) \in E$ *weakly converges to* $a \in E$, *then a sequence of convex combinations of* $x(n)$ *strongly converges to* a.

This theorem is proved with the aid of the separability theorems which follow from the Hahn-Banach theorem. The proof simplifies if any linear combinations are considered instead of convex.

The property of weak compactness for a space dual to the separable normed space is expressed by a theorem similar to the classical Weierstrass convergent subsequence theorem.

THEOREM (Alaoglu). *If a sequence of continuous linear functionals on a separable normed space is bounded, then it has a subsequence convergent at every point.*

The general case is handled by a theorem on the weak compactness of closed bounded sets in dual spaces, which follows from Tychonoff's theorem on the product of compact sets (Ref. [2], Sect. 5.4).

The Mazur and Alaoglu theorems are applied to solving linear operator equations. Passing over to coordinates in a bounded sequence of approximate solutions permits one to pick a subsequence which converges weakly to a generalized solution. Since there exists a sequence of linear combinations of approximate solutions which converges strongly to this generalized solution, one can obtain a solution meeting the desired conditions.

3.6. Compact Linear Operators

Bounded linear operators map bounded sets onto bounded sets. Outstanding among bounded operators are *compact* ones which map bounded sets onto parts of compact sets. Such operators are also said to be *completely continuous*. In separable Hilbert space compact operators are approximated by degenerate operators (having finite-dimensional images). This explains many nice properties of compact linear operators.

The classical examples of compact linear operators are integral operators. Compact linear operators are useful in defining Fredholm operators. Equations with these operators will be discussed in the second part of this book.

3.6.1. Examples of Compact Operators

Consider two normed spaces E and F.

DEFINITION. *A linear operator* $T: B \to F$ *is called compact if the image* $T(B)$ *of each bounded set* $B \subseteq E$ *is contained in a compact set* $C \subseteq F$.

It should be noted that the condition $T(B) \subseteq C$ is tantamount to the compactness of the closure $\overline{T(B)}$ of the image $T(B)$ of a bounded set B.

3.6 Compact Linear Operators

Since T is linear, then in place of an arbitrary bounded set B it will suffice to take the unit ball $B(0, 1) \subseteq E$ and its image $T(B(0, 1) \subseteq F$.

A linear operator $T: E \to F$ is compact if and only if, for each bounded sequence $x_n \in E$, the sequence $Tx_n = y_n \in F$ has a convergent subsequence. This immediately follows from the definitions.

An operator $T: E \to F$ whose image $T(E)$ is finite-dimensional will be called *degenerate*. Thus, by definition, degenerate operators map E onto finite-dimensional subspaces of the space F. These subspaces are also called *degenerate*. Since in a finite-dimensional space the closure of any bounded set is compact, degenerate operators are compact.

It is straightforward to verify that a normed space is finite-dimensional if and only if the closed unit ball in it is compact, that is, the identity operator on this space is compact.

EXAMPLE 1. Let $E = \mathbb{R}^n$, $F = \mathbb{R}^m$. Then every linear operator $T: E \to F$ is degenerate and so compact. Linear mappings of finite-dimensional spaces are known as *matrix operators*.

EXAMPLE 2. Let $E = F = l^2$ be Banach spaces of number sequences with summable squares. Every bounded numerical sequence $\{\alpha_n\}$ defines a linear operator $T: E \to F$ transforming a sequence $x = \{\xi_n\}$ into the sequence $Tx = y = \{\eta_n\}$ of numbers $\eta_n = \alpha_n \xi_n$. By analogy with matrices such operators are called *diagonal*.

The operator T is compact if and only if $\alpha_n \to 0$.

EXAMPLE 3. Let $E = F = \mathscr{C}[a, b]$, and $T: E \to F$ be given by

$$Tf = g, \quad g(y) = \int_a^b k(y, x) \cdot f(x) \cdot dx \quad (y \in [a, b])$$

for $f \in \mathscr{C}[a, b]$, $k \in \mathscr{C}([a, b] \times [a, b])$. Clearly T is linear. It is called an integral operator with continuous kernel. Since for any $\varepsilon > 0$ and some $\delta > 0$

$$|g(y)| \leq (b - a) \cdot \|k\| \cdot \|f\|,$$
$$|g(y) - g(z)| \leq \varepsilon \cdot (b - a) \cdot \|f\| \quad (|y - z| \leq \delta),$$

then $g \in \mathscr{C}[a, b]$ and the closure of the set $T(B(0, 1)) = \{Tf = g : \|f\| < 1\}$ is uniformly bounded and equicontinuous. By Ascoli's theorem it is compact, and so T is compact.

EXAMPLE 4. Let $E = F = \mathscr{C}[a, b]$, and $T: E \to F$ be defined by

$$Tf = g,$$

$$g(y) = \int_a^y k(y, x) \cdot f(x) \cdot dx \quad (y \in [a, b]),$$

where $f \in \mathscr{C}[a, b]$, $k \in \mathscr{C}([a, b] \times [a, b])$. Like in Example 3, it is not hard to check that T is a compact linear operator known as a *Volterra's integral operator*.

EXAMPLE 5. Let $E = F = \mathscr{L}^2([a, b], dx)$ and the operator be defined by

$$Tf = g,$$

$$g(y) = \int_a^b k(y, x) \cdot f(x) \cdot dx \quad (y \in [a, b]),$$

where $f \in \mathscr{L}^2([a, b], dx)$, $k \in \mathscr{L}^2([a, b] \times [a, b], dxdy)$. According to Fubini's theorem

$$k(y, \cdot) \in \mathscr{L}^2([a, b], dx), \quad k(y, \cdot) \cdot f \in \mathscr{L}([a, b], dx)$$

for almost every $y \in [a, b]$. In other points of $[a, b]$ the function g may be deemed equal to zero. Using Cauchy's inequality it is easy to check that $g \in \mathscr{L}^2([a, b], dy)$. Clearly T is a linear operator.

Using Riesz's compactness criterion for spaces of integrable functions (see 3.3.3), it is not hard to prove that the operator T is compact. It is called an *integral operator with Hilbert-Schmidt kernel*.

3.6.2. Properties of Compact Operators

In this section we shall consider compact operators mapping a normed space into themselves. They make up a closed two-sided ideal in the normed ring of bounded operators. This is stated in the following

THEOREM. (i) *A linear combination of compact operators is also a compact operator.*

(ii) *The product of a compact operator by a bounded one is a compact operator.*

(iii) *The limit of a convergent sequence of compact operators is a compact operator.*

PROOF. For example we prove the statement (ii). Let T be a compact linear operator, L be a bounded linear operator, and B be a bounded set in a normed space E. Then $L(B)$ is bounded and $TL(B)$ is contained in some compact set $C \subseteq E$.

In exactly the same manner, $T(B)$ is contained in some compact set, as well as $LT(B)$. Thus, the products TL and LT are compact operators, as stated. ∎

The statement (ii), valid also for $L : T(E) \to E$, gives rise to the following important

COROLLARY. *In a finite-dimensional normed space, an operator inverse to the injective compact operator is unbounded*

3.6. Compact Linear Operators

PROOF. This assertion is true, otherwise the identity operator for an infinite-dimensional Banach space would be compact. ∎

Consider a normed space E, a linear compact operator $T: E \to E$, and an equation $Tx = y$. This equation cannot be well posed: we have just proved that the uniqueness of its solution excludes its continuous dependence on the right-hand side.

Compact operators transform a weakly convergent sequence into a strongly convergent one. In a reflexive Banach space, every bounded linear operator endowed with such a property is compact. Using the property of compact operators to render weak convergence into strong one can prove that in a separable Hilbert space each compact operator is the limit of a sequence of degenerate operators. These properties along with statement (iii) of the above theorem characterize compact operators in such spaces.

PROPOSITION. *In a separable Hilbert space H a linear operator is compact if and only if there exists a sequence of degenerate operators S_n in H such that $\|T - S_n\| \to 0$.*

This proposition is helpful in verifying the condition of compactness for the diagonal operator in Example 2 of Section 3.6.1.

3.6.3. Adjoint Operators

We shall consider normed spaces E, F, a bounded linear operator $T: E \to F$, and the adjoint spaces E', F'. T maps a point $x \in E$ onto $Tx = y \in F$. The equalities

$$x' = T'y' = y'T, \quad y'T(x) = y'(Tx)$$

define a bounded linear operator $T': F' \to E'$ which maps the coordinate $y' \in F'$ of the new point $Tx = y$ onto the coordinate $x' = T'y'$ of x. The linearity and boundedness of T' are readily verified. It is called the *Banach adjoint* of T.

EXAMPLE. Let $E = F = l^1$ and $T(\xi_1, \xi_2, ...) = (0, \xi_1, \xi_2, ...)$. Then $E' = F' = l^\infty$ and $T'(\eta_1, \eta_2, ...) = (\eta_2, \eta_3, ...)$. Observe that $\|T\| = \|T'\| = 1$.

Using the Hahn-Banach theorem it is not hard to prove that $\|T\| = \|T'\|$. The mapping $T \to T'$ is a linear isometry of the space $\mathcal{B}(E, F)$ into $\mathcal{B}(F', E')$. Furthermore, if T is invertible, so is T' with $(T^{-1})' = (T')^{-1}$.

Let $E = F = H$ be a Hilbert space and $E' = F' = H^*$ its dual space. Then according to Riesz's representation theorem there is a conjugate linear isometry $C: H \to H^*$ which maps $c \in H$ into a linear functional $x \to c \cdot x$ ($x \in H$). The operator

$$T^* = C^{-1}T'C : H \to H$$

is said to be (Hilbert) adjoint for the operator $T : H \to H$. It is bounded and linear along with T.

The adjoint operator T^* is defined by the identity

$$Tx \cdot y = x \cdot T^*y \quad (x, y \in H).$$

Indeed,

$$y \cdot Tx = Cy(Tx) = T'Cy(x) = C^{-1}T'Cy \cdot x = T^*y \cdot x.$$

EXAMPLE 1. Let $H = \mathbb{C}^n$ and the linear operator T have the matrix $A = [a_{ij}]$ in the standard basis. Then the adjoint operator T^* has the matrix $A^* = (\bar{a}_{ji})$, which is the transpose and complex-conjugate of A.

EXAMPLE 2. Let $H = \mathscr{L}^2([a, b], dx)$ and T be an integral operator with Hilbert-Schmidt kernel k. Then the adjoint operator T^* is a Hilbert-Schmidt operator with kernel $k^*(y, x) = \overline{k(x, y)}$, the transpose and complex-conjugate of k.

It is easy to verify that, for any bounded linear operators S and T in H,

$$(ST)^* = T^* \cdot S^* \quad (T^*)^* = T, \quad \|T^*T\| = \|T\|^2.$$

In addition, if T is invertible, so is T^* and $(T^{-1})^* = (T^*)^{-1}$.

Using the properties of compact operators, the criterion of compactness for spaces of continuous mapping, and the embedding into the second dual space, we can prove the statement (Ref. [4], Sect. 10.4) of the following

THEOREM. *A bounded linear operator $T = E \to F$ is compact if and only if the adjoint operator $T' : F' \to E'$ is compact.*

Since the product of a compact operator by a bounded one is also a compact operator, from this theorem and the identity $T^* = C^{-1}T'C$ follows

COROLLARY. *A bounded linear operator $T : H \to H$ is compact if and only if its adjoint $T^* : H \to H$ is compact.*

This result allows one to pass over from operators to their adjoints without losing compactness.

3.6.4. Fredholm Operators

Consider normed spaces E, F and a bounded linear operator $A = E \to F$. The equation

$$Ax = y \quad (x \in E, y \in F)$$

is a natural generalization of systems of linear equations. For some classes

3.6. Compact Linear Operators

of operators, the theory of linear operator equations is well developed, in particular for compact operators and their close associates – Fredholm operators.

In this section we shall always consider one and the same normed space E and linear operators $A : E \to E$. The identity operator for E will be denoted by I.

DEFINITION. *An operator equal to the difference of the identity operator and a compact operator is called Fredholm.*

Thus, an operator $A = I - T$ is Fredholm if T is compact. Since $-T$ is compact as well as T, we can treat A as the sum $A = I + (-T)$.

EXAMPLE. Let T be the integral linear operator of Examples 3–5 in Section 3.6.1. Then the operator $A = I - T$ is Fredholm.

The equation $Af = h$ for this operator is the Fredholm integral equation

$$f(y) + \int_a^b k(y, x) \cdot f(x) \cdot dx = h(y).$$

In Example 4, k is replaced with another kernel with the property $k(y, x) = 0$ for $y < x$. This allows integration over the whole interval $[a, b]$ instead of $[a, y]$.

A matrix injective operator is necessarily covering. It turns out that so is the Fredholm. This follows from Riesz's theorem on the injective Fredholm operator. Before we begin proving it, let us formulate some auxiliary statements regarding the properties of Fredholm operators.

Let T be compact and $A = I - T$ and Fredholm operators, respectively, in a normed space E, and L, M be closed subspaces of E such that $L \subset M$, $L \neq M$ and $AM \subseteq L$.

LEMMA 1. *For all $x \in L$ and some $b \in M$ with the norm $\|b\| = 1$, $\|Tb - Tx\| \geq 1/2$.*

PROOF. Since $L \subset M$, $L \neq M$ and L is closed, there is $a \in M$ for which

$$\text{dist}(a, L) = \inf\{\|a - x\| : x \in L\} = \alpha > 0$$

and so there is $u \in L$, for which $\alpha \leq \|a - u\| \leq 2\alpha$. Pick $b = \|a - u\|^{-1}(a - u)$, $\|b\| = 1$. Observe that $\|b - x\| = \|a - u\|^{-1} \|a - (u + \|a - u\| \cdot x)\|$ $\geq \alpha \cdot (2\alpha)^{-1} = 2^{-1}$ ($x \in L$), because $u + \|a - u\| \cdot x \in L$ for $x \in L$. Consequently, $\|Tb - Tx\| = \|(I - A)b - Tx\| = \|b - (Ab + Tx)\| \geq 1/2$ ($x \in L$), in view of $Ab \in L$ and so $Tx = x - Ax \in L$, $Ab + Tx \in L$ for $x \in L$. ∎

NOTE. Using this lemma, one can determine inductively a sequence $\{b_n\}$ inside a closed unit ball in an infinite-dimensional normed space, such that $\|b_m - b_n\| \geq 1/2$ for $m \neq n$. This implies noncompactness of the ball.

Let T be compact and $A = I - T$ injective Fredholm operators, respectively, in a normed space E.

LEMMA 2. *The operator A maps any bounded set $X \subseteq E$ into the closed set $AX = Y \subseteq E$.*

PROOF. Pick a sequence of $y_n \in Y$ converging to $\bar{y} \in E$. To prove that $\bar{y} \in Y$ we consider the sequence of $x_n = A^{-1} y_n \in X$ and demonstrate that it is bounded. Suppose that it is not so and there is a subsequence $\{ x_{r(n)} \}$ for which $\| x_{r(n)} \| \geq n$.

Note that $Au_n = \| x_{r(n)} \|^{-1} \cdot y_{r(n)} \to 0 \cdot \bar{y} = 0$ for some $u_n = \| x_{r(n)} \|^{-1} \cdot x_{r(n)}$. Consider $v_n = Tu_n$. Since T is compact and $\| u_n \| = 1$, there is a subsequence $v_{s(n)}$ converging to some $v \in E$. Thus $u_{s(n)} = I \cdot u_{s(n)} = (A + T) u_{s(n)} = Au_{s(n)} + v_{s(n)} \to 0 + v = v$. Since T is continuous, $v_{s(n)} = Tu_{s(n)}$ and $v_{s(n)} \to v$, $u_{s(n)} \to v$, $v = Tv$ and $Av = v - Tv = 0$. Consequently, $v = A^{-1} 0 = 0$, on the other hand $\| v \| = \| \lim u_{s(n)} \| = \lim \| u_{s(n)} \| = 1$. Supposing $\{ x_n \}$ is unbounded leads us to contradiction.

The boundedness of the sequence $\{ x_n \}$ and the compactness of T imply that there exists a subsequence $z_{p(n)} = Tx_{p(n)}$ converging to some $\bar{z} \in E$. Therefore,

$$x_{p(n)} = I \cdot x_{p(n)} = (A + T) x_{p(n)} = y_{p(n)} + z_{p(n)} \to \bar{y} + \bar{z}.$$

Since under the conditions of the lemma the set X is closed and $x_{p(n)} \in X$, $x = \bar{y} + \bar{z} \in X$. Consequently,

$$\bar{y} = \lim y_{p(n)} = \lim A x_{p(n)} = A (\lim x_{p(n)}) = Ax \in Y.$$

Thus, Y is closed, as stated. ∎

This lemma has an important

COROLLARY. *The operator inverse to a one-one Fredholm operator is continuous.*

PROOF. By what we have proved, for such an A^{-1}, the preimage $(A^{-1})^{-1} X = AX = Y$ of each closed set X is also closed. Hence, A^{-1} is continuous, as required. ∎

Now we are in a position to formulate and prove the central theorem of this section.

THEOREM (Riesz). *If a Fredholm operator A in a normed space E is injective, then A maps E onto itself and A^{-1} is continuous.*

PROOF. (i) We will prove that $AE = E$. Consider the sequence of operators

$$A^n = (I - T)^n = I - (\alpha_1 T^1 + \ldots + \alpha_n T^n) = I - T_n.$$

3.6. Compact Linear Operators

Here $T_n = \alpha_1 T^1 + ... + \alpha_n T^n$ are compact along with $T = I - A$ and so all A^n are Fredholm. Let also $A^0 = I$.

The operators A^n define the sets $Y_n = A^n E$. By Lemma 2 these are closed subspaces of E. Furthermore, $Y_{n+1} = A^n(AE) \subseteq A^n E = Y_n$ ($n = 0, 1, 2, ...$). There is a number $n(0)$ such that $Y_n = Y_{n(0)}$ for all $n \geq n(0)$. Indeed, otherwise a strictly decreasing subsequence of the subspaces $Z_n = Y_{r(n)}$ would exist, and by Lemma 1 we could pick a sequence of $z_n \in Z_n$ such that $\|z_n\| = 1$ and $\|Tz_n - Tz_p\| \geq 1/2$ for $p > n$. This contradicts the compactness of T. (For the subspaces $L = Z_{n+1} = Y_{r(n+1)}$, $M = Z_n = Y_{r(n)}$ and the operator $A^{r(n+1) - r(n)}$ the conditions of Lemma 1 are fulfilled.)

Among the numbers $n(0)$ such that $Y_n = Y_{n(0)}$ for all $n \geq n(0)$ there is the minimum one $m \geq 0$. We will show that $m = 0$. Indeed, if $m > 0$ then $Y_{m-1} \supset Y_m = Y_{m+1}$, $Y_{m-1} \neq Y_m$. At the same time $Y_{m-1} = A^{-1} Y_m = A^{-1} Y_{m+1} = Y_m$. Consequently, $m = 0$ and $E = Y_0 = Y_1 = ...$, so $AE = AY_0 = Y_1 = E$. The operator A maps E onto itself. The first part of the theorem is established.

(ii) By the Corollary to Lemma 2, the operator A^{-1} inverse to A is continuous. ∎

We emphasize that the Riesz theorem on the injective Fredholm operator has been proved for normed spaces which are not necessarily complete.

3.6.5. Fredholm's Theorems

These theorems state the conditions of well-posedness of Fredholm's linear operator equation and the existence and uniqueness of its solution.

Consider a Fredholm operator $A = I - T$ in a Banach space E and its adjoint $A' = I' - T'$ in the Banach space E'. Since T' is compact along with T, A' is Fredholm along with A. These operators define the equations

$$Ax = y, \qquad (1)$$

$$Az = 0 \quad (x, y, z \in E) \qquad (2)$$

$$A'x' = y', \qquad (1')$$

$$A'z' = 0 \quad (x', y', z' \in E') \qquad (2')$$

We will say that equation (1) is *well-posed* (or *correctly solvable* if in the Russian literature) if it has a unique solution $x = A^{-1} y \in E$ for every $y \in E$, and the operator A^{-1} is continuous.

From Riesz's theorem follows immediately

THEOREM 1. *Either equation* (1) *is well-posed or equation* (2) *has a nonzero solution.*

PROOF. By Riesz's theorem, the well-posedness (or correct solvability) of equation (1) is equivalent to bijection property of A. On the other hand, the assertion that equation (2) has a nonzero solution is equivalent to saying that A is not one-one. Thus, Theorem 1 is equivalent to the statement of the operator A being either one-one or not. ∎

Theorem 1 is known as the *Fredholm alternative*. It is tantamount to Riesz's theorem.

THEOREM 2. *Equation* (1) *has a solution for a given* $y \in E$ *if and only if* $z'(y) = 0$ *for each solution* z' *of equation* (2').

PROOF. If $Ax = y$ and $A'z' = 0$, then

$$z'(y) = z'(x) - z'(Tx) = z'(x) - T'z'(x) = A'z'(x) = 0.$$

Conversely, if for $A'z' = 0$, $z'(y) = 0$, then, since, for $y \notin Y$, $AE = Y$ is closed, the Hahn-Banach theorem implies the existence of a continuous linear functional φ on E such that $\varphi(y) = 1$ and $\varphi(z) = 0$ for each $z \in Y$. Thus

$$0 = \varphi(x - Tx) = \varphi(x) - \varphi(Tx) = \varphi(x) - T'\varphi(x) = A'\varphi(x)$$

for all $x \in E$ and $A'\varphi = 0$. This contradicts to the equality $\varphi(y) = 1$. Hence, $y \in Y = AE$ and equation (1) has a solution, as stated. ∎

The third theorem specifies the dimensionality of the spaces of the solutions.

THEOREM 3. *Equations* (1) *and* (1') *have the same finite number of linear independent solutions.*

Consider the kernels

$$\text{Ker } A = \{z : Az = 0\}, \quad \text{Ker } A' = \{z' : A'z' = 0\}$$

of A and A'. Theorem 3 states that

$$\dim(\text{Ker } A) = \dim(\text{Ker } A') < \infty.$$

The proof relies on the properties of compact operator spectra which will be described later. The solution is unique if and only if the kernels are singular (zero).

If the space $E = H$ is Hilbert, then instead of equations (1'), (2') one must consider

$$A^* x^* = y^* \tag{1*}$$

$$A^* z^* = 0 \quad (x^*, y^*, z^* \in H). \tag{2*}$$

All the three Fredholm theorems remain valid under such a change. In this case the condition of the existence of a solution uniqueness in Theorem 2 means that the right-hand side of equation (1) is orthogonal to each solution of equation (2^*).

NOTE. The Fredholm alternative does not hold for all bounded linear operators. In particular, the Volterra integral operator with kernel $k = 1$ maps the Banach space $E = \mathscr{C}[a, b]$ one-one to its proper part and has a discontinuous inverse operator.

3.7. Self-Adjoint Operators

Of all operators in Hilbert space, Hermitian and normal are of major interest. They are defined in terms of adjoint operators. By some of their properties, Hermitian operators are like real numbers, while normal operators are like complex numbers.

This section deals with complex separated spaces.

3.7.1. Banach Space Adjoints

Let E and F be normed spaces, A be a subspace dense in E, $T: A \to F$ be a linear operator, A^*, E^*, and F^* be their algebraic adjoints, and A', E' and F' be their topological adjoints. Since $\overline{A} = E$, every functional $a' \in A'$ is extended by continuity to $x' \in E'$, and this extension is unique.

The identity

$$S^* y^* = y^* T \quad (y^* \in F^*)$$

defines a linear operator $S^* = F^* \to A^*$. It is the algebraic adjoint of T. Consider a subspace

$$B' = F' \cap (S^*)^{-1} A'$$

consisting of all $y' \in F'$ for which $S^* y' = a' \in A'$. If $y' \in B'$, then by replacing $a' \in A'$ with its continuous extension $x' \in E'$ we obtain a linear operator $T': B' \to E'$. It is said to be the Banach space adjoint of T and defined by

$$x' = T' y' \quad (y' \in B),$$
$$x'(x) = y'(Tx) \quad (x \in A).$$

If $A = E$ and the operator T is bounded, then $B' = F'$ and its adjoint T' is bounded too. We considered such operators in Section 3.6.3.

The adjoint T' of any densely defined operator T is necessarily closed, but its domain B' may be trivial, e.g., to consist of a zero functional on F alone. The properties of Banach space adjoints are described in detail in the book of Kato [7] having numerous illustrative examples.

3.7.2. Hilbert Space Adjoints

We define Hilbert space adjoints independently of Banach space adjoints, so it will be easier to understand. Consider Hilbert spaces E and F, their vector subspaces A and B, linear operators $T : A \to F$ and $S : B \to E$. If

$$Tx \cdot y = x \cdot Sy \quad (x \in A, \ y \in B),$$

we will say that the operators T and S are adjoints of one another. For example, the zero operator $T = 0$ on the zero subspace $A = \{0\}$ is the adjoint of any linear operator $S : B \to E$.

If A is dense in E then a functional $a' \in A'$ with values $a'(x) = x \cdot Sy$ ($x \in A$) is extended by continuity to the functional $x' \in E'$, $x'(x) = x \cdot Sy$ ($x \in E$). This allows us to find the largest operator among all the adjoints S of T. Pick a set A^* of all $y^* \in F$ for which

$$Tx \cdot y^* = x \cdot x^* \quad (x \in A)$$

for some $x^* \in E$. It is a vector subspace of the Hilbert space F. Since $\overline{A} = E$, the element we are looking for is unique: from $x \cdot x_1^* = x \cdot x_2^* \ (x \in A)$ it follows that $x \cdot x_1^* = x \cdot x_2^*$ ($x \in E$) and $x = x_2^*$. Thus, the above identity defines an operator $T^* : A^* \to E$ which maps $y^* \in A^*$ to $x^* = T^* y^* \in E$. It can be easily verified T^* that is linear:

$$Tx \cdot (\beta_1 y_1^* + \beta_2 y_2^*) = \beta_1 (Tx \cdot y_1^*) + \beta_2 (Tx \cdot y_1^*)$$
$$= \beta_1 (x \cdot x_1^*) + \beta_2 (x \cdot x_2^*) = x \cdot (\beta_1 x_1^* + \beta_2 x_2^*)$$

for $y_j^* \in A^*$, $x_j^*, = T^* y_j^* \ \beta_j \in \mathbb{F}$ ($j = 1, 2$). From the definitions given it follows that T^* is the adjoint of T, and any operator $S : B \to E$ adjoint of T is a contraction of $T^* \cdot T^*$ is called the *Hilbert adjoint* of T.

Thus, by definition, the Hilbert adjoint $T^* : A^* \to E$ of a densely defined operator $T : A \to F$ acts only on those $y^* \in F$ for which there exists an $x^* \in E$ such that $Tx \cdot y^* = x \cdot x^*$ for all $x \in A$; then $T^* y^* = x^*$. It should be noted that the kernel of T^* equals the orthogonal complement of the range of T:

$$\operatorname{Ker}(T^*) = (\operatorname{Ran} T)^{\perp}.$$

If T^* is also densely defined ($\overline{A}^* = F$), then its adjoint can be

3.7. Self-Adjoint Operators

defined $T^{**} : A^{**} \to F$ ($A^{**} \subseteq E$). It is not hard to verify that T^{**} extends T and $A \subseteq A^{**}$. Moreover, $A^{**} = E$ if T is bounded.

The properties of operators are closely connected with the properties of their graphs. It can be readily verified that

$$G(T^*) = W(T)^\perp, \quad W(T) = \{(-Tx, x) : x \in A\}.$$

Indeed,

$$W(T)^\perp = \{(y, y^*) : (y, y^*) \cdot (-Tx, x)$$
$$= -y \cdot Tx + y^* \cdot x = 0 (x \in A)\}.$$

The graph $G(T^*)$ is closed, like any orthogonal complement. If T is injective and has a dense range ($\overline{\operatorname{Ran} T} = F$) then T^* is also injective and $(T^*)^{-1} = (T^{-1})^*$.

The Hilbert adjoints are described in detail in the book of Weidmann [8]. In particular, the reader can find the proofs of all the mentioned properties there. Henceforth we will refer to Hilbert adjoints briefly as *adjoints*.

EXAMPLE 1 (*multiplication operator*). Let $E = F = \mathcal{L}^2(\mathbb{R}, dx)$ and $Tf = g$, $g(x) = x \cdot f(x)$ ($x \in \mathbb{R}$). T is said to be the operator of multiplication by an independent variable. Its domain is defined by the condition $g \in \mathcal{L}^2$. It is easy to check that $\overline{A} = \mathcal{L}^2$, T is unbounded, and $T^* = T$.

EXAMPLE 2 (*differentiation operator*). Let $E = F = \mathcal{L}^2([0, 1] dx)$ and $Tf = g$, $g(x) = i^{-1} f'(x)$ ($x \in [0, 1]$). The derivatives are computed almost everywhere and the domain A of T is considered to be the set of all absolutely continuous functions f with the derivatives $f' \in \mathcal{L}^2$ and subject to the conditions $f(0) = f(1) = 0$. It can be proved that T is densely defined and unbounded, and T^* is its extension to the set A^* of all absolutely continuous functions f with the derivatives $f' \in \mathcal{L}^2$ (without boundary conditions): $T^* \supseteq T$, $T^* \neq T$.

If one takes $\mathcal{L}^2(\mathbb{R}, dx)$ instead of $\mathcal{L}^2([0, 1], dx)$, then the identity $T^* = T$ will hold.

It is worth noting that since the theory has been developed for separated spaces, in the above examples, the points are classes of functions equivalent with respect to the Lebesgue measure. One could also form a theory considering inseparated spaces by making necessary stipulations and replacing identities with equivalencies.

3.7.3. Hermitian and Normal Operators

This and the two subsequent sections will be concerned with linear operators $T : A \to H$ on a dense vector subspace A of a Hilbert space H and with their adjoints $T^* : A^* \to H$ acting on a subspace A^* of H.

If T^* extends T, then T is said to be *symmetric*. For such operators,

$$Tx \cdot y = x \cdot Ty \quad (x, y \in A)$$

but it is possible that $A^* \neq A$, as in Example 2 of Section 3.7.2. If $T^* = T$ then T is said to be *self-adjoint*. The same identity holds for it and $A^* = A$. The self-adjoint operator T is closed along with T^*. In particular, the multiplication and differentiation operators in Examples 1 and 2 of Section 3.7.2 are closed. If a one-one operator T is self-adjoint, then so is its inverse T^{-1}. This fact follows from the properties of the associated graphs. If T is closed then TT^* and T^*T are self-adjoint.

If $A = H$ then a self-adjoint operator $T : A \to H$ is called *Hermitian*. Thus, by definition, a Hermitian operator is a symmetric, everywhere defined operator, that is, an operator $T : H \to H$ such that

$$Tx \cdot y = x \cdot Ty \quad (x, y \in H).$$

For example, a matrix diagonal operator having real eigenvalues is Hermitian. By the closed graph theorem, any Hermitian operator is bounded.

An operator T is said to be *normal* if it is closed and $T^*T = TT^*$. The definitions imply that

$$\|Tx\| = \|T^*x\| \quad (x \in \text{Dom}(T^*T) = \text{Dom}(TT^*)).$$

It can be proved that

$$\text{Ker } T = \text{Ker } T^* = \text{Dom}(T^*T) = \text{Dom}(TT^*)$$

(see Kato [7], Sect. 5.3), whence

$$\|Tx\| = \|T^*x\| \quad (x \in \text{Dom } T = \text{Dom } T^*).$$

Consider Hermitian operators $R : H \to H$, $S : H \to H$, the operator $R + iS = T$ and its adjoint $R - iS = T^*$. Obviously, $TT^* = T^*T$ if and only if $RS = SR$. Consequently, commutability of the real and imaginary parts of an operator is a criterion of its normality. The analogy between Hermitian and normal operators on the one side and real and complex numbers on the other is rather deep.

3.7. Self-Adjoint Operators

3.7.4. Unitary Operators

A bounded linear operator $T: H \to H$ is said to be *isometric* if it preserves the inner product:

$$Tx \cdot Ty = x \cdot y \quad (x, y \in H).$$

Instead of the inner product we can require preservation of the norm

$$\|Tx\| = \|x\| \quad (x \in H).$$

Clearly, this identity follows from the previous one for $x = y$. The fact that they are equivalent follows from the polarization identity (3.1.2).

An isometric operator $T: H \to H$ is said to be *unitary* if it maps a complex Hilbert space H onto itself, $T(H) = H$. In view of the separatedness of the Hilbert space the isometricity of T implies that it is one-one: from $Tx_1 = Tx_2$ follows

$$\|x_1 - x_2\| = \|T(x_1 - x_2)\| = \|Tx_1 - Tx_2\| = 0.$$

A bounded linear operator is unitary if and only if its adjoint is equal to its inverse: $T^* = T^{-1}$. Indeed, if T is unitary then T^{-1} is defined on H, and from the condition for the inner product it follows that $T^*T = TT^* = I$. Thus, $T^* = T^{-1}$. Conversely, if $T^* = T^{-1}$ then $T^*T = I$ and the identity for the inner product is valid. Moreover, in this case $H = \text{Ran}(T) = \text{Dom}(T^{-1}) = \text{Dom}(T^*)$, and, consequently, T is unitary.

Note that since $T^*T = TT^* = I$ the unitary operator is normal.

3.7.5. Positive Operators

Each operator $T: H \to H$ defines a functional with values

$$\varphi(x) = Tx \cdot x \quad (x \in H).$$

T is said to be *positive* if φ is positive: $\varphi(x) \geq 0$ for $x \in H$.

Since

$$T^*Tx \cdot x = Tx \cdot Tx = \|Tx\|^2 \quad (x \in H),$$

the operator T^*T is positive for any bounded linear T.

That a Hermitian T is positive will be written as $T \geq O$. This property allows us to order the Hermitian operators $S: H \to H$ and $R: H \to H$. Like with real numbers, $S \geq R$ means that $T = S - R \geq O$ that is,

$$(S - R)x \cdot x \geq 0 \quad (x \in H).$$

This order for Hermitian operators has many habitual properties which are easily verifiable.

Applying the operator Newton algorithm, similar to the conventional Newton algorithm of root extraction one can show that for any Hermitian $T \geqslant O$ there is a unique Hermitian $S \geqslant O$ such that $T = S^2$. Furthermore, S commutes with any bounded R that commutes with T. The operator S is said to be the *square root* of T, denoted by $T^{1/2}$.

Now we can define the absolute value of the bounded operator T:

$$|T| = (T^*T)^{1/2}.$$

A bounded operator $U : H \to H$ is said to be *partially isometric* if

$$\|Ux\| = \|x\| \quad (x \in (\text{Ker } U)^\perp).$$

For any bounded linear operator $T : H \to H$ there exists a unique partially isometric $U : H \to H$ such that

$$T = U \cdot |T|, \quad \text{Ker } U = \text{Ker } T.$$

This identity is called the *polar decomposition* of T. A necessary and sufficient condition for T to be normal is the commutability of $|T|$ and U, that is $|T| \cdot U = U \cdot |T|$.

Thus, with the aid of Hermitian operators we obtained the Cartesian and polar decompositions of bounded linear operators on a Hilbert space. Normal operators are of special interest for the commutability of the expansion terms.

3.8. Operator Spectra

The spectrum of a matrix operator consists of its eigenvalues. Many operator properties are convenient to describe in terms of operator spectra. In the general case, the definition and analysis of operator spectra are more complicated than the matrix operator procedures. However, for some operator classes, the spectrum approach proves to be quite efficient.

The central task of spectral operator theory is solving the equation $\lambda x - Tx = y$, where λ is a scalar, x and y are vectors, T is a linear operator. Spectral theory examines the conditions of existence and uniqueness of the solution x to this equation and its continuous dependence on the right-hand side y.

3.8.1. Classification of Spectra

Let E be a complex Banach space, X be a vector subspace dense in it, $T : X \to E$ be a linear operator, $I : E \to E$ be the identity operator, and $\lambda \in \mathbb{C}$. The contraction of the identity operator I to X will be also denoted by I. As usual, we suppose that E is nonzero and the norm separates its points.

3.8. Operator Spectra

The equation
$$\lambda x - Tx = y \quad (x \in X, \ y \in E) \tag{1}$$
leads quite naturally to the operator
$$S(\lambda, T) = \lambda I - T : X \to E$$
and to the inverse relation
$$R(\lambda, T) = (\lambda I - T)^{-1} : Y \to E,$$
where
$$Y = \operatorname{Ran} S(\lambda, T) = S(\lambda, T) X.$$
By definition, the image
$$R(\lambda, T) y = \{x : \lambda x - Tx = y\}$$
consists of the solutions x of the equation $\lambda x - Tx = y$. Therefore, the correspondence $R(\lambda, T)$ is said to be the *resolvent*.

The type of solvability of equation (1) leads us naturally to a partitioning of the complex numbers λ into classes; first, into the basic two. If equation (1) is well-posed (correctly solvable), that is, if $S(\lambda, T)$ *injectively* maps X onto a subspace Y dense in E and the inverse $R(\lambda, T)$ is continuous, then λ belongs to the *resolvent set* $P(T)$ of T. Otherwise λ belongs to the *spectrum* $\Sigma(T)$ of T. Thus, by definition,
$$\Sigma(T) = \mathbb{C} \setminus P(T).$$
We emphasize that the continuity of T, and hence that of $S(\lambda, T)$, has not been assumed. The operator $R(\lambda, T)$ may be continuous while $S(\lambda, T)$ may not.

The well-posedness (correct solvability) of equation (1) is understood here in a generalized sense: the solution can exist not for any, but only for those y which belong to a certain dense subspace. This generalization is justified by the fact that, in many important cases, this subspace coincides with the whole space.

Observe that, for $\lambda \neq 0$, equation (1) is equivalent to
$$x - \lambda^{-1} Tx = \lambda^{-1} y \quad (x \in X, \ y \in E). \tag{2}$$
If T is compact, this is a Fredholm equation. In this context, equation (1) with a compact T is called a *Fredholm equation* too, and the operator $S(\lambda, T) = \lambda I - T$ is said to be *Fredholm*.

The spectrum $\Sigma(T)$ of T breaks into several parts. The first one is the point spectrum $\Sigma_p(T)$ that consists of the eigenvalues λ of T, for which at $y = 0$ equation (1) has a nonzero solution, $x \neq 0$, that is, such $\lambda \in \mathbb{C}$ that $Tx = \lambda x$ for some $x \neq 0$, $x \in X$. In the general case the definition of eigenvalues is the same as for matrix operators. From the definitions and the properties of matrix operators it follows that the matrix operator T has a point spectrum $\Sigma(T) = \Sigma_p(T)$.

The subspace H of E,

$$H(\lambda, T) = \{x : Tx = \lambda x\},$$

is said to be the *proper subspace* of an operator T defined by the eigenvalue $\lambda \in \Sigma_p(T)$. This subspace consists of the eigenvectors X of T corresponding to λ. In some cases, it is convenient to consider $H(\lambda, T) = \{0\}$ for $\lambda \notin \Sigma_p(T)$. The dimensionality $\dim H(\lambda, T)$ is said to be the *multiplicity* of the eigenvalue λ of T. This multiplicity may be infinite.

Since equation (1) is linear, then $\lambda \notin \Sigma_p(T)$ means that, for any $y \in Y = \mathrm{Ran}\, S(\lambda, T)$, equation (1) has exactly one solution. If $\overline{Y} = E$ and $\lambda \notin \Sigma_p(T)$, $\lambda \notin P(T)$, λ is said to belong to the *continuous spectrum* of T, written $\lambda \in \Sigma_c(T)$, conversely if $\overline{Y} \neq E$, then the number $\lambda \notin \Sigma_p(T)$, belongs to the *residual spectrum* $\Sigma_r(T)$ of T.

Thus,

$$P(T) + \Sigma(T) = C,$$

$$\Sigma(T) = \Sigma_p(T) + \Sigma_c(T) + \Sigma_r(T).$$

The classification of operator T spectra is summarized in the table below, which presents types of spectra along with the defining properties of the resolvent correspondence. For example, if $R(\lambda, T)$ is densely defined, injective and continuous, then $\lambda \in P(T)$. Note that other classifications may be found in the literature.

$R(\lambda, T)$	Injective		Noninjective
	continuous	discontinuous	
densely defined	$P(T)$	$\Sigma_c(T)$	$\Sigma_p(T)$
nondensely defined	$\Sigma_r(T)$		

Let us illustrate the point with several examples.

3.8. Operator Spectra

EXAMPLE 1. Let $T = O$, then $S(\lambda, T) = \lambda \cdot I$ and
$$\Sigma(T) = \Sigma_p(T) = \{0\}.$$
Let $T = I$, then $S(\lambda, T) = (\lambda - 1) \cdot I$ and
$$\Sigma(T) = \Sigma_p(T) = \{1\}.$$

EXAMPLE 2. Let $E = \mathbb{C}^n$. Then T, $S(\lambda, T)$ are matrix operators and
$$\Sigma(T) = \Sigma_p(T) = \{\lambda : \det S(\lambda, T) = 0\},$$
because all matrix operators are continuous, and injective matrix operators with nonzero determinant are necessarily covering.

EXAMPLE 3. Let $X = E = \mathscr{C}[0, 1]$ and $Tf = g$, $g(u) = u \cdot f(u)$ ($u \in [0, 1]$). Clearly, T is bounded. It is not hard to check that, for $\lambda \notin [0, 1]$, $S(\lambda, T)$ is a homeomorphism of E on E. For $\lambda \in [0, 1]$, it is injective, has a nondense image and a discontinuous inverse, that is, for $\lambda \in [0, 1]$, the resolvent relation $R(\lambda, T)$ is nondensely defined, injective and discontinuous. Thus, $\Sigma(T) = \Sigma_r(T) = [0, 1]$.

Let now $X = E = \mathscr{L}^2([0, 1], du)$ and, as before, T be the operator of multiplication by an independent variable. It is bounded again and has the same spectrum $\Sigma(T) = [0, 1]$. However, in the new space the resolvent operator $R(\lambda, T)$ is densely defined and the spectrum of T becomes continuous, $\Sigma(T) = \Sigma_c(T)$.

Let $E = \mathscr{L}^2(\mathbb{R}, du)$, and the operator T is defined on a subspace X dense in E
$$X = \{f : \int |u \cdot f(u)|^2 du < \infty\},$$
like in Example 1 of Section 3.7.2. T is unbounded. It can be easily verified that, for $\lambda \notin \mathbb{R}$, $S(\lambda, T)$ is a homeomorphism of X on E. Furthermore, for $\lambda \in \mathbb{R}$ it is one-one, has a dense image, and its inverse is discontinuous. That is, for $\lambda \in \mathbb{R}$, the resolvent relation $R(\lambda, T)$ is densely defined, injective and discontinuous. Thus,
$$\Sigma(T) = \Sigma_c(T) = \mathbb{R}.$$

EXAMPLE 4. Let $E = \mathscr{C}[0, 1]$, $X = \mathscr{C}^1[0, 1]$, and T be the differentiation operator $Tf = g$, $g(u) = f'(u)$ ($u \in [0, 1]$). Let $\lambda \in \mathbb{C}$. To each $h \in E$ and $R(\lambda, T)$ there corresponds a solution $f \in X$ of the equation $\lambda f - f' = h$ given by
$$f(u) = c e^{\lambda u} - \int_0^u e^{\lambda(u-t)} h(t) dt.$$

Here $c = f(0)$ is an arbitrary complex number. Hence, for any $\lambda \in \mathbb{C}$, the resolvent correspondence $R(\lambda, T)$ is defined on the whole E and is not one-one. Thus, $\Sigma(T) = \Sigma_p(T) = \mathbb{C}$ and $P(T) = O$. In this case T has a purely point spectrum and its resolvent set is empty.

Let, as before, $E = \mathscr{C}[0, 1]$, but the differentiation operator be defined on a smaller subspace X which consists of the functions $f \in \mathscr{C}^1[0, 1]$ having $f(0) = 0$. Now the resolvent $R(\lambda, T)$ is defined on the whole E and is one-one. For the values of the solution f in the interval $[0, 1]$, the boundedness of the exponent under the integral sign implies the continuity of $R(\lambda, T)$. Thus, now $P(T) = \mathbb{C}$ and $\Sigma(T) = O$. The spectrum of T is empty.

If we take $E = \mathscr{L}^2([0, 1], du)$ and define the differentiation operator T on the space X of absolutely continuous functions, then everything will be the same: without the condition $f(0) = 0$ the resolvent set $P(T)$ will be empty, and under this condition the spectrum $\Sigma(T)$ will be empty. This agrees with the common knowledge: the solution of this differential equation without additional conditions is not unique, whereas subject to this condition this equation is well-posed. It should be noted that since $f \in X$ is continuous the condition $f(0) = 0$ makes sense also in the space $E = \mathscr{L}^2([0, 1], du)$ because continuous functions equivalent with respect to the Lebesgue measure are equal. Therefore each class of equivalent functions has at most one continuous.

Let now $E = \mathscr{L}^2(\mathbb{R}, du)$, and the differentiation operator T with the factor i^{-1}, as in Example 2 of Section 3.7.2, be defined on the subspace X, of all functions absolutely continuous on each finite interval. In this case, the spectrum of T is identical to that of the operator of multiplication by an independent variable: $\Sigma(T) = \Sigma_c(T) = \mathbb{R}$. The reason is that these operators of differentiation and multiplication are unitarily equivalent: they are obtained from one another by the Fourier-Plancherel unitary transformation.

EXAMPLE 5. Let $X = E = \mathscr{L}^2([0, 1], du)$ and $Tf = g$, where

$$g(v) = \int_0^v f(u)\,du \quad (v \in [0, 1]).$$

The operator T is injective and compact, and, for $\lambda \neq 0$, the operator $S(\lambda, T) = \lambda I - T$ is one-one. By the Riesz theorem about injective Fredholm operators, the inverse $R(\lambda, T)$ is defined on the whole E and continuous. On the other hand, the operator $R(0, T)$ inverse to the compact operator $-T = S(0, T)$ is discontinuous. This is the differentiation operator and it is defined on the dense in E subspace of absolutely continuous functions equal to zero at 0. Hence, $\Sigma(T) = \Sigma_c(T) = \{0\}$.

3.8. Operator Spectra

If $E = \mathscr{C}\,[\,0,\,1\,]$, the operator image $S\,(0,\,T)$ will not be dense, and the continuous spectrum becomes residual: $\Sigma\,(T) = \Sigma_r\,(T) = \{\,0\,\}$.

These examples demonstrate that the spectrum depends essentially on the choice of space over which the operator acts.

3.8.2. Spectrum of a Closed Operator

If T is closed, then for any number $\lambda \in P\,(T)$, the resolvent relation $R\,(\lambda,\,T)$ is a bounded operator defined on the whole E. This allows us to define on the resolvent set $P = P\,(T)$ an operator-valued function $R = R\,(\,\cdot\,,\,T)$ with the range in the Banach algebra $\mathscr{B} = \mathscr{B}\,(E,\,E)$ of bounded linear operators in E. The function $R\,:\,P \to \mathscr{B}$ is called the *resolvent* of T.

LEMMA. *If T is closed and $\lambda \in P$, then* $\mathrm{Ran}\,S\,(\lambda,\,T) = E$.

PROOF. By the definition of the resolvent set P, for $\lambda \in P$, the image $S(\lambda,\,T) = \lambda\mathrm{I} - T$ is dense in E, and to prove the statement it suffices to show that $\mathrm{Ran}\,S(\lambda,\,T)$ is closed. This closeness follows from the fact that T is closed. ∎

By the lemma and the definition of the resolvent correspondence, $R\,(\lambda) = R\,(\lambda,\,T) \in \mathscr{B}$ for $\lambda \in P$. Hence, R really maps P into $\mathscr{B}$.

The function R is said to be analytic if its composition $x'\,R\,:\,P \to C$ with any functional $x' \in E'$ is an analytical function. This is tantamount to saying that the set $P \subseteq \mathbb{C}$ is open and in the neighborhood of each $\lambda \in P$, R expands in a power series with the coefficients from $\mathscr{B}$ (see Reed and Simon [9], Theorem VI.4). Operator-valued analytic functions have many properties in common with numerical-valued analytic functions.

THEOREM. *The resolvent set $P = P\,(T)$ of a closed operator T is open, and its resolvent $R = R\,(\,\cdot\,,\,T)$ is an analytic function.*

PROOF. Pick $\lambda \in P$ and choose $\mu \in \mathbb{C}$ so that the norm of $U = (\lambda - \mu) \cdot R\,(\lambda)$ would be strictly less than unity, $\|\,U\,\| < 1$. To this end it suffices to have $|\,\lambda - \mu\,| \cdot \|\,R\,(\lambda)\,\| < 1$. Then, by the operator-progression sum formula, the operator $\mathrm{I} - U$ has an inverse $(\mathrm{I} - U)^{-1} \in \mathscr{B}$. Thus,

$$R(\mu) = (\mu\mathrm{I} - T)^{-1} = (\mu\mathrm{I} - \lambda\mathrm{I} + \lambda\mathrm{I} - T)^{-1}$$
$$= (\lambda\mathrm{I} - T)^{-1} \cdot [\,\mathrm{I} - (\lambda - \mu)(\lambda\mathrm{I} - T)^{-1}\,]^{-1} = R(\lambda) \cdot (\mathrm{I} - U)^{-1} \in \mathscr{B},$$

that is, $\mu \in P$. Thus, the disc $B\,(\lambda,\,r)$ about λ with radius $r \in \,]\,0,\,\|\,R(\lambda)\,\|^{-1}\,[$ is contained in P, i.e., the set P is open.

For $|\,\mu - \lambda\,| < r$, the operator-progression sum formula yields

$$R(\mu) = R(\lambda + (\mu - \lambda) = \Sigma(-1)^n\,R^{n+1}(\lambda) \cdot (\mu - \lambda)^n.$$

Hence, the function R is analytic. ∎

Since the complement of an open set in the complex plane is closed, the proved theorem leads to

COROLLARY. *The spectrum of a closed operator is a closed set.*

Take the numbers λ, μ and the closed operators T, U defined on one and the same subspace A dense in E. Denote I_A and I_E the identical transformations of A and E, respectively. We have

$$R(\lambda, T) - R(\mu, U) = R(\lambda, T) \cdot I_E - I_A \cdot R(\mu, U)$$
$$= R(\lambda, T) \cdot [S(\mu, U) \cdot R(\mu, T)] - [R(\lambda, T) \cdot S(\lambda, T)] \cdot R(\mu, U)$$
$$= R(\lambda, T) \cdot [S(\mu, U) - S(\lambda, T)] \cdot R(\mu, U).$$

For $U = T$ and $\mu = \lambda$ these identities give the *resolvent equations*

$$R(\lambda, T) - R(\mu, T) = -(\lambda - \mu) \cdot R(\lambda, T) \cdot R(\mu, T),$$
$$R(\lambda, T) - R(\lambda, U) = R(\lambda, T) \cdot (T - U) \cdot R(\lambda, U).$$

The first of them is known as the *Hilbert identity*. It implies that

$$R(\lambda) \cdot R(\mu) = R(\mu) \cdot R(\lambda)$$

and

$$R'(\lambda) = \lim_{\mu \to \lambda} (\mu - \lambda)^{-1} \cdot [R(\mu) - R(\lambda)] = -R^2(\lambda).$$

There is a simple relation between a closed operator T and its Banach adjoint T':

$$\Sigma(T') = \Sigma(T),$$
$$R(\lambda, T') = R(\lambda, T)',$$

for any number λ from the common resolvent set of T and T'. If the space E is Hilbert and T^* is the Hilbert adjoint of T, then

$$\Sigma(T^*) = \Sigma(T)^*,$$
$$R(\lambda, T^*) = R(\lambda^*, T)^*$$

for any $\lambda^* \in P(T)$. Here $\Sigma(T)^* = \{\lambda^* : \lambda \in \Sigma(T)\}$ is mirror image of the set $\Sigma(T)$ in the real line.

Extensively treated examples of resolvents for differential operators may be found in Section 3.6 of the book by Helgason [6].

3.8. Operator Spectra

3.8.3. Spectrum of a Bounded Operator

The bounded operator is closed. Therefore, everything that has been said about the spectrum of the closed operator applies to the spectrum of the bounded operator. However, the spectrum of a bounded operator is nonempty and bounded.

Let $T \in \mathcal{B} = \mathcal{B}(E, E)$ be a bounded linear operator over a Banach space $E \neq \{0\}$. By the operator-progression sum formula, for $|\lambda| > \|T\|$,

$$R(\lambda, T) = (\lambda I - T)^{-1} = \lambda^{-1}(I - \lambda^{-1}T)^{-1} = \lambda^{-1}\Sigma \lambda^{-n} T^n \quad (n \geq 0).$$

Consequently, $\lambda \in P(T)$ for $|\lambda| > \|T\|$, and the spectrum $\Sigma(T)$ is contained in a closed disc about 0 with radius $\|T\|$. Outside the disc the resolvent is expanded in the above Laurent series.

Taking the inverse $z = \lambda^{-1}$ and using the formula for the radius of convergence of a power series, it is easily checked that the resolvent series with the terms $\lambda^{-n} T^n = T^n z^n$ has a sum for $|\lambda| > \rho(T)$, and does not have a sum for $|\lambda| < \rho(T)$, where

$$\rho(T) = \overline{\lim} \, (\|T^n\|^{1/n}).$$

This number $\rho(T)$ is called the *spectral radius* of an operator T. It is possible that $\rho(T) = 0$, as for the Volterra type operators. It can be proved that the sequence $\|T^n\|^{1/n}$ converges and so

$$\rho(T) = \lim (\|T^n\|^{1/n}).$$

The spectrum of a bounded operator is characterized by

THEOREM. *The spectrum $\Sigma(T)$ of a bounded operator T is a nonempty, bounded, closed set contained in the disc $\overline{B}(0, \rho(T))$.*

PROOF. The boundedness of $\Sigma(T)$ has been proved, $\Sigma(T) \subseteq \overline{B}(0, \|T\|)$. The inclusion $\Sigma(T) \subseteq \overline{B}(0, \rho(T))$ follows from the definition of spectral radius $\rho(T)$. That $\Sigma(T)$ is closed follows from the closedness of T. It remains to prove that $\Sigma(T)$ is not empty.

Note that for $|\lambda| > \|T\|, \|R(\lambda, T)\| \leq (|\lambda| - \|T\|)^{-1}$, thus $\|R(\lambda, T)\| \to 0$ as $|\lambda| \to \infty$. If $\Sigma(T) = O$, then $P(T) = \mathbb{C}$ and the resolvent is an entire analytic function. By Liouville's theorem, which holds also for operator-valued functions of a complex variable, the resolvent $R(\cdot, T)$ equals some constant $C \in \mathcal{B}$. Since $\|R(\lambda, T)\| \to 0$ as $|\lambda| \to \infty$, C is the zero operator on E. By the assumption $E \neq \{0\}$, and the zero operator on E is not one-one. Therefore, it cannot be a value of the resolvent of T. Consequently, $\Sigma(T)$ cannot be empty. ∎

It is not hard to check that, for the spectral radius, the following identities hold

$$\rho(T) = \inf\{\|T^n\|^{1/n} : n \geq 1\} = \sup\{|\lambda| : \lambda \in \Sigma(T)\},$$

which are useful in computing the radius.

Pick out numbers $a_0, a_1, \ldots, a_n$ and consider the operator polynomial

$$p(T) = a_0 I + a_1 T + \ldots + a_n T^n.$$

It is checked easily that

$$\Sigma(p(T)) = p(\Sigma(T)) = \{p(\lambda) : \lambda \in \Sigma(T)\},$$

where

$$p(\lambda) = a_0 + a_1 \lambda + \ldots + a_n \lambda^n.$$

This result is a particular case of the general spectral mapping theorem.

3.8.4. Spectrum of a Compact Operator

In many respects compact operators are similar to matrix. The spectra of compact operators also resemble those of matrix operators. The spectra of compact operators are described by

THEOREM (Riesz-Schauder). *Let T be a compact operator on a Banach space E. Then*

(i) *every number $\lambda \neq 0$ from the spectrum $\Sigma(T)$ is an eigennumber of T;*

(ii) *outside any disc about 0 of radius $r > 0$ in the complex plane $\mathbb{C}$ there are only a finite number of points belonging to the spectrum $\Sigma(T)$;*

(iii) *every eigennumber $\lambda \neq 0$ of T is of finite multiplicity.*

The finite set of points outside a disc in $\mathbb{C}$ can, in particular, be empty. This is the case, e.g., for matrix operators.

That the point 0 plays a special role is obvious for infinite-dimensional diagonal operators in the space of sequences $E = l^2$. Such an operator is compact if and only if its diagonal elements tend to zero. Therefore, these elements cannot include an infinitely recurring nonzero number, while zero can recur any number of times.

By the Riesz-Schauder theorem, the only nonisolated point of a compact operator spectrum $\Sigma(T)$ can be 0. Any $\lambda \neq 0$ of $\Sigma(T)$ has a neighborhood which contains no other points of $\Sigma(T)$.

The Riesz-Schauder theorem is closely related to Fredholm's theorems. The proper subspaces $H(\lambda, T)$ are finite-dimensional, so for $\lambda \neq 0$ one can reduce Fredholm equations to matrix.

A detailed proof of the Riesz-Schauder theorem may be found in the book of Kato [7, Ch. 3, theorem 6.2.6].

3.8. Operator Spectra

3.8.5. Spectrum of a Self-Adjoint Operator

In this section we shall consider a nonzero complex Hilbert space H, a vector subspace A dense in it, a linear operator $T : A \to H$, the identity operator $I : H \to H$, a number $\lambda \in \mathbb{C}$, an operator $S(\lambda, T) = \lambda I - T : A \to H$, the Hilbert space adjoints $T^* : A^* \to H$ and $S^*(\lambda, T) = S(\lambda^*, T^*) = \lambda^* I - T^* : A^* \to H$.

If T is one-one and has a dense image Ran $T = T(A)$, then T^* is also one-one, and $(T^*)^{-1} = (T^{-1})^*$. This assertion is easily proved.

Theorem about the Spectrum of a Self-Adjoint Operator. The spectrum of a self-adjoint operator is described by

THEOREM. *Let $T = T^*$. Then*

$$\Sigma(T) = \{ \lambda : \operatorname{Ran} S(\lambda, T) \neq H \} \subseteq \mathbb{R}.$$

Furthermore,

$$\Sigma_p(T) = \{ \lambda : \operatorname{Ran} S(\lambda, T) \neq H, \overline{\operatorname{Ran} S(\lambda, T)} \neq H \}, \quad (1)$$

$$\Sigma_c(T) = \{ \lambda : \operatorname{Ran} S(\lambda, T) \neq H, \overline{\operatorname{Ran} S(\lambda, T)} = H \}, \quad (2)$$

$$\Sigma_r(T) = O. \quad (3)$$

PROOF. Observe that for $T = T^*$, $Tu = \lambda u$, $\|u\| = 1$ we have

$$\lambda = u \cdot \lambda u = u \cdot Tu = Tu \cdot u = (\lambda u) \cdot u = \lambda^*,$$

hence, $\Sigma_p(T) \subseteq \mathbb{R}$.

Let $\Lambda = \{ \lambda : \overline{\operatorname{Ran} S(\lambda, T)} \neq H \}$. We will prove that $\Sigma_p(T) = \Lambda$. Indeed, by the theorem about orthogonal projection, $\lambda \in \Lambda$ is tantamount to $B = \overline{\operatorname{Ran} S(\lambda, T)}^{\perp} \neq \{ 0 \}$, therefore, $S(\lambda, T)x \cdot b = 0$ $(x \in A)$ for some $b \neq 0$, $b \in B$. Hence, $x \cdot S^*(\lambda, T) b = 0$ $(x \in A)$, $S(\lambda^*, T^*) b = 0$, $\lambda^* \in \Sigma_p(T)$ and $\lambda = \lambda^* \in \Sigma_p(T)$ since $\Sigma_p(T) \subseteq \mathbb{R}$. The identity (1) is established.

By definition, $\Sigma_r(T) \subseteq \Lambda$ and $\Sigma_r(T) \cap \Sigma_p(T) = O$. For $\Sigma_p(T) = \Lambda$, this is possible if and only if $\Sigma_r(T) = O$. The statement (3) is established.

Now we prove that $\Sigma(T) \subseteq \mathbb{R}$, i.e., that $\mathbb{C} \setminus \mathbb{R} \subseteq P(T)$. Let $\lambda = \mu + i\nu$; $\mu, \nu \in \mathbb{R}$. Notice that for $z = (\mu I - T) x$ and $T = T^*$,

$$z \cdot x = (\mu I - T) x \cdot x = x \cdot (\mu I - T) x = x \cdot z,$$

$$\|(\lambda I - T) x\|^2 = \|z + i\nu x\|^2$$
$$= \|z\|^2 + i\nu(z \cdot x) - i\nu(x \cdot z) + \nu^2 \cdot \|x\|^2$$
$$= \|z\|^2 + \nu^2 \cdot \|x\|^2 \quad (x \in A)$$

Consequently,

$$\|S(\lambda, T)x\| \geq |\nu| \cdot \|x\| \quad (x \in A).$$

If $\lambda \in \mathbb{R}$, $\nu \neq 0$, then this $S(\lambda, T)$ is injective. Since $\Lambda = \Sigma_p(T) \subseteq \mathbb{R}$ (just proved), then $\lambda \notin \Lambda$ and $\overline{\operatorname{Ran} S(\lambda, T)} = H$. Thus, for $\lambda \in \mathbb{R}$ the resolvent correspondence is densely defined and one-one. Since

$$\|R(\lambda, T)y\| \leq |\nu|^{-1} \cdot \|y\| \quad (y = S(\lambda, T)x, \ x \in A),$$

then $R(\lambda, T)$ is continuous. Therefore $\lambda \in P(T)$.

Let us prove the identity (2). Denote its right-hand side by M. We need to show that $\Sigma_c(T) \subseteq M$. Let $\lambda \in \Sigma_c(T)$. Then, by the definition of continuous spectrum, $\overline{\operatorname{Ran} S(\lambda, T)} = H$ and $R(\lambda, T)$ is discontinuous. Since $\Sigma(T) \subseteq \mathbb{R}$, then $\lambda = \lambda^*$ and $R(\lambda, T)^* = R(\lambda, T)$ for $T = T^*$. Consequently, the operator $R(\lambda, T)$ is closed. By the closed graph theorem, if $R(\lambda, T)$ is unbounded then $\operatorname{Ran} S(\lambda, T) \neq H$. Hence, $\lambda \in M$ and $\Sigma_c(T) \subseteq M$.

Now we prove the inverse inclusion. Let $\lambda \in M$. Then, $\lambda \notin \Sigma_p(T)$ and $R(\lambda, T)$ is a densely defined operator. It remains to show that $R(\lambda, T)$ is unbounded. The self-adjoint operator $T = T^*$ is closed. If $R(\lambda, T)$ is bounded, then $\lambda \in P(T)$ and, by the theorem about the closed-operator resolvent, $\operatorname{Ran} S(\lambda, T) = H$, which contradicts the assumption $\lambda \in M$. Hence, $R(\lambda, T)$ is a densely defined discontinuous operator, $\lambda \in \Sigma_c(T)$ and $M \subseteq \Sigma_c(T)$. The identity (2) is established, which completes the proof of the theorem. ∎

Another characteristic of a self-adjoint operator may be given by defining the limit spectrum.

A number $\lambda \in \mathbb{C}$ for which there exist associated vectors $u_n \in A$, $\|u_n\| = 1$ such that $\lambda u_n - T u_n \to 0$ is said to be the *limit eigennumber* of a linear operator $T: A \to H$. Denote the set of all limit eigennumbers of T by $\Sigma_l(T)$ and call it the *limit spectrum* of T. By definitions, $\Sigma_p(T) \subseteq \Sigma_l(T)$, i.e., each eigennumber is also the limit eigennumber.

Using the estimates for the norms of the operators $S(\lambda, T)$ and $R(\lambda, T)$ obtained in the proof of the theorem about the self–adjoint operator spectrum, it is not hard to prove

CRITERION (Weyl). *Let $T = T^*$. Then $\Sigma(T) = \Sigma_l(T)$.*

Alternatively, a number belongs to the spectrum of a self-adjoint operator if and only if it is its limit eigennumber.

Theorem about the Spectrum of a Hermitian Operator. Self-adjoint operators defined on a whole Hilbert space will be called Hermitian. By the closed graph theorem, if an operator $T = T^*: H \to H$ is closed, then it is bounded. The latter implies the boundedness of the spectrum $\Sigma(T)$ of a Hermitian operator T.

3.8. Operator Spectra

Using the Riesz representation theorem, it is not difficult to verify that the identity

$$B(z, x) = Tx \cdot z \quad (x, z \in H)$$

puts into a one-to-one correspondence the Hermitian operators on H and the Hermitian forms on $H \times H$. The polarization identity $B(z, x) = 4^{-1} [B(z + x, z + x) - B(z - x, z - x) - iB(z + ix, z + ix) + iB(z - ix, z - ix)]$ sets up a one-to-one correspondence between the Hermitian forms $B(z, x)$ and the quadratic forms

$$Q(x) = B(x, x) = Tx \cdot x \quad (x \in H)$$

on the complex Hilbert space H. This fact allows one to identify Hermitian forms with the quadratic forms induced by them, and to describe the properties of Hermitian operators and their spectra with the aid of quadratic forms. Since $T = T^*$,

$$Q(x) = Tx \cdot x = x \cdot Tx = \overline{Tx \cdot x} \quad (x \in H)$$

and all the values of the quadratic form Q are real, $Q(x) \in \mathbb{R}$. Observe that the quadratic form Q is determined by values it takes on the unit sphere

$$Q(x) = B(x, x) = \|x\|^2 \cdot B(u, u) = \|x\|^2 \cdot Q(u)$$

for $u = \|x\|^{-1} \cdot x$, $x \neq 0$. Furthermore, $Q(0) = B(0, 0) = 0$. The boundedness of T implies that, for $\|u\| = 1$, Q is bounded on the unit sphere S:

$$|Q(x)| = |Tu \cdot u| \leq \|Tu\| \leq \|T\|.$$

Therefore we can consider the bounds of the values of $Q(u)$ and $|Q(u)|$ on the sphere S:

$$a = \inf \{Tu \cdot u : \|u\| = 1\},$$
$$b = \sup \{Tu \cdot u : \|u\| = 1\},$$
$$c = \sup \{|Tu \cdot u| : \|u\| = 1\}.$$

Notice that

$$|Q(x)| \leq c \cdot \|x\|^2 \quad (x \in H).$$

LEMMA. $\|T\| = \max \{|a|, |b|\} = c$.

PROOF. It has been shown that $|Q(u)| \leq \|T\|$ for $\|u\| = 1$, and so $c \leq \|T\|$. Let us prove the reverse inequality. Let $\|u\| = 1$, $Tu \neq 0$ and

$z = \|Tu\|^{-1} \cdot Tu$. Then

$$B(z, u) = Tu \cdot z = \|Tu\|^{-1} \cdot \|Tu\|^2 = \|Tu\|.$$

In view of the polarization identity, the real-valuedness of $Q(x) = B(x, x)$, the inequality $|Q(x)| \leq c \cdot \|x\|^2$, and the parallelogram law, we obtain

$$\|Tu\| = B(z, u) = 4^{-1}|Q(z + u) - Q(z - u)|$$
$$\leq 4^{-1} \cdot (|Q(z+u)| + |Q(z-u)|) \leq 4^{-1} c \cdot (\|z+u\|^2 + \|z-u\|^2)$$
$$\leq 2^{-1} c \cdot (\|z\|^2 + \|u\|^2) = 2^{-1} c \cdot 2 = c.$$

Consequently,

$$\|T\| = \sup\{\|Tu\| : \|u\| = 1\} \leq c.$$

Thus, $\|T\| = c$.

The identity $\max\{|a|, |b|\} = c$ follows from the definitions of the numbers a, b, c. This is readily verified by considering $0 < a \leq b$, $a \leq b \leq 0$, and $a \leq 0 \leq b$. In the first case, $c = b$, in the second $c = -a$, in the third $c = \max\{-a, b\}$. ∎

The spectrum $\Sigma(T)$ of a Hermitian operator T is characterized by

THEOREM. $\{a, b\} \subseteq \Sigma(T) \subseteq [a, b]$.

PROOF. First, consider a positive Hermitian operator A and prove that $\|A\| \in \Sigma(A)$. Since the spectrum of a self-adjoint operator coincides with its limit spectrum, it will suffice to show that $\mu = \|A\|$ is a limit eigennumber of A.

Since $A \geq 0$, then $|Au \cdot u| = Au \cdot u \geq 0$, and by the lemma,

$$\mu = \sup\{|Au \cdot u| : \|u\| = 1\} = \sup\{Au \cdot u : \|u\| = 1\}.$$

Thus, there are $u_n \in H$, $\|u_n\| = 1$ such that $Au_n \cdot u_n > \mu - 1/n$. Consequently,

$$\|(\mu 1 - A)u_n\|^2 = \mu^2 - 2\mu \cdot Au_n \cdot u_n + \|Au_n\|^2$$
$$\leq \mu^2 - 2\mu(\mu - 1/n) + \mu^2 = 2\mu/n \quad (n \geq 1),$$
$$\|S(\mu, A) \cdot u_n\| \leq (2\mu/n)^{1/2} \to 0 \quad (n \to \infty).$$

Hence, $\mu \in \Sigma_l(A)$.

Consider the operators $A = T - aI$, $B = bI - T$. Since $a \leq Tu \cdot u \leq b$ ($\|u\| = 1$), so $A \geq O$, $B \geq O$. It is an easy matter to check that $\|A\| = \|B\| = b - a$, by what we have proved $b - a \in \Sigma(A)$, $b - a \in \Sigma(B)$. From the definitions it follows that $\mu \in \Sigma(A)$ is equivalent

3.8. Operator Spectra

to $\lambda = \mu + a \in \Sigma(T)$, and $\nu \in \Sigma(B)$ is equivalent to $\lambda = b - \nu \in \Sigma(T)$. Therefore, $a = b - (b - a) \in \Sigma(T)$, $b = (b - a) + a \in \Sigma(T)$ and hence, $\{a, b\} \subseteq \Sigma(T)$.

Since the operators A and B are self-adjoint and bounded, their spectra are real and bounded by the numbers $-(b - a)$ and $b - a$. Since $\Sigma(T) = \Sigma(A) + a = b - \Sigma(B)$, the spectrum of T is bounded by the numbers $a = b - (b - a)$, $b = (b - a) + a$. Indeed, if $\lambda = \mu + a = b - \nu$, $\mu \in \Sigma(A)$, and $\nu \in \Sigma(B)$, then $\lambda \leq (b - a) + a$, $\mu \leq b - a$, $-(b - a) \leq \nu$ and $\lambda = \mu + a \leq (b - a) + a = b$, $\lambda = b - \nu \geq b - (b - a) = a$. Thus, $\Sigma(T) \subseteq [a, b]$. ∎

This theorem has several important consequences.

COROLLARY 1. *A Hermitian operator is positive if and only if its spectrum consists of positive numbers alone.*

PROOF. Let T be Hermitian and $T \geq O$. Then, by definition, $Tu \cdot u \geq 0$ ($\|u\| = 1$), thus $a = \inf\{Tu \cdot u : \|u\| = 1\} \geq 0$. Consequently, by the theorem, $\Sigma(T) \subseteq [a, \infty[\subseteq [0, \infty[$. Conversely, if $a \geq 0$, then $Tx \cdot x = \|x\|^2 (Tu \cdot u) \geq a \cdot \|x\|^2 \geq 0$ for $u = \|x\|^{-1} \cdot x$, $x \neq 0$, $x \in H$. Hence, $T \geq O$, as desired. ∎

COROLLARY 2. *The spectral radius $\rho(T)$ of a Hermitian operator T equals its norm $\|T\|$.*

PROOF. Since $\|T^n\| \leq \|T\|^n$, then $\rho(T) = \overline{\lim}(\|T^n\|^{1/n}) \leq \|T\|$. By the theorem and the lemma, $\|T\| = \max\{|a|, |b|\}$ and both $a \in \Sigma(T)$ and $b \in \Sigma(T)$, therefore $\rho(T) = \sup\{|\lambda| : \lambda \in \Sigma(T)\} \geq \|T\|$. ∎

Consider an n-dimensional complex space $H = \mathbb{C}^n$, a matrix operator T, its conjugate T^*, their product $S = T^* T$. Notice that S is a positive Hermitian operator:

$$S^* = (T^*T)^* = T^* \cdot T^{**} = T^* \cdot T = S,$$
$$Sx \cdot x = T^*Tx \cdot x = Tx \cdot Tx = \|Tx\|^2 \geq 0 \quad (x \in H).$$

The eigennumbers of S are said to be the *singular numbers* of T. Denote them by σ_k ($1 \leq k \leq n$). By Corollary 1, $S \geq 0$ implies $\sigma_k \geq 0$. Pick the largest number $\sigma = \max\{\sigma_k : 1 \leq k \leq n\}$.

COROLLARY 3. *The Euclidean norm of a matrix operator equals the square root of its largest singular number.*

PROOF. Indeed, $\Sigma(S) = \{\sigma_k : 1 \leq k \leq n\}$ and $\sigma_k \geq 0$. By Corollary 2, this implies that $\|S\| = \rho(S) = \max\{\sigma_k : 1 \leq k \leq n\} = \sigma$. By the lemma $\|S\| = \sup\{|Su \cdot u| : \|u\| = 1\} = \sup\{Su \cdot u : \|u\| = 1\} = \sup\{\|Tu\|^2 : \|u\| = 1\} = \|T\|^2$. Consequently, $\|T\|^2 = \sigma$ and $\|T\| = \sigma^{1/2}$. ∎

In particular, if a matrix operator T is Hermitian, then $T^* = T$,

$S = T^2$, $\sigma_k = \lambda_k^2$ where λ_k are the eigennumbers of T, and so $\|T\| = \max\{\lambda_k : 1 \leq k \leq n\}$. The norm of the matrix Hermitian operator equals the largest absolute value of its eigennumbers.

Theorem about the Spectrum of a Compact Hermitian Operator. Amidst the eigenvectors of a matrix Hermitian operator on a finite dimensional Euclidean space, one can pick vectors whose collection will be an orthonormal basis of this space. The same is true for a compact Hermitian operator on a separable Hilbert space.

Suppose in addition that the Hilbert space H under consideration is separable, that is, it contains a countable orthonormal basis (h_k) consisting of mutually orthogonal normed vectors, $h_j \perp h_k$ ($j \neq k$), $\|h_k\| = 1$, and for every $x \in H$ there exists a unique coordinate family $\xi_k \in C$, such that $x = \Sigma \, \xi_k \, h_k$. (Here we assume that finite sums converge in norm in H).

The spectrum of a compact Hermitian operator is given by the famous

THEOREM (Hilbert-Schmidt). *For any compact Hermitian operator T on a complex separable Hilbert space H there is a countable orthonormal basis consisting of eigenvectors h_k of T.*

PROOF. Let $A = \overline{\cup H(\lambda, T)}$ ($\lambda \in \Sigma(T)$) and $B = A^\perp$. Here $H(\lambda, T) = \{0\}$, for $\lambda \notin \Sigma_p(T)$. Clearly
$$TA = T[\overline{\cup H(\lambda, T)}] \subseteq \overline{T[\cup H(\lambda, T)]} \subseteq \overline{\cup TH(\lambda, T)}$$
$$= \overline{\cup H(\lambda, T)} = A.$$

Since $T^* = T$, and $z \cdot y = 0$ ($z \in A$, $y \in B$) implies $x \cdot Ty = Tx \cdot y = 0$ ($x \in A$, $y \in B$) for $TA \subseteq A$, then $Ty \in B$ for $y \in B$. From $TA \subseteq A$ it follows $TB \subseteq B$.

Since T is compact, by the Riesz-Schauder theorem, for $\lambda \neq 0$, the dimensionality of the proper subspace (eigenspace) $H(\lambda, T)$ is finite. The dimensionality of $H(0, T)$ is countable since H is assumed to be separable. (In particular, it may also be finite.) Since T is Hermitian, the subspaces $H(\lambda, T)$ are mutually orthogonal. Therefore, by picking an orthonormal basis in each $H(\lambda, T) \neq \{0\}$ we obtain an orthonormal basis for A. It remains to demonstrate that $B = \{0\}$, so $A = H$.

Let $U = T_B$ be the contraction of T to the subspace B. Since $TB \subseteq B$, then $U : B \to B$ is a compact Hermitian operator, for such is $T : H \to H$. We will prove that U has no eigennumbers: if $\lambda \in \Sigma_p(U)$ then $Ty = Uy = \lambda y$ for some $y \neq 0$, $y \in B$, therefore $\lambda \in \Sigma_p(T)$. Hence, $y \in H(\lambda, T) \subseteq A$. Consequently, $y \in AB = \{0\}$ and $y = 0$, but by definition $y \neq 0$. We have arrived at a contradiction which shows that $\Sigma_p(U) = O$. Since U is compact, by the theorem on the spectrum of a bounded operator and the Riesz-Schauder theorem, this implies

3.8. Operator Spectra

$\Sigma(U) = \{0\}$, whence $\|U\| = |0| = 0$ and $U = O$ because U is Hermitian.

Therefore, $Uy = Oy = 0 = 0 \cdot y$ for each $y \in B$. If $y \neq 0$, then $0 \in \Sigma_p(U)$, which is impossible since, as we have proved, the operator U has no eigennumbers. Thus $B = \{0\}$. ∎

Among the basis vectors h_k we pick out the vectors g_j which constitute the basis of the subspace $H(0, T) = \text{Ker } T$. The rest of the vectors will be denoted by f_i. We will refer to the family (f_i) as the *fundamental family* of eigenvectors of T. The vectors f_i are defined by the identities $Tf_i = \lambda_i f_i$ for the eigennumbers $\lambda_i \neq 0$ of T.

The Hilbert-Schmidt theorem enables us to represent the values of a compact Hermitian operator in the form of linear combinations of vectors f_i from the fundamental family of T. Thus, the Hilbert-Schmidt theorem implies a generalization of the statement that any matrix Hermitian operator can be reduced to diagonal form.

COROLLARY 1. $Tx = \Sigma \alpha_i f_i$ ($\alpha_i = f_i \cdot Tx$, $x \in H$).

PROOF. Since the vectors f_i, g_j, constitute a basis for H, then $x = \Sigma \xi_i f_i + \Sigma \eta_i g_i$ for some numbers ξ_i and η_i. Whence $Tx = \Sigma \xi_i Tf_i + \Sigma \eta_j Tg_j$ since T is linear and continuous. But $Tf_i = \lambda_i f_i$, $Tg_j = 0 \cdot g_j = 0$ and $\xi_i \lambda_i = (f_i \cdot x) \lambda_i = (\lambda_i f_i) \cdot x = Tf_i \cdot x = f_i \cdot Tx = \alpha_i$ since T is Hermitian and $\lambda_i = \lambda_i^* \in \mathbb{R}$. Hence, $\xi_i \cdot Tf_i = \xi_i \lambda_i \cdot f_i = \alpha_i f_i$, as stated. ∎

The product $S = T^*T$ of any compact operator T by the adjoint T^* is a positive compact Hermitian operator. The eigennumbers σ_k of S are said to be the *singular numbers* of T. By the Hilbert-Schmidt theorem there is an orthogonal basis of H consisting of the eigenvectors b_k of S. Let $\sigma_i > 0$, $\mu_i = \sigma_i^{1/2}$, and $e_i = \mu_i^{-1} \cdot Tb_i$. Then it is easy to check that $\sigma_i = \|Tb_i\|^2$, $\mu_i = \|Tb_i\|$, the family (e_i) is orthonormal, and

$$Tx = \Sigma \mu_i (b_i x) e_i \quad (x = \Sigma (b_k x) b_k \in H).$$

This is the *canonical form* of writing compact operator values.

The proper basis (h_k) of a compact Hermitian operator permits one to get an easy solution of the equation $\lambda x - Tx = y$.

COROLLARY 2. *Let* $\lambda \notin \Sigma_p(T)$ *and* $x = \Sigma \xi_k h_k$, $y = \Sigma \eta_k h_k$. *Then the equation* $\lambda x - Tx = y$ *has a unique solution. It is a vector* x *with the coordinates* $\xi_k = (\lambda - \lambda_k)^{-1} \eta_k$, $\lambda_k \in \Sigma_p(T)$.

PROOF. Multiplying both sides of $\lambda x - Tx = y$ by h_k and using the relations $Tx = \Sigma \xi_k \cdot Th_k = \Sigma \xi_k \cdot \lambda_k h_k$ we obtain

$$\lambda \cdot \xi_k - \xi_k \cdot \lambda_k = h_k \cdot \lambda x - h_k \cdot Tx = h_k \cdot y = \eta_k.$$

For $\lambda \neq \lambda_k$, it follows that $\xi_k = (\lambda - \lambda_k)^{-1}\eta_k$. Thus, the solution x is unique.

Such an x is a solution. Indeed, if $\xi_k = (\lambda - \lambda_k)^{-1}\eta_k$, then $\lambda\xi_k - \xi_k\lambda_k = \eta_k$ and

$$\lambda \cdot \xi_k h_k - T(\xi_k h_k) = \lambda\xi_k \cdot h_k - \xi_k \lambda_k \cdot h_k = \eta_k h_k,$$

$$\lambda x - Tx = \lambda\Sigma\xi_k h_k - T(\Sigma\xi_k h_k)$$

$$= \Sigma[\lambda\xi_k h_k - T(\xi_k h_k)] = \Sigma \eta_k h_k = y.$$

Hence, $x = \Sigma \xi_k h_k$ is the desired solution. ∎

COROLLARY 3. *Let* $\lambda \in \Sigma_p(T)$ *and* $x = \Sigma\xi_k h_k$, $y = \Sigma\eta_k h_k$. *Then the equation* $\lambda x - Tx = y$ *has a solution only if* $\eta_k = 0$ *for* $\lambda_k = \lambda$. *In this case, any vector* x *with the coordinates* $\xi_k = (\lambda - \lambda_k)^{-1}\eta_k$ *for* $\lambda_k \neq \lambda$, $\lambda_k \in \Sigma_p(T)$ *is a solution.*

PROOF. Using the argument of Corollary 2, we obtain for the solution x the identity $\lambda\xi_k - \xi_k\lambda_k = \eta_k$. If $\lambda_k = \lambda$ then $\eta_k = 0$.

To check that under this condition the vector x with these coordinates is a solution we may again argue as in Corollary 2. If $\lambda_k = \lambda$, the coordinate ξ_k may be arbitrary. ∎

Corollaries 2 and 3 refine Fredholm's theorems for this particular case. Notice that the condition $\eta_k = 0$ for $\lambda_k = \lambda$ is equivalent to $y \perp H(\lambda, T)$. Since $\lambda = \lambda^*$, $T = T^*$, this condition means that y is orthogonal to any solution z^* of the equation $\lambda^* z^* - T^* z^* = 0$.

3.9. The Spectral Theorem

The Hilbert-Schmidt theorem implies that the values of a compact Hermitian operator may be presented as sums of vectors from the fundamental family with some scalar coefficients. The spectral theorem generalizes this result to the case of self-adjoint operators: their values may be represented by integrals of some numerical-valued functions with respect to projection-valued measures.

3.9.1. Projection-Valued Measures

In this section we will introduce first projection operators (or projectors), then projection-valued measures, and, finally, integral sums and integrals with respect to these measures.

We will consider orthogonal projectors alone, i.e., Hermitian idempotents on a complex Hilbert space H. By definition, a linear operator $P : H \to H$ such that $P^2 = P$ and $P^* = P$ is called a projector (onto

3.9. The Spectral Theorem

the subspace PH). Denote the set of all projectors $P : H \to H$ by $\mathcal{P} = \mathcal{P}(H)$. Let $P, Q \in \mathcal{P}$.

It is easy to verify that PQ, $P + Q$, $P - Q$ are projectors if and only if $PQ = $ respectively QP, $PQ = 0$, and $PQ = QP = Q$. If $PQ = O$, P and Q are said to be *orthogonal*, written $P \perp Q$. (In this case, $QP = O$ as well.)

Since $Pf \cdot f = P^2 f \cdot f = P^*Pf \cdot f = Pf \cdot Pf = \|Pf\|^2 \geq 0$ ($f \in H$), the projection operator P is positive, $P \geq O$. This allows us to define the partial order for projection operators: $P \geq Q$ is equivalent to $P - Q \geq O$, that is, $PQ = QP = Q$.

For projection operator sequences $\{P_n\}$, the fact that the weak convergence $P_n f \cdot g \to Pf \cdot g$ ($f, g \in H$) is equivalent to the strong convergence $P_n f \to Pf$ ($f \in H$) is readily verifiable. The following statement is true.

THEOREM. *An increasing (decreasing) sequence of projection operators converges strongly to its upper (lower) bound.*

Notice that O and i are projection operators, with $O = \min \mathcal{P}$, $I = \max \mathcal{P}$. They bound any projection operator sequence from below and from above, $O \leq P_n \leq I$.

The theorem tells us that a series of mutually orthogonal projection operators has a strong sum equal to its upper bound.

Projection-Valued Measures. Consider a set X, an algebra $\mathcal{R}$ of some parts of X, a Hilbert space H over a field of complex numbers $\mathbb{C}$, the algebra $\mathcal{P}$ of projectors $H \to H$. We will call the sets from $\mathcal{R}$ simple. Suppose that there exist $B_n \in \mathcal{R}$ constituting a sequence $B_n \uparrow X$. In particular, one can choose $X = \mathbb{R}$ and the algebra $\mathcal{R}$ of simple sets $A = \Sigma[a_i, b_i[$ constituted by a finite number of mutually disjoint intervals $[a_i, b_i[$ ($-\infty < a_i \leq b_i < \infty$).

A projection-valued measure is defined as an additive mapping $P : \mathcal{R} \to \mathcal{P}$ strongly continued above at zero and below at unity. Thus,

$$P(A + B) = P(A) + P(B) \quad (A, B, A + B \in \mathcal{R}),$$
$$P(A_n) \to O \quad (A_n \downarrow 0, A_n \in \mathcal{R}),$$
$$P(X_n) \to I \quad (X_n \uparrow X, X_n \in \mathcal{R}).$$

If $X = \mathbb{R}$, it is convenient to specify projection-valued measures in terms of projection-valued distribution functions. These are defined as normed increasing functions $F : \mathbb{R} \to \mathcal{R}$ strongly continuous on the left. Thus,

$$F(s) \leq F(t) \quad (s \leq t), \quad F(t) \to F(u) \quad (t \uparrow u);$$
$$F(s) \to O \quad (s \downarrow -\infty), \quad F(t) \to I \quad (t \uparrow \infty).$$

We emphasize that the convergence is strong everywhere (at each point).

Instead of the continuity on the left one can require continuity on the right, as is frequently done.

The identities

$$P([s, t[) = F(t) - F(s),$$
$$F(t) = \lim_{s \downarrow -\infty} P([s, t[) \quad (-\infty < s \leq t < \infty)$$

establish a one-to-one correspondence between projection-valued distribution functions and measures. If $X = \mathbb{R}$, this correspondence allows one to consider in place of projection-valued measures the corresponding distribution functions.

It is straightforward to verify that a projection-valued measure is multiplicative as well as additive:

$$P(AB) = P(A) \cdot P(B) \quad (A, B \in \mathcal{R}).$$

The continuity above at zero implies that the projection-valued measure is continuous and countably additive.

There is a close analogy between projection-valued and probabilistic measures.

A projection-valued measure $P : \mathcal{R} \to \mathcal{P}$ is quite naturally associated with the families of scalar- and vector-valued measures defined by

$$\mu_f(A) = P(A)f \cdot f,$$
$$m_f(A) = P(A)f \quad (A \in \mathcal{R})$$

for each $f \in H$. The necessary and sufficient conditions for a family of positive number-valued measures $\mu_f : \mathcal{R} \to \mathbb{R}$ ($f \in H$) to be induced by a projection-valued measure $P : \mathcal{R} \to \mathcal{P}$ are known. Therefore, the properties of projection-valued measures can be deduced from the known properties of number-valued measures.

In addition to positive measures μ_f, the projection-valued measure P is associated with the family of number-valued measures μ_{fg} ($f, g \in H$) defined with the aid of the polarization identity

$$\mu_{fg} = 4^{-1}(\mu_{f+g} - \mu_{f-g} - i\mu_{f+ig} + i\mu_{f-ig}),$$

which is equivalent to

$$\mu_{fg}(A) = P(A)f \cdot g \quad (A \in \mathcal{R}).$$

EXAMPLE 1. Let $X = \mathbb{R}$, $H = l^2$, $P_n f = \xi_n e_n$ for $f = (\xi_m) \in H$ and $e_n = (\delta_{mn})$, $\delta_{mn} = 1$ for $m = n$ and $\delta_{mn} = 0$ for $m \neq n$. That is, P_n projects

3.9. The Spectral Theorem

$H = l^2$ onto the nth coordinate axis ($n = 1, 2, \ldots$). The identity

$$P(A) = \sum_{n \in A} P_n \quad (A \in \mathcal{R})$$

defines a discrete projection-valued measure.

EXAMPLE 2. Let $X = \mathbb{R}$, $H = \mathcal{L}^2(\mathbb{R}, dx)$. For every function $f \in H$ and any number $t \in \mathbb{R}$ we define a function $g =]-\infty, t[\cdot f$ with the values $g(x) = f(x)$ for $x < t$ and $g(x) = 0$ for $t \leq x$. Observe that $g \in H$. The identity $F(t)f = g$ defines a projection-valued distribution function $F : \mathbb{R} \to \mathcal{P}$ with values $F(t) \in \mathcal{P}$ ($t \in \mathbb{R}$). This situation is easily verified using the Lebesgue theorem on passage to the limit under the integral sign. It is an easy matter to check that $F(t)f \to 0$ ($t \downarrow -\infty$) and $F(t)f \to f(t \uparrow \infty)$.

The function F induces a projection-valued measure P given by

$$P(A)f = A \cdot f \quad (A \in \mathcal{R}, f \in H).$$

The projection operator $P(A)$ performs multiplication of a function f by the set A (i.e., by its indicator).

Integral Sums. In order to simplify the formulas, we will identify sets with their indicators and use the same notation for the both.

Let

$$u = \sum a_i X_i \quad (a_i \in \mathbb{C}, X_i \in \mathcal{R})$$

be a simple number-valued function on X, and $P : \mathcal{R} \to \mathcal{P}$ be a projection-valued measure. The operator

$$S(u, P) = \sum a_i P(X_i)$$

is called the integral sum of u with respect to measure P. It is straightforward to verify that the value of the integral sum is independent of the method by which the simple function is represented as a linear combination of simple sets:

$$\sum a_i P(X_i) = \sum b_j P(Y_j) \quad (\sum a_i X_i = \sum b_j Y_j)$$

where $a_i, b_j \in \mathbb{C}$ and $X_i, Y_j \in \mathcal{R}$.

Let S be the algebra of simple functions. When it is clear which measure is meant by P, we will write $S(u)$ instead of $S(u, P)$. Denote by $\mathcal{L} = \mathcal{L}(H)$ the algebra of all linear operators $T : H \to H$, and by $\mathcal{B} = \mathcal{B}(H)$ the algebra of bounded linear operators.

These definitions suggest that an integral sum S with values $S(u)$ is a bounded linear operator

$$S(\alpha u + \beta v) = \alpha S(u) + \beta S(v) \quad (\alpha, \beta \in \mathbb{C}; \ u, v \in \mathcal{S}).$$

Thus, $S: \mathcal{S} \to \mathcal{B}$.

Since the measure P is multiplicative, so is the integral sum S:

$$S(uv) = S(u) \cdot S(v) \quad (u, v \in \mathcal{S}).$$

The value of an integral sum $S(u)$ needs not necessarily be a Hermitian operator, as the value $P(A)$ of a projection-valued measure P. But $S(u)$ is a normal operator

$$S^*(u) \cdot S(u) = S(u) \cdot S^*(u) \quad (u \in \mathcal{S}).$$

Indeed, since the values of the measure P are Hermitian,

$$S(u^*) = S^*(u) \quad (u \in \mathcal{S}).$$

Along with the fact that S is multiplicative this yields

$$S^*(u) \cdot S(u) = S(u^*) \cdot S(u) = S(u^*u)$$
$$= S(|u|^2) = S(uu^*) = S(u) \cdot S(u^*).$$

It is an easy matter to verify the inner product rule for values of S

$$S(v)g \cdot S(u)f = \int v^* u \cdot d\mu_{gf} \quad (f, g \in H; \ u, v \in \mathcal{S}).$$

Here, the number-valued measure μ_{gf} is given by

$$\mu_{gf}(A) = P(A)g \cdot P(A)f = P(A)g \cdot f = g \cdot P(A)f \quad (A \in \mathcal{R}).$$

It equals a linear combination of positive number-valued measures, and an integral with respect to it equals the respective linear combination of the integrals by these measures

$$\int w \cdot d\mu_{gf} = 4^{-1} \Big(\int w \cdot d\mu_{g+f} - \int w \cdot d\mu_{g-f}$$
$$- i\int w \cdot d\mu_{g+if} + i\int w \cdot d\mu_{g-if} \Big)$$

for every function w integrable with respect to the measures $\mu_{g+f}, \mu_{g-f}, \mu_{g+ig}, \mu_{g-if}$.

By analogy with positive number-valued measures the integral of the product v^*u of functions v, u can be considered to be their inner product denoted by $\langle v, u \rangle_{gf}$. Then the inner-product rule will be written as

$$\langle S(v)g, S(u)f \rangle = \langle v, u \rangle_{gf} \quad (f, g \in H, u, v \in \mathcal{S}).$$

3.9. The Spectral Theorem

For $g = f$ and $v = u$, the inner-product rule yields
$$\|S(u)f\|^2 = \|u\|_f^2 = \int |u|^2 d\mu_f \quad (u \in S, \ f \in H).$$
Here, the positive number-valued measure μ_f has the values
$$\mu_f(A) = P(A)f \cdot P(A)f = \|P(A)f\|^2 \quad (A \in \mathcal{R}, \ f \in H).$$

The identity $\|S(u)f\| = \|u\|_f$ enables the proof of an important statement which lies behind the definition of the integral with respect to a projection-valued measure. Let $\mathcal{L}_f^2 = \mathcal{L}^2(\mathbb{R}, \mu_f)$ be the Hilbert space of μ_f-measurable number-valued functions on $\mathbb{R}$, square-integrable with respect to measure μ_f for a given $f \in H$. The norm in $\mathcal{L}_f^2$ will be denoted by $\|\cdot\|_f$. Clearly, $S \subseteq \mathcal{L}_f^2$ for each $f \in H$. Thus, in the above identities $\|u\|_f$ denotes the norm of $u \in S$ in $\mathcal{L}_f^2$.

LEMMA. *If a sequence of functions $u_n \in S$ converges in $\mathcal{L}_f^2$, then the sequence of vectors $g_n = S(u_n)f$ converges in H.*

PROOF. In view of the linearity of the integral sum and the identity for norms we have
$$\|g_m - g_n\| = \|S(u_m - u_n)f\| = \|u_m - u_n\|_f \to 0$$
as $m, n \to \infty$. ∎

Note that the set S of simple functions is dense in $\mathcal{L}_f^2$ for each $f \in H$.

For real-valued simple functions, the values $S(u)$ of the sum S are Hermitian operators. If $u \geq 0$, then $S(u) \geq O$. This follows from the fact that the values of the projection-valued measure are Hermitian and positive.

3.9.2. Integrals of Bounded Functions

Consider the algebra $\mathcal{M}$ of functions $u : X \to \mathbb{C}$ measurable with respect to every measure μ_f ($f \in H$). We will conventionally say that these functions are measurable with respect to P, or, briefly, measurable. The algebra $\mathcal{M}$ is Borel, i.e., closed under the pointwise convergence of sequences of functions, and thus $\mathcal{M}$ contains the algebra S of simple functions and its Borel closure, $\overline{S}$, as well as the Borel closure $\overline{\mathcal{R}}$ of the algebra of simple sets. For example, when $X = \mathbb{R}$ and $\mathcal{R}$ is generated by intervals, then $\mathcal{M}$ contains all Borel, specifically continuous, functions $\mathbb{R} \to \mathbb{C}$.

Denote by $\mathcal{M}_0$ the algebra consisting of bounded functions $u : X \to \mathbb{C}$ measurable with respect to P. Notice that $S \subseteq \mathcal{M}_0 \subseteq \mathcal{M}$. We define the integral by a projection-valued measure for bounded measurable functions and investigate its properties. It is a continuous extension of the integral sum S from the algebra $\mathcal{S}$ to $\mathcal{M}_0$.

Preliminaries. First we consider a positive, bounded, number-valued measure $\mu : \mathcal{R} \to \mathbb{R}$, the Hilbert space $\mathcal{L}^2 = \mathcal{L}^2(x, \mu)$, and its part $\mathcal{L}_0^2 = \mathcal{L}_0^2(x, \mu)$ consisting of bounded functions $u \in \mathcal{L}^2$.

To every function $u : X \to \mathbb{C}$ measurable with respect to μ we can associate a sequence of functions $\varphi_n : X \to \mathbb{R}$ given by

$$\varphi_n(x) = 1 \quad (x : |u(x)| \leq n), \quad \varphi_n(x) = 0 \quad (x : |u(x)| > n),$$

and the sequence of functions $\bar{u}_n = \varphi_n u$ with values

$$\bar{u}_n(x) = u(x) \quad (x : |u(x)| \leq n), \quad u_n(x) = 0 \quad (x : |u(x)| > n).$$

Clearly,

$$|\bar{u}_n(x)| \leq |u(x)|, \quad |\bar{u}_n(x)| \leq n \quad (x \in X).$$

The functions φ_n, $\bar{u}_n$ are measurable with respect to μ along with u.

By a more or less standard argument, we can prove the following statements which are used in defining the integrals with respect to a projection-valued measure and in analyses of its properties.

LEMMA 1. *Any constant $c : X \to \mathbb{C}$ is locally integrable with respect to a locally bounded measure μ.*

LEMMA 2. *The functions $\bar{u}_n \in \mathcal{L}_0^2$ and $\bar{u}_n \to u$ at each point.*

LEMMA 3. *Let $u, v, u_n \in \mathcal{L}^2$ and $|u_n| \leq |v|$ $(n = 1, 2, \ldots)$. Then $u_n \to u$ (μ) implies $u_n \to u$ $(\mathcal{L}^2)$.*

Here $u_n \to u$ (μ) means convergence almost everywhere with respect to μ, and $u_n \to u(\mathcal{L}^2)$ means that $\|u_n - u\| \to 0$. Lemmas 1 and 3 together yield

LEMMA 4. *Let $u, u_n \in \mathcal{L}^2$ and $|u_n| \leq c$ $(n = 1, 2, \ldots)$ for some $c > 0$. Then, $u_n \to u$ (μ) implies $u_n \to u$ $(\mathcal{L}^2)$.*

From Lemmas 2 and 3 readily follows

PROPOSITION 1. *The set $\mathcal{L}_0^2$ is dense in the space $\mathcal{L}^2$.*

Using this proposition together with Lemma 4 it can be proved that S is dense in $\mathcal{L}_0^2$. Thus,

PROPOSITION 2. *The set $\mathcal{S}$ is dense in $\mathcal{L}^2$.*

These statements have several important sequels.

3.9. The Spectral Theorem

COROLLARY 1. *For any $f \in H$, $\mathcal{M}_0$ is contained in $\mathcal{L}_f^2$.*

COROLLARY 2. *For any $u \in \mathcal{M}_0$ and $f \in H$, there is a sequence of $u_n^f \in S$, which converges to u in $\mathcal{L}_f^2$.*

COROLLARY 3. *Let $f \in H$, $u \in \mathcal{L}_f^2$, $u_n^f \in S$ and $u_n^f \to u$ ($\mathcal{L}_f^2$). Then there exists $g \in H$ such that $S(u_n^f)f \to g$ (H).*

COROLLARY 4. *Let $f \in H$, $u \in \mathcal{L}_f^2$, $u_n^f \in S$, $v_n^f \in S$ and $u_n^f \to u$, $v_n^f \to u$ ($\mathcal{L}_f^2$). Then $\lim S(u_n^f)) f = \lim S(u_n^f) f$ (H).*

Thus, for every $u \in \mathcal{M}_0$ and each vector $f \in H$ there is a sequence of simple functions u_n^f approximating u in the space $\mathcal{L}_f^2$, whereas the sequence of vectors $S(u_n^f)$ has a limit $g \in H$, one and the same for all sequences of simple functions which approximate u. It is quite natural to define the integral of a function u with respect to the projection-valued measure P in terms of this limit.

Definition of the Integral. Denote the algebra of all transformations (linear and nonlinear) of space H by $\mathcal{F}_0 = \mathcal{F}_0(H)$. We will refer to these transformations as operators acting on H.

DEFINITION. *The integral of a function $u \in \mathcal{M}_0$ with respect to measure P is the operator $E_0(u) \in \mathcal{F}_0$ defined by*

$$E_0(u) f = \lim S(u_n^f) f$$

for every $f \in H$ and a sequence of simple functions $u_n^f \in S$, $u_n^f \to u$ ($\mathcal{L}_f^2$).

By the integral of a bounded measurable function with respect to a projection-valued measure we mean a mapping $E_0 : \mathcal{M}_0 \to \mathcal{F}_0$ carrying a function $u \in \mathcal{M}_0$ to the operator $E_0(u) \in \mathcal{F}_0$.

Clearly, the integral E_0 extends the integral sum S, that is,

$$E_0(u) = S(u) \quad (u \in S).$$

If $u \in S$, as an approximating sequence for each $f \in H$ we may choose the constant sequence $u_n^f = u$ ($n = 1, 2, \dots$).

EXAMPLE 1. Consider a set $X = \mathbb{N}$, the algebra $\mathcal{R}$ of all its finite parts, the Hilbert space $H = l^2$, and its standard basis $e_n = (\delta_n(m))$. Define P_n to be the projection on the coordinate axis $\mathbb{C}e_n$

$$P_n f = f(n) \cdot e(n) \quad (f = (f(m)) \in H).$$

Define a projection-valued measure P by
$$P(A) = \sum_{n \in A} P_n \quad (A \in \mathcal{R}).$$
Every bounded sequence $u : \mathbb{N} \to \mathbb{C}$ is measurable with respect to P. For each $f \in H$, it is approximated in $\mathcal{L}_f^2$ by a sequence of simple functions
$$u_n = \sum_{m \leq n} u(m) \cdot \{m\},$$
with values $u_n(m) = u(m)$ for $m \leq n$, and $u_n(m) = 0$ for $m > n$. Indeed,
$$\| u - u_n \|_f^2 = \sum_{m > n} |u(m)|^2 \cdot |f(m)|^2 \leq c^2 \cdot \sum_{m > n} |f(m)|^2 \to 0.$$
Since
$$S(u_n) = \sum_{m \geq n} u(m) \cdot P_m, \quad S(u_n)f = \sum_{m \leq n} u(m) \cdot f(m) \cdot e_m,$$
then
$$E_0(u)f = \lim S(u_n)f = \Sigma u(m) \cdot f(m) \cdot e_m,$$
or
$$E_0(u)f = uf = (u(m) \cdot f(m))_m.$$
The integration of a sequence thus means multiplying by this sequence.

EXAMPLE 2. Let X be a set, $\mathcal{R}$ be the algebra of some of its parts, μ be a positive number-valued measure on $\mathcal{R}$ (not necessarily bounded), $H = \mathcal{L}^2 = \mathcal{L}^2(X, \mu)$ be a Hilbert space, and P be a projection-valued measure on $\mathcal{R}$ defined by
$$P(A)f = Af \quad (A \in \mathcal{R}, f \in H).$$
Here, as usual, $Af(x) = f(x)$ for $x \in A$, and $Af(x) = 0$ for $x \notin A$.

Observe that
$$S(u)f = \Sigma a_i P(X_i)f = \Sigma a_i X_i f = uf \quad (u = \Sigma a_i X_i \in S, f \in H).$$
Since
$$\mu_f(A) = \| P(A)f \|^2 = \| Af \|^2 = \int A |f|^2 \cdot d\mu,$$
then
$$d\mu_f / d\mu = |f|^2, \quad d\mu_f = |f|^2 d\mu.$$

3.9. The Spectral Theorem

Using this fact, and changing the variables, yields

$$\| uf - u_n^f f \|^2 = \int | uf - u_n^f f |^2 d\mu$$
$$= \int | u - u_n^f |^2 \cdot | f |^2 d\mu = \int | u - u_n^f |^2 d\mu_f = \| u - u_n^f \|_f^2.$$

Thus, $u_n^f \to u$ (H) for $u_n^f \to u$ ($\mathscr{L}_f^2$). Therefore,

$$E_0(u) f = \lim S(u_n^f) f = \lim (u_n^f f) = uf.$$

In this example the integration with respect to projection-valued measure again reduces to multiplication. The first example is a special case of the second when $X = N$ and the measure μ is counting.

Properties of the Integrals. The value $E_0(u)$ of the integral E_0 for any bounded measurable function $u \in \mathcal{M}_0$ is an everywhere defined, bounded, normal operator on H. An essential role in the proof of this and other statements regarding properties of the integrals is played by the following

LEMMA. *For any $f, g \in H$ there exists $h \in H$ such that $\mu_f, \mu_g \ll \mu_h$.*

Here, as before, $\mu \ll \nu$ means that measure μ is continuous with respect to measure ν. The lemma is proved by a rather lengthy but elementary argument (*cf.* Weidmann [8], Theorem 7.13). From this lemma it is not hard to deduce

COROLLARY. *Let $u \in \mathcal{M}_0$, $|u| < c$, and $f_i \in H$ ($i = 1, ..., m$). Then there exist $u_n \in S$ such that $u_n \to u$ (μ_{f_i}; $i = 1, ..., m$), $u_n \to u$ ($\mathscr{L}_{f_i}^2$; $i = 1, ..., m$) and $|u_n| < c$ ($n = 1, 2, ...$).*

Applying this corollary and using the appropriate properties of integral sums, it is an easy matter to prove the statement, formulated earlier in this section, which characterizes the values of the integral E_0 and their class.

THEOREM 1. *For a function $u \in \mathcal{M}_0$, the value $E_0(u)$ of the integral E_0 is an everywhere defined, bounded, normal, linear operator on H.*

PROOF. That $E_0(u)$ is linear follows from the identities

$$E_0(u)(af + bg) = \lim S(u_n)(af + bg) = \lim (aS(u_n)f + bS(u_n)g)$$
$$= a \lim S(u_n) f + b \lim S(u_n) g = a E_0(u) f + b E_0(u) g$$
$$(a, b \in \mathbb{C}; f, g \in H),$$

which hold for $u_n \in S$, $u_n \to u$ ($\mathscr{L}_f^2, \mathscr{L}_g^2, \mathscr{L}_{af+bg}^2$).

The boundedness of the linear operator $E_0(u)$ follows from the relations

$$\| E_0(u)f \| = \| \lim S(u_n)f \| = \lim \| S(u_n)f \| = \lim \| u_n \|_f \leq c \cdot \| f \|,$$

which hold for $f \in H$, $u_n \in S$, $u_n \to u$ ($\mathscr{L}_f^2$), $|u_n| < c$ ($n = 1, 2, \ldots$).

The normality of $E_0(u)$ results from

$$E_0(u^*) = E_0^*(u), \quad E_0(uv) = E_0(u) \cdot E_0(v) \quad (u, v \in \mathcal{M}_0).$$

Indeed,

$$E_0(u) \cdot E_0^*(u) = E_0(u) \cdot E_0(u^*) = E_0(uu^*)$$
$$= E_0(|u|^2) = E_0(u^*u) = E_0^*(u) \cdot E_0(u)$$

for every $u \in \mathcal{M}_0$. ∎

Let $\mathscr{L}_0 = \mathscr{L}_0(H)$, rather than $\mathscr{B} = \mathscr{B}(H)$, denotes the algebra of bounded linear operator on H. The integral $E_0: \mathcal{M}_0 \to \mathscr{L}_0$ is a homeomorphism of the algebra $\mathcal{M}_0$ into the algebra $\mathscr{L}_0$. It preserves all the operations of addition, multiplication, multiplication by a scalar and conjugation.

THEOREM 2. *For any $a, b \in \mathbb{C}$ and $u, v \in \mathcal{M}_0$,*

$$E_0(au + bv) = aE_0(u) + bE_0(v) \tag{L}$$

$$E_0(uv) = E_0(u) \cdot E_0(v), \tag{M}$$

$$E_0(u^*) = E_0^*(u). \tag{N}$$

These identities are obtained by passage to the limit in the similar identities for integral sums.

COROLLARY. *For any real-valued $u \in \mathcal{M}_0$ the value $E_0(u)$ of E_0 is a Hermitian operator.*

PROOF. If $u^* = u$, then by the identity (N)

$$E_0^*(u) = E_0(u^*) = E_0(u)$$

and $E_0(u)$ is Hermitian, as stated. ∎

Passing to the limit in the inner product rules for integral sums yields a similar rule for the integral.

THEOREM 3. *For any $u, v \in \mathcal{M}_0$ and $f, g \in H$,*

$$\langle E_0(v)g, E_0(u)f \rangle = \langle v, u \rangle_{gf}, \tag{C}$$

3.9. The Spectral Theorem

where

$$\langle v, u \rangle_{gf} = \int v^* u \, d\mu_{gf},$$
$$\mu_{gf}(A) = \langle P(A)g, P(A)f \rangle \quad (A \in \mathcal{R}).$$

For $g = f$ and $v = u$, from the identity (C) of Theorem 3 follows

COROLLARY. *For any $u \in \mathcal{M}_0$ and $f \in H$*

$$\| E_0(u) f \| = \| u \|_f. \tag{I}$$

The letter I in this designation stands for "isometry."

By the assumption involved in the definition of a projection-valued measure P, there is a sequence of simple sets $X_n \in \mathcal{R}$ such that $X_n \uparrow X$ and $P(X_n) \to I$, where I is the identity transformation on the space H.

THEOREM 4. *The integral E_0 is normed:*

$$E_0(1) = I.$$

PROOF. Clearly, the constant $1 \in \mathcal{M}_0$, and the sequence of indicators of sets $X_n \uparrow X$ approximates it in $\mathcal{L}_f^2$ for each $f \in H$.

Thus

$$E_0(1) f = \lim S(X_n) f = \lim P(X_n) f = f.$$

Hence $E_0(1) = I.$ ∎

Since the integral E_0 is linear, this theorem yields immediately

COROLLARY. *For any constant c, $E_0(c) = c \cdot I$.*

(Here, as before in this chapter, c denotes both a constant function and its value).

The order properties of an integral sum S, as well as its algebraic properties, are also carried over to the integral E_0.

THEOREM 5. *The integral E_0 is positive and monotonic for real-valued functions $u, v \in \mathcal{M}_0$.*

$$E_0(u) \geq 0 \quad (u \geq 0),$$
$$E_0(u) \leq E_0(v) \quad (u \leq v).$$

PROOF. By the Corollary to Theorem 2, for $u \geq 0$, the operator $E_0(u)$ is Hermitian. Using identity (C) of Theorem 3 for $v = 1$, $g = f$ and identity $E_0(1) = I$ we obtain

$$f \cdot E_0(u) f = E_0(1) f \cdot E_0(u) f = \int u \, d\mu_f \geq 0 \quad (u \geq 0).$$

Thus, $E_0(u) \geq 0$ for $u \geq 0$, i.e., E_0 is positive.

That E_0 is monotonic follows from its positivity and linearity:
$$E_0(v) - E_0(u) = E_0(v - u) \geq 0,$$
$$E_0(u) \leq E_0(v)$$
for $u \leq v$ and $u, v \in \mathcal{M}_0$. ∎

Using the fact that the integral E_0 is isometric and linear, one can easily prove for it the limit theorem which describes the convergence of sequences of its values. Since for some f a sequence of functions $u_n \in \mathcal{M}_0$ can converge in $\mathcal{L}_f^2$ to a function $u \notin \mathcal{M}_0$, the limit theorem consists of two statements, for the cases of $u \in \mathcal{M}_0$ and $u \notin \mathcal{M}_0$, respectively. As a spin-off we prove that the limit is independent of the choice of the approximating sequence.

THEOREM 6. *Let* $u_n, v_m \in \mathcal{M}_0$ *and* $f \in H$, $u \in \mathcal{L}_f^2$, $u_n, v_n \to u \ (\mathcal{L}_f^2)$. *Then there is a* $g \in H$ *such that*
$$\lim E_0(u_n) = \lim E_0(v_n) = g,$$
and $g = E_0(u)$ *if* $u \in \mathcal{M}_0$.

PROOF. Indeed,
$$\| E_0(v_p) f - E_0(u_n) f \| = \| E_0(v_p - u_n) f \| = \| v_p - u_n \|_f.$$
For $v_p = u_p$, this implies that $g_n = E_0(u_n) f$ converges to some $g \in H$, and for $v_p \neq u_p$, we obtain the required equality of the limits.

In exactly the same fashion
$$\| E_0(u) f - E_0(u_n) f \| = \| u - u_n \|_f,$$
therefore,
$$\lim E_0(u_n) f = E_0(u) f$$
for $u \in \mathcal{M}_0$. ∎

Theorems 1–6 indicate that the integrals of bounded measurable functions with respect to projection-valued measures have many valuable properties.

3.9.3. Integrals of Unbounded Functions

The values of the integrals of unbounded functions are unbounded densely defined linear operators. Therefore, the definition and examination of the properties of such integrals is more complicated than for the integrals of bounded functions. However, both integrals have much in common.

3.9. The Spectral Theorem

Definition. Pick out a measurable function $u \in \mathcal{M}$ and consider the set

$$D(u) = \{f \mid u \in \mathcal{L}_f^2\} \subseteq H.$$

Thus, by definition, $f \in D(u)$ is equivalent to $u \in \mathcal{L}_f^2$. This ensures the existence of a sequence of bounded measurable functions $u_n^f \in \mathcal{M}_0$ which for any $f \in D(u)$ converges to a function $u \in \mathcal{M}$ in $\mathcal{L}_f^2$. As such approximating bounded functions we may choose simple functions $u_n^f \in S$ (by Proposition 2 of Section 3.9.2). By Theorem 6 about convergence, the sequence of vectors $g_n = E_0(u_n^f)f$ has a limit $g \in H$, one and the same for all sequences of functions from $\mathcal{M}_0$ approximating $u \in \mathcal{M}$. It is natural to define with this limit the integral of a function $u \in \mathcal{M}$ with respect to the projection-valued measure P.

Denote $\mathcal{F} = \mathcal{F}(H)$ to be the set of all partial transformations of the space H (mappings of parts of H into H). We will call these mappings partial operators acting in H, or briefly, operators, like conventional mappings defined on the entire H.

DEFINITION. *We say the integral of a function $u \in \mathcal{M}$ by measure P about the partial operator $E(u) \in \mathcal{F}$ defined by the identity*

$$E(u)f = \lim E_0(u_n^f)f$$

for each $f \in D(u)$ and some sequence of bounded functions $u_n^f \in \mathcal{M}_0$, $u_n^f \to u$ $(\mathcal{L}_f^2)$.

By the integral of a function measurable with respect to a projection-valued measure we mean a mapping $E: \mathcal{M} \to \mathcal{F}$ carrying a function $u \in \mathcal{M}$ into a partial operator $E(u) \in \mathcal{F}$.

Clearly, the integral E extends the integral E_0 and the integral sum S, that is

$$E(u) = S(u) \quad (u \in S),$$
$$E(u) = E_0(u) \quad (u \in \mathcal{M}_0).$$

In either of these cases, we may take the corresponding constant sequence as an approximating sequence.

EXAMPLE 1. In Example 1 of Section 3.9.2, in place of the bounded sequence we take an arbitrary sequence $u: \mathbb{N} \to \mathbb{C}$. It is not hard to check that

$$D(u) = \{f: \Sigma \mid u(n) \cdot f(n) \mid^2 < \infty\},$$
$$E(u)f = uf \quad (f \in D(u)).$$

EXAMPLE 2. In Example 2 of Section 3.2.9, we take in place of a bounded function a measurable function $u \in \mathcal{M}$. It can be readily verified that

$$D(u) = \{ f : \int |uf|^2 d\mu < \infty \},$$
$$E(u)f = uf \quad (f \in D(u)).$$

Again, the integration of a function with respect to a projection-valued measure reduces to multiplication by this function.

Properties. The value $E(u)$ of the integral E for any measurable function $u \in \mathcal{M}$ is a densely defined, normal, linear operator in H. Because we are concerned with the partially defined operators, the proofs of most statements regarding the properties E are complicated. Therefore, we will merely formulate them as the generalizations of Theorems 1–6 about the properties of E_0.

LEMMA. *For the integral E of a function $u \in \mathcal{M}$, the domain $D(u)$ of the value $E(u)$ is a dense vector subspace in H.*

THEOREM 1. *The value $E(u)$ of the integral E of a function $u \in \mathcal{M}$ is a densely defined normal linear operator in H.*

THEOREM 2. *For any $a, b \in \mathbb{C}$ and $u, v \in \mathcal{M}$ the following inclusions hold*:

$$E(au + bv) \supseteq aE(u) + bE(v), \qquad (L)$$
$$E(uv) \supseteq E(u) \cdot E(v), \qquad (M)$$
$$E(u^*) = E^*(u). \qquad (N)$$

These inclusions mean that the left-hand sides are extensions of the right-hand sides. We emphasize that by the partial operator rules the domains of results equal the intersection of the domains of relevant operators. For example, $D(uv) = D(u) \cap D(v)$.

COROLLARY. *For each real-valued $u \in \mathcal{M}$ the value $E(u)$ of the integral E is a self-adjoint operator.*

THEOREM 3. *For any $u, v \in \mathcal{M}$, $f \in D(u)$, and $g \in D(v)$,*

$$\langle E(v) g, E(u) f \rangle = \langle v, u \rangle_{gf}; \qquad (C)$$

where

$$\langle v, u \rangle_{gf} = \lim \int \bar{v}_n^* \cdot \bar{u}_n \cdot d\mu_{gf}.$$

COROLLARY. *For any $u \in \mathcal{M}$ and $f \in D(u)$, $\| E(u) f \| = \| u \|_f$.*

(The bounded functions $\bar{v}_n, \bar{u}_n$ which approximate v and u have been defined in 3.9.2.)

3.9. The Spectral Theorem

THEOREM 4. *The integral E is normed and, for any constant c, $E(c) = c \cdot I$.*

THEOREM 5. *The integral E of real-valued functions $u, v \in \mathcal{M}$ is positive and monotonic*

$$E(u) \geq 0 \quad (u \geq 0), \quad E(u) \leq E(v) \quad (u \leq v).$$

THEOREM 6. *Let $u_n, u \in \mathcal{M}$ and $f \in D(u) \cap D(u_n)$ $(n = 1, 2, \ldots)$, $u_n \to u$ $(\mathcal{L}_f^2)$. Then*

$$\lim E(u_n) f = E(u) f.$$

Theorems 4 and 5 are identical to those for the integral E_0. They are given here merely to preserve numeration.

The integral $E(u)$ of a function $u \in \mathcal{M}$ by a projection-valued measure is denoted also like the integral by a number-valued measure

$$E(u) = \int u dP.$$

If the measure is induced by a distribution function F, then an alternative designation is

$$E(u) = \int u dF.$$

Where needed, the variable x and the domain X of u are also indicated.

3.9.4. The Spectral Theorem

The integral of a real-valued measurable function with respect to a projection-valued measure is a self-adjoint operator. A natural question to ask is then whether or not every self-adjoint operator is representable as an integral of a function with respect to some measure? The spectral theorem asserts that the question should be answered in the affirmative. This theorem is discussed in depth in Chapter 7 of the book by Weidmann [8].

Preliminaries. In the proof of the spectral theorem we shall use the Stieltjes inversion formula which can be readily verified.

Let $\rho : \mathbb{R} \to \mathbb{R}$ be a function of bounded variation continuous on the left (difference of two increasing functions), and $\rho(-\infty) = 0$. For each $z \in \mathbb{C}$, $\operatorname{Im} z \neq 0$, we consider a function $r(z, \cdot) : \mathbb{R} \to \mathbb{C}$ given by

$$r(z, x) = (z - x)^{-1} \quad (x \in \mathbb{R}).$$

The function $r(z, \cdot)$ is integrable with respect to a numerical-valued measure defined by ρ. If $\rho = \sigma - \tau$, where σ and τ are increasing positive distribution functions, then

$$\int r(z, \cdot) d\rho = \int r(z, \cdot) d\sigma - \int r(z, \cdot) d\tau.$$

Consider also a function φ given by

$$\varphi(z) = \int_{-\infty}^{\infty} (z - x)^{-1} d\rho(x) \quad (\operatorname{Im} z \neq 0).$$

The inversion formula holds

$$\rho(x) = \lim_{\delta \downarrow 0} \lim_{\varepsilon \downarrow 0} (2\pi i)^{-1} \int_{-\infty}^{x+\delta} [\varphi(s - i\varepsilon) - \varphi(s + i\varepsilon)] ds.$$

Recall also that the resolvent

$$R(z, T) = (zI - T)^{-1} \quad (z \in P(T))$$

of a closed operator T is a bounded everywhere defined operator, $R(\cdot, T)$ is an analytical function on the resolvent set $P(T)$, and that a self-adjoint operator is closed and its spectrum is contained in $\mathbb{R}$.

Theorem Formulation. Consider a complex Hilbert space H, a vector subspace A dense in H, the algebra $\mathcal{P}$ of projectors on H, the real line $\mathbb{R}$, the algebra $\mathcal{R}$ of simple sets in $\mathbb{R}$ induced by the intervals $[a, b[$ ($-\infty < a \leq b < \infty$), and a projection-valued measure $P: \mathcal{R} \to \mathcal{P}$.

In order to simplify notations, we will denote by the letter x an arbitrary point of $\mathbb{R}$ and the identity function on $\mathbb{R}$.

THEOREM. *For every self-adjoint operator $T: A \to H$ there exists a unique projection-valued measure $P: \mathcal{R} \to \mathcal{P}$ such that*

$$T = \int x \, dP.$$

The measure P is defined by

$$\| P([a, b[) f \|^2$$
$$= \lim_{\delta \downarrow 0} \lim_{\varepsilon \downarrow 0} (2\pi i)^{-1} \int_{a+\delta}^{b+\delta} \langle f \cdot [R(x - i\varepsilon, T) - R(x + i\varepsilon, T)] f \rangle dx$$

for all $f \in H$ and $-\infty < a \leq b < \infty$.

Here under the integral sign we have the inner product of a vector f by the vector $g = [R(x - i\varepsilon, T) - R(x + i\varepsilon, T)] f$. Since $\varepsilon > 0$, $x - i\varepsilon$,

3.9. The Spectral Theorem

$x + i\varepsilon \notin \mathbb{R}$. Consequently $x - i\varepsilon, x + i\varepsilon \in P(T)$ and the integrand is defined for each $f \in H$.

The measure P defined in this theorem is called the *spectral measure* of an operator T.

It would be rather cumbersome to prove that a self-adjoint operator T has the spectral measure P, whereas the uniqueness of this measure is simple to prove. We choose the latter to clarify the statement of the theorem.

PROOF. Let $T = E(x)$, $D(x) = A$. Then $E(z - x) = zI - T : A \to H$ since E is linear and normed. Pick the functions

$$u = z - x, \quad v = r(z, x) = (z - x)^{-1} \quad (\text{Im } z \neq 0).$$

Notice that v is bounded and $uv = vu = 1$. Since E is multiplicative, then

$$E(u) \cdot E(v) = I, \quad E(v) \cdot E(u) = I_A,$$

where I and I_A are the identity mappings of H and A. Since $E(u) = zI - T$, these relations yield

$$E(v) = E^{-1}(u) = R(z, T) : H \to H.$$

By the inner product rule

$$\varphi(z) = f \cdot E(v) f = \int v \, d\mu_f = \int_{-\infty}^{\infty} (z - x)^{-1} d\mu_f(x).$$

Applying the inversion formula we obtain

$$\mu_f([a, b[) = \lim_{\delta \downarrow 0} \lim_{\varepsilon \downarrow 0} \frac{1}{2\pi i} \int_{a+\delta}^{b+\delta} [\varphi(x - i\varepsilon) - \varphi(x + i\varepsilon)] dx$$

for $-\infty < a \leq b < \infty$. It remains to note that for each $f \in H$

$$\mu_f([a, b[) = \|P([a, b[) f\|^2,$$

$$\varphi(x - i\varepsilon) - \varphi(x + i\varepsilon) = f \cdot [R(x - i\varepsilon, T) - R(x + i\varepsilon, T)] f,$$

and that the family of positive number-valued measures μ_f ($f \in H$) uniquely defines a projection-valued measure P by the above equality.

Indeed, let $Q : \mathcal{R} \to \mathcal{P}$ be a projection-valued measure for which

$$f \cdot Q([a, b[) f = \|Q([a, b[) f\|^2 = \|P([a, b[) f\|^2$$
$$= f \cdot P([a, b[) f$$

for $f \in H$ and $-\infty < a \leq b < \infty$. Then, by the polarization formula,
$$g \cdot Q([a, b[)f = g \cdot P([a, b[)f \quad (g \in H),$$
and so
$$Q([a, b[)f = P([a, b[)f \quad (f \in H),$$
$$Q([a, b[) = P([a, b[) \quad (-\infty < a \leq b < \infty).$$
Hence $P = Q$. The uniqueness of spectral measure is established. ∎

EXAMPLE 1. Consider a compact Hermitian operator $T : H \to H$, sequences of its eigennumbers $\lambda_n \neq 0$ and projectors P_n on the corresponding proper subspaces (eigenspaces) $H_n = \text{Ker}(\lambda_n I - T)$, $n = 1, 2, \ldots$, and the projector P_0 on $\text{Ker }T$. Let $\lambda_0 = 0$.

Pick a projection-valued measure $P : \mathcal{R} \to \mathcal{P}$ with values
$$P([a, b[) = \sum_{a \leq \lambda_n < b} P_n \quad (-\infty < a \leq b < \infty).$$

Here $n = 0, 1, 2, \ldots$, and the sum is strong. By the Hilbert-Schmidt theorem we have
$$E(x)f = \Sigma \lambda_n P_n f = Tf \quad (f \in H),$$
that is,
$$T = \Sigma \lambda_n P_n = \int x \, dP,$$
and P is the spectral measure of T.

Observe that $\lambda_n \to 0$ if $\dim H = \infty$. This follows from the Riesz-Schauder theorem. (Otherwise the sequence of eigenvalues $\lambda_n \neq 0$ would have a limit point $\lambda \neq 0$, or else one of the eigenspaces corresponding to them would be infinite-dimensional.)

EXAMPLE 2. Let
$$H = \mathcal{L}^2(R, dx), \quad A = \{f \in H : \int |xf|^2 \, dx < \infty\},$$
and $T = M$ be the operator of multiplication by an independent variable x (the identity function on $\mathbb{R}$):
$$Tf = x \cdot f \quad (f \in A).$$

Pick a projection-valued measure $P : \mathcal{R} \to \mathcal{P}$ with values
$$P(x)f = X \cdot f \quad (X \in \mathcal{R}, f \in A).$$

3.9. The Spectral Theorem

It is not too difficult to verify, following the argument of Example 2 in Section 3.9.2, that

$$E(x)f = x \cdot f \quad (f \in A).$$

Thus

$$T = \int x \, dP$$

and P is the spectral measure of $T = M$.

The spectral theorem suggests that every self-adjoint operator is unitarily equivalent to the multiplication operator

$$T = U^{-1}MU,$$

where U is a unitary operator. This operator also relates the spectral measures of T and M (see Theorem 7.18 in Ref. [8]).

NOTE. Define $\mathbb{C}$ to be the complex plane, $\mathcal{R}$ the algebra of simple sets in $\mathbb{C}$ induced by the rectangles $[a, b[\times [c, d[$ ($-\infty < a \leq b, c \leq d < \infty$), $\mathcal{B}$ the algebra of bounded linear operators on H, and $P: \mathcal{R} \to \mathcal{B}$ an operator-valued measure.

By analogy with the integral with respect to projection-valued measure one can define the integral with respect to an operator-valued measure for functions of a complex variable, and to prove the spectral theorem for normal operators (see Theorem 7.32 in Weidmann's book [8]).

3.9.5. Operator-Valued Functions

The spectral theorem and the notion of integration of measurable functions by projection-valued measures enables us to define functions of self-adjoint operators.

Let $T: A \to H$ be a self-adjoint operator, $P: \mathcal{R} \to \mathcal{P}$ be its spectral measure, $u: \mathbb{R} \to \mathbb{C}$ be a function measurable with respect to P. Then, by definition,

$$u(T) = \int u \, dP.$$

In particular, this identity is valid for any Borel function u.

EXAMPLE 1. Let

$$u(x) = \sum_{0 \leq m \leq n} a_m x^m \quad (a_m \in \mathbb{C}).$$

Then it is easy to verify, using the multiplicativity of the integral with respect to a projection-valued measure, that

$$u(T) = \sum_{0 \leq m \leq n} a_m T^m \quad (T^0 = I).$$

EXAMPLE 2. For any natural number $n \geq 1$ and a positive self-adjoint operator T, there exists a unique positive self-adjoint operator $T^{1/n}$ such that $(T^{1/n})^n = T$. $T^{1/n}$ is defined by

$$T^{1/n} = \int x^{1/n}\, dP,$$

where P is the spectral measure of T. If T is compact then so is $T^{1/n}$.

Observe that since T is positive and self-adjoint, its spectrum $\Sigma(T)$ lies on the positive semi-axis of $\mathbb{R}$. Therefore the spectral measure P of T is zero on sets from the negative semi-axis. The values of the integrand $x^{1/n}$ on the negative semi-axis do not matter and may be chosen arbitrarily.

NOTE. Using the spectral theorem for normal operators one can prove that for any natural number $n \geq 1$ and a normal operator T there exists a normal operator $T^{1/n}$ such that $(T^{1/n})^n = T$.

The spectral measure of an operator T will be denoted by the same letter E as the integral with respect to this measure, written

$$T = \int \lambda\, dE(\lambda), \quad u(T) = \int u(\lambda)\, dE(\lambda).$$

In particular, a function power of a positive self-adjoint operator is defined by

$$T^{m/n} = \int \lambda^{m/n}\, dE(\lambda).$$

This integral representation is used to define the fractional powers T^α ($0 < \alpha < 1$) for a wider class of operators (see Kato [7], Section 5.3).

3.10. Operator-Valued Exponential

Exponentials are outstanding among operator-valued functions. The spectral theorem suggests that they may be defined as integrals of conventional exponential functions. Operator-valued exponentials are used efficiently in solving differential equations with operator coefficients.

3.10. Operator-Valued Exponential

3.10.1. Problem Formulation

Consider a complex-valued exponential $f: [0, \infty[\to \mathbb{C}$ with the exponential coefficient $a \in \mathbb{C}$

$$f(t) = e^{at} \quad (t \geq 0).$$

It is a continuous homeomorphism from the additive semigroup of positive numbers onto the multiplicative semigroup of some complex numbers:

$$f(s + t) = e^{a(s+t)} = e^{as} e^{at} = f(s) \cdot f(t)$$

for all $s, t \geq 0$. In particular, for $a = 2\pi i$, the exponential maps $[0, \infty[$ onto the unit circle in the complex plane $\mathbb{C}$.

Each value $f(t)$ of the exponential f determines a linear transformation $U(t)$ of $\mathbb{C}$, mapping a point $z \in \mathbb{C}$ into the point

$$U(t)z = e^{at}z.$$

Like the values of $f(t)$, the transformations $U(t)$ constitute a multiplicative semigroup:

$$U(s + t) = U(s) \cdot U(t) \tag{1}$$

for all $s, t \geq 0$. Thus, the exponential may be related to a semigroup of linear transformations.

If $a = i\alpha$ ($\alpha \in \mathbb{R}$), then

$$U(0) = 1, \quad U^*(t) = U(-t) = U^{-1}(t) \quad (t \in \mathbb{R}).$$

The transformations $U(t)$ are unitary and constitute a group.

Observe also that

$$e^{-at} = \lim_{n \to \infty} \left[\left(1 + \frac{t}{n} a \right)^{-n} \right]. \tag{2}$$

The coefficient a of the exponential f equals its derivative at zero, $\dot{f}(0)$,

$$a = \lim_{t \downarrow 0} [t^{-1}(e^{at} - 1)], \tag{3}$$

and f is the unique solution of the differential equation.

$$\dot{f} = af \tag{4}$$

obeying the condition $f(0) = 1$.

Finally, the exponential is an analytic function:

$$e^{at} = \sum_{n \geq 0} \frac{(at)^n}{n!} \quad (n \geq 0), \tag{5}$$

and the series

$$\sum \frac{|at|^n}{n!} < \infty$$

converges absolutely.

Now we formulate a problem: define an exponential with an operator coefficient, which will have similar properties.

For a bounded operator T on a Banach space X, this goal is achieved easily by using the analytic operator-valued functions

$$e^{tT} = \sum_{n \geq 0} \frac{(tT)^n}{n!} \quad (n \geq 0).$$

Since

$$\sum \frac{\|tT\|^n}{n!} < \infty,$$

the right-side series defines a bounded linear operator on X. If T is unbounded, this definition is no longer valid and to define the operator exponential one has to use its other properties.

3.10.2. Operator Semigroup

It seems fairly natural to define an operator-valued exponential by analogy with equation (2) as the limit of a special sequence of operators. We need to prove only that the limit exists for a sufficiently wide class of operators. Observe that the negative exponents in the chosen sequence allow us to use the boundedness of the values of the resolvent for closed operators.

It can be proved (see Kato [7], Section 9.1) that if T is a densely defined closed operator on a Banach space X, the resolvent set $\mathcal{R}(-T)$ for operator $-T$ contains the negative semiaxis, and for $\lambda > 0$, $\|(\lambda I + T)^{-1}\| \leq \lambda^{-1}$, then for each $t \geq 0$ there exists the limit

$$U(t) = \lim_{n \to \infty} \left[\left(I + \frac{t}{n}T\right)^{-n}\right]$$

in the strong sense. The operators $U(t)$ defined by this identity constitute

3.10. Operator-Valued Exponential

a semigroup:

$$U(s + t) = U(s) \cdot U(t)$$

for all $s, t \geq 0$. This is a contraction semigroup, i.e., $\| U(t) \| \leq 1$. The function U with the values $U(t)$ is strongly continuous. It will be called the exponential with the coefficient $-T$. Its values are written in the habitual form

$$U(t) = e^{-tT} \quad (t \geq 0).$$

For each $x \in \text{Dom } T$,

$$-T(x) = \lim_{t \downarrow 0} [t^{-1}(e^{-tT} - I)x],$$

where I denotes the identity transformation on X. In this context $-T$ is known as the *infinitesimal generator* of the semigroup $(U(t))$. Different infinitesimal generators generate different semigroups.

3.10.3. The Laplace Transform

The Laplace transform of an exponential generated by the operator $-T$ yields the resolvent of this operator.

Consider a Borel function $f : \mathbb{R} \to \mathbb{C}$ such that

$$|f(t)| \leq ce^{\alpha t} \quad (t \geq 0) \quad \text{and} \quad f(t) = 0 \quad (t < 0)$$

for some $c > 0$ and $\alpha \in \mathbb{R}$. Observe that for any $z \in \mathbb{C}$ with $x = \text{Re } z > \alpha$, the function $g : \mathbb{R} \to \mathbb{C}$ given by

$$g(t) = e^{-tx} f(t) \quad (t \in \mathbb{R})$$

is integrable by Lebesgue's measure dt since $x - \alpha > 0$ and

$$|g(t)| \leq ce^{-t(x-\alpha)} \quad (t \geq 0), \; g(t) = 0 \quad (t < 0).$$

By the Laplace transform of a function f we mean the function h given by

$$h(z) = \int_0^\infty e^{-tz} f(t)\, dt \quad (\text{Re } z > \alpha).$$

The Laplace transform of f is related to the Fourier transform of the functions $\exp(ty) \cdot f(t)$ for $y = \text{Im } z < -\alpha$. This fact can be verified with the aid of the inversion formula (see Kolmogorov and Fomin [1], Section 8.6).

For continuous operator-valued functions, there is a standard way to define the proper and improper Riemann integrals. These integrals pave the way for the definition of the Laplace transform of an operator-valued exponential U in a similar manner to that of a numerical-valued function f.

Under the conditions formulated in Section 3.10.2,

$$(z\mathrm{I} + T)^{-1} = \int_0^\infty e^{-tz} U(t) \, dt \quad (\operatorname{Re} z > 0).$$

Indeed, differentiating vector-valued functions F, G given by

$$F(t) = U(t)v, \quad G(t) = e^{-tz} U(t)v \quad (t \geq 0, \, v \in X)$$

yields

$$\dot{F}(t) = -U(t) \cdot Tv,$$
$$\dot{G}(t) = -e^{-tz} U(t) \cdot (z\mathrm{I} + T)v.$$

Integrating, we obtain

$$(z\mathrm{I} + T)^{-1} w = \int_0^\infty e^{-tz} U(t) w \, dt, \quad w = (z\mathrm{I} + T)v,$$

which provides the required identity for $R(z, -T)$.

3.10.4. Stone's Theorem

If instead of a Banach space X we deal with a Hilbert space H, then in order to define operator-valued exponentials we can use integrals with respect to a projection-valued measure.

Let A be a vector subspace dense in H, $T : A \to H$ be a self-adjoint operator, and P be its spectral measure. Then the identity

$$U(t) = e^{itT} = E(e^{itx}) = \int e^{itx} \, dP \quad (t \in \mathbb{R})$$

defines an exponential with the coefficient iT. The operators $U(t)$ constitute a group

$$U(0) = \mathrm{I}, \quad U(s + t) = U(s) \cdot U(t) \quad (s, t \in \mathbb{R}).$$

The function U with values $U(t)$ is strongly continuous. Finally, $U(t) f \in A$ for all $f \in A$ and $t \in \mathbb{R}$. All these properties follow from the properties of the integrals.

Since the imaginary exponential is bounded, the operators $U(t)$ are bounded and defined on the whole H. Furthermore

$$U^*(t) = (e^{itx}) = E^*(e^{-itx}) = U(-t) = U^{-1}(t).$$

The operator $U(t)$ is unitary.

3.10. Operator-Valued Exponential

It is not hard to prove also that

$$iTf = \lim_{t \to 0} [t^{-1}(e^{-itT} - I)f] \quad (f \in A).$$

In this context, iT is called the infinitesimal generator of the group $(U(t))$.

Consider also the algebra $\mathcal{B} = \mathcal{B}(H)$ of bounded linear operators on H. The following theorem holds (see Weidmann [8], Section 7.6).

THEOREM (Stone). *If a function $U : \mathbb{R} \to \mathcal{B}$ is strongly continuous, its values $U(t)$ are unitary, and*

$$U(0) = I, \quad U(s+t) = U(s) \cdot U(t) \quad (s, t \in \mathbb{R}),$$

then there exists a unique densely defined self-adjoint operator T on H such that

$$U(t) = e^{itT} \quad (t \in \mathbb{R}).$$

The content of this section can be briefly put in the following statement: unitary operators constitute a one-parameter unitary group if and only if they are values of an imaginary operator-valued exponential.

3.10.5. Evolution Equations

Stone's theorem enables an easy solution of the operator differential equation

$$\dot{U} = iT \cdot U \quad (1)$$

with the coefficient iT, subject to the condition $U(0) = I$. The only solution to meet these conditions is the exponential

$$U(t) = e^{itT} \quad (t \in \mathbb{R}).$$

Indeed, using the definitions and the rules of integration with respect to a projection-valued measure, we obtain

$$\dot{U}(t) = \lim_{\Delta \to 0} (\Delta^{-1}[U(t+\Delta) - U(t)]) = \lim_{\Delta \to 0} E(\Delta^{-1}[e^{i(t+\Delta)x} - e^{itx}])$$

$$= E(ixe^{itx}) = E(ix) \cdot E(e^{itx}) = iT \cdot U(t) \quad (t \in \mathbb{R}).$$

In addition, $U(0) = I$.

Let V be an arbitrary solution of equation (1) and $V(0) = I$. Then $W = U - V$ is also a solution and $W(0) = O$. Consider $f \in H$ and a function

$\varphi : \mathbb{R} \to \mathbb{R}$ given by

$$\varphi(t) = \|W(t)f\|^2 = W(t)f \cdot W(t)f \quad (t \in \mathbb{R}).$$

Observe that

$$\Delta^{-1}[\varphi(t + \Delta) - \varphi(t)] = W(t + \Delta)f \cdot \Delta^{-1}[W(t + \Delta) - W(t)]f$$
$$+ \Delta^{-1}[W(t + \Delta) - W(t)]f \cdot W(t)f$$

for $\Delta \neq 0$. Therefore,

$$\dot{\varphi}(t) = 2\operatorname{Re}[i(g \cdot Tg)], \quad g = W(t)f \quad (t \in \mathbb{R}).$$

Since T is self-adjoint, the product $g \cdot Tg$ is real-valued, and $\dot{\varphi}(t) = 0$ identical. Thus, φ is constant. By the condition, $\varphi(0) = 0$, so $\varphi(t) = 0$ and $U(t) - V(t) = W(t) = O$ for all $t \in \mathbb{R}$. The required solution is unique.

Observe that since $U(A) \subseteq A$, the right-hand side of equation (1) is defined on $A = \operatorname{Dom} T$. Let $f \in A$ and $u(t) = U(t)f$. Then the operator equation (1) is equivalent to the vector equation

$$\dot{u}(t) = iT \cdot u(t), \qquad (2)$$

and the condition $U(0) = I$ is equivalent to $u(0) = f$.

EXAMPLE 1. Let $H = \mathcal{L}^2(\mathbb{R}, dx)$ and T be the operator of multiplication by independent variable x

$$Tf = x \cdot f \quad (f, xf \in H).$$

Then, it is easy to verify that

$$e^{itT}f = e^{itx}f \quad (t \in \mathbb{R}, f \in H).$$

Thus, the exponential is a miltiplication operator.

EXAMPLE 2. Let $H = \mathcal{L}^2(\mathbb{R}, dx)$ and T be the differential operator

$$Tf = i^{-1} \cdot f' \quad (f, f' \in H).$$

Then it is easy to verify that

$$e^{itT}f(x) = f(x + t) \quad (t, x \in \mathbb{R}; f \in H).$$

The exponential is a translation operator.

The operator exponentials are an efficient tool in solving the abstract

3.10. Operator-Valued Exponential

Cauchy problem for a wide class of *evolution equations*

$$\dot{U}(t) = T(t) \cdot U(t) \tag{3}$$

with variable operator coefficients $T(t)$ (see Yošida [4], Chapters 14, and Balakrishnan [10], Chapter 4).

Consider a Banach space Y, an interval $[0, 1] \subseteq \mathbb{R}$, a vector subspace A dense in Y, and a family of closed linear operators $T(t) : A \to Y$ ($t \in [0, 1]$). Rewrite equation (3) in vector form

$$\dot{u}(t) = T(t) \cdot u(t) \quad (0 \leq t \leq 1). \tag{4}$$

Here $u(t) = U(t) \cdot f$ for a vector f from the common domain of operator in the right-hand side of equation (3). Sometimes the solution of equation (4) is sought only for $t \neq 0$. This constraint allows solutions satisfying the condition $u(0) = f$ for any $f \in Y$, not necessarily from the domain of the operators under consideration.

We now formulate a theorem about the solution of Cauchy's problem for equation 4.

Let $f \in Y$, $T(0) = T$ and $\mathcal{R}(t) = \mathcal{R}(T(t))$,

$$R(\lambda, t) = R(\lambda, T(t)) \quad (\lambda \in \mathcal{R}(t)).$$

Define the sum F of the iterated kernels K_n ($n = 1, 2, \ldots$):

$$K_1(t, r) = [T(t) - T(r)]e^{(t-r)T(r)} \quad (r \leq t),$$
$$K_1(t, r) = 0 \quad (t < r);$$
$$K_n(t, r) = \int_0^1 K_1(t, s) \cdot K_{n-1}(s, r) \, ds \quad (n \geq 2);$$
$$F(t, s) = \Sigma K_n(t, s) \quad (n \geq 1).$$

Make some additional assumptions:

(a) $Z = \{0\} \cup \{z : |\arg z| < \theta\} \subseteq \mathcal{R}(t)$ for all $t \in [0, 1]$ and some $\theta > \pi/2$;

(b) $R(z, \cdot)$ is strongly continuous uniformly relative to z on each compact set in the sector Z, and there are $\alpha > 0$, $\beta > 0$ such that

$$\|R(z, t)\| \leq \alpha(|z| - \beta)^{-1} \quad (z \in Z, |z| > \beta, t \in [0, 1]);$$

(c) for some $\gamma > 0$ and all $r, s, t \in [0, 1]$,

$$\|T(t) \cdot T^{-1}(r) - T(s) \cdot T^{-1}(r)\| \leq \gamma |t - s|.$$

Observe that the condition $0 \in \mathcal{R}(t)$ guarantees the existence of the inverse operator

$$T^{-1}(r) = R(0, r) : Y \to A.$$

The boundedness of the operator

$$S(t, r) = T(t) \cdot T^{-1}(r) : Y \to Y$$

follows from the closed graph theorem. The condition (c) means that the function $S(\cdot, r)$ satisfied Lipschitz's condition uniformly relative to r on $[0, 1]$.

THEOREM. *Under the above assumptions, the identity*

$$u(t) = e^{tT} f + \int_0^t e^{(t-s)T(s)} F(s, 0) f \, ds \tag{5}$$

for $0 < t \leq 1$ determines the unique solution of equation (4), *which meets the condition $u(0) = f$.*

A detailed proof of this theorem may be found in the book by Yošida [4]. It is based on the Fredholm method and consists in the justification of the transition from the differential equation

$$\dot{u}(t) = T(t) \cdot u(t) = T \cdot u(t) + (T(t) - T) \cdot u(t)$$

to the integral equation

$$u(t) = e^{tT} f + \int_0^t e^{(t-s)T} (T(s) - T) \cdot u(s) \, ds.$$

This equation is solved by the method of successive approximations. The result is equation (5).

NOTE. When solving evolution equations, one frequently has to differentiate and integrate vector-valued functions. The differentiation of such functions has been described in Chapter 1. Relevant proofs omitted in this text may be found in the book of Savel'ev [11, Part 4].

The integration of number-valued functions has been discussed in Chapter 2, and the skipped proofs may be found in Savel'ev [12]. The integration of vector-valued functions is reduced to the integration of scalar functions by a passage to coordinates (continuous linear functionals).

Part Two

Ill-Posed Problems

Chapter 4. CLASSICAL PROBLEMS

4.1. Mathematical Description of Physical Laws

A mathematical description of physical phenomena consists in associating with each point in space (or in a spatial region) a scalar or vector corresponding to the properties of the matter at this point. If a local property of the matter is described by one number then the resultant field is said to be scalar, while if the property is given by a vector then the result is a vector field. Examples of scalar fields may be density, temperature, and pressure. Examples of vector fields are velocities of particles in moving fluids, strength of a force field, in particular gravitational, magnetic, or electrical.

Physical laws relate fields and their derivatives. Stated differently, the fields satisfy differential equations or systems of differential equations. Some physical laws lead to integrodifferential equations.

Let us now consider examples of basic physical fields.

Gravitational Field. The equation of a gravitational field follows from Newton's law of gravitation which states that two particles of mass m and $\mathcal{M}$ separated by a distance r exert a gravitational force $\mathcal{F}$ of mutual attraction on one another,

$$\mathcal{F} = k \frac{m \mathcal{M}}{r^2}, \tag{1}$$

where k is the gravitational constant.

From equation (1) it follows that a mass distributed in space with density ρ induces a gravitational field which has a potential

$$\mathscr{F} = \operatorname{grad} \varphi,$$

and this potential φ satisfies Poisson's equation,

$$\Delta \varphi = 4\pi k \rho. \tag{2}$$

Electrostatic and Magnetostatic Fields. An electrostatic field, like a gravitational field, may be represented as a gradient of a scalar potential function u,

$$E = \operatorname{grad} u,$$

and u satisfies Poisson's equation,

$$\Delta u = \frac{4\pi}{\varepsilon} \rho. \tag{3}$$

Here ρ is the spatial electric charge density, and ε is the dielectric permittivity of the medium.

A static magnetic field obeys the following set of equations

$$\operatorname{curl} H = \frac{4\pi}{c} j,$$
$$\operatorname{div} H = 0, \tag{4}$$

where j is the vector of current density, and c is the speed of light.

In regions with no current, the magnetic field H may be written in terms of potential v obeying Laplace's equation,

$$H = \operatorname{grad} v, \quad \Delta v = 0.$$

Electromagnetic Field. Time-varying electromagnetic fields are described by Maxwell's field equations

$$\operatorname{curl} H = \frac{4\pi}{c} j + \frac{1}{c} \frac{\partial}{\partial t} \mathscr{D},$$
$$\operatorname{div} \mathscr{D} = 4\pi\rho,$$
$$\operatorname{curl} E = -\frac{1}{c} \frac{\partial}{\partial t} B, \tag{5}$$
$$\operatorname{div} B = 0,$$
$$j = \sigma(E + E').$$

4.1. Mathematical Description of Physical Laws

Here $B = \mu H$ and $\mathcal{D} = \varepsilon E$ are the magnetic and electric induction vectors, respectively, μ is the magnetic permeability of the medium, σ is the electrical conductivity, c is the speed of light, E' is the electric field due to external sources independent of H and E.

Observe that when ε and μ are constant and the fields are time-invariant, equations (3) and (4) follow from the system (5).

In the general case, the system (5) is quite complicated.

In a free space, in vacuum or in air, the medium parameters are $\mu = 1$, $\varepsilon = 1$, $\sigma = 0$. If we suppose also that the medium contains no spatial electric charges, that is, that $\rho = 0$, the system becomes

$$\operatorname{curl} H = \frac{1}{c} \frac{\partial E}{\partial t},$$

$$\operatorname{div} E = 0,$$

$$\operatorname{curl} E = -\frac{1}{c} \frac{\partial H}{\partial t}, \quad (6)$$

$$\operatorname{div} H = 0.$$

The last equation suggests that H has a vector potential A such that

$$H = \operatorname{curl} A. \quad (7)$$

Substituting equation (7) into the third line in the system (6) we obtain

$$\operatorname{curl}\left(\frac{\partial A}{\partial t} + cE\right) = 0 \quad (8)$$

whence follows

$$\frac{\partial A}{\partial t} + cE = -\operatorname{grad} \varphi. \quad (9)$$

Substituting equation (9) into the second equation of the system (6), we get

$$\Delta \varphi + \frac{\partial}{\partial t} \operatorname{div} A = 0. \quad (10)$$

The vector potential A is known to be determined by H up to a gradient of an arbitrary scalar potential.

If we pick out A, subject to the condition, such that

$$\text{div } A = -\frac{1}{c^2}\frac{\partial \varphi}{\partial t}, \tag{11}$$

then we obtain the scalar potential φ and the vector potential A which satisfy the wave equations:

$$\frac{1}{c^2}\frac{\partial^2 \varphi}{\partial t^2} = \Delta\varphi, \tag{12}$$

$$\frac{1}{c^2}\frac{\partial^2 A}{\partial t^2} = \Delta A.$$

Fluid Dynamics. Fluid dynamics deals with motions of gases and fluids. Denote V to be a field of local velocities of fluid particles, p the local pressure, and ρ the fluid density. If the fluid is ideal (no viscosity) and the flow is adiabatic (no heat transfer between different fluid elements), then the vector velocity field and the scalar pressure and density fields satisfy the following system of equations:

$$\frac{\partial V}{\partial t} + (V, \nabla)V = -\frac{1}{\rho}\text{grad } p, \tag{13}$$

$$\frac{\partial \rho}{\partial t} + \text{div}(\rho V) = 0, \tag{14}$$

$$p = a\rho^n. \tag{15}$$

Equation (13) is known as Euler's equation, it follows from the conservation law for the momentum. Equation (14) is called the equation of continuity. It is in charge of mass conservation. Equation (15) is the equation of state of the fluid.

Equation (13) may be recasted into a form involving the field V alone

$$\frac{\partial}{\partial t}\text{curl } V = \text{curl}[V \times \text{curl } V]. \tag{16}$$

By way of example we consider two cases when these equations of fluid dynamics reduce to simpler formulations.

(1) In incompressible fluids

$$\rho = \rho_0 = \text{const},$$

4.1. Mathematical Description of Physical Laws

and the velocity field is of the potential type

$$V = \text{grad } \varphi.$$

Equation (16) is satisfied automatically, and from equation (14) follows

$$\Delta \varphi = 0, \tag{17}$$

that is, the potential satisfies Laplace's equation.

(2) Small vibrations set in a fluid or gas are called *sound waves*. Their smallness may be represented by

$$p = p_0 + p_1,$$
$$\rho = \rho_0 + \rho_1,$$

where p_0, ρ_0 are constant, and p_1, ρ_1 are sufficiently small perturbations. Assume further that the velocity field is also small, so that in equations (13), (14), (15) we may neglect the square terms of the quantities V, p, ρ, their product, and the products of their derivatives. In this case equations (13)–(15) take the form

$$\frac{\partial V}{\partial t} = -\frac{1}{\rho_0} \text{grad } p,$$

$$\frac{\partial \rho_1}{\partial t} + \rho_0 \text{div } V = 0, \tag{18}$$

$$p_1 = a\rho_0^{n-1} \rho_1.$$

From these equations it follows that

$$\frac{\partial p_1}{\partial t} + a\rho_0^n \text{div } V = 0.$$

Suppose that the velocity field has a potential,

$$V = \text{grad } \varphi.$$

Then by equation (18) the function φ satisfies the wave equation

$$\frac{\partial^2 \varphi}{\partial t^2} + c^2 \Delta \varphi,$$

where $c = \sqrt{a\rho_0^{n-1}}$ is the velocity of sound.

Heat Equation. Let u be a temperature distribution in a medium, c be the specific heat capacity of that medium, and ρ be its density. The profile of the temperature field in the medium obeys Fourier's law:

$$W = -\lambda \, \text{grad} \, u,$$

where W is the vector of heat flux density, and λ is the thermal conductivity.

By Fourier's law, the function u satisfies the equation

$$\frac{\partial u}{\partial t} = \frac{1}{c\rho} \, \text{div} \, (\lambda \, \text{grad} \, u). \tag{19}$$

4.2. Classification of Differential Equations of the Second Order

Consider a linear differential equation of the second order in an n-dimensional space R^n

$$\sum_{j,\,k=1}^{n} a_{jk} \frac{\partial^2 u}{\partial x_j \partial x_k} + \sum_{j=1}^{n} b_j \frac{\partial u}{\partial x_j} + cu = f. \tag{1}$$

Introduce new independent variables ($y_1, ..., y_n$) related to ($x_1, ..., x_n$) by

$$y_k = \sum_{j=1}^{n} b_{kj} x_j.$$

The matrix $[b_{kj}]$ is assumed to be nonsingular.

In the new variables equation (1) takes the form

$$\sum_{p,\,q=1}^{n} a'_{pq} \frac{\partial^2 u}{\partial y_p \partial y_q} + \sum_{p=1}^{n} b'_p \frac{\partial u}{\partial y_p} + cu = f, \tag{1'}$$

where

$$a'_{pq} = \sum_{j,\,k=1}^{n} b_{jp} b_{kp} a_{jk},$$

$$b'_p = \sum_{j=1}^{n} b_{pj} b_j.$$

Define also the quadratic form

$$\sum_{j,\,k=1}^{n} a^0_{jk} \xi_j \xi_k, \tag{2}$$

4.2. Classification of Differential Equations

whose coefficients are the coefficients of equations a_{jk} taken at some point $(x_1^0, \ldots, x_n^0)$.

With the new variables

$$\xi_k = \sum_{j=1}^{n} b_{kj} \eta; \quad k = 1, \ldots, n,$$

this quadratic form becomes

$$\sum_{p,q} a_{pq}^{01} \eta_p \eta_q,$$

where

$$a_{pq}^{01} = \sum_{j,k=1}^{n} b_{jp} b_{kq} a_{jk}^{0}.$$

Thus, a linear change of variables alters the coefficients of the second derivatives in equation (1) at point $(x_1^0, \ldots, x_n^0)$. along with the coefficients of the associated quadratic form (2).

It is known that a change of variables can reduce a quadratic form to its canonical form, having

$$\begin{aligned} a_{pq}^{01} &= 0, \quad p \neq q, \\ a_{pp}^{01} &= 1, \quad p = 1, \ldots, m-1, \\ a_{pp}^{01} &= -1, \quad p = m, \ldots, n_1 - 1, \\ a_{pp}^{01} &= 0, \quad p = n_1, \ldots, n. \end{aligned} \quad (3)$$

The relations of parameters in the set (3) that result in popular equations at point $(x_1^0, \ldots, x_n^0)$. are summarized below.

if the set (3) is such that	then eq (1) is
(a) $m = n + 1$ or $m = 1, n_1 = n + 1$	elliptic
(b) $m = n, n_1 = n + 1$ or $m = 2, n_1 = n + 1$	hyperbolic
(c) $2 < m < m, n_1 = n + 1$	ultrahyperbolic
(d) $n_1 \leq n$	parabolic

If equation (1) is elliptic (hyperbolic, parabolic) at each point of a domain, then it is said to be elliptic (hyperbolic, parabolic) within the domain.

4.3. Elliptic Equations

Elliptic equations describe, first of all, stationary (time-invariable) fields. The classical elliptic equations are those associated with the names of Laplace,

$$\Delta u = 0 \tag{1}$$

and Poisson,

$$\Delta u = f. \tag{2}$$

Stationary fields in a medium with spatial gradients obey equations with variable coefficients. For example, the concentration field in a stationary diffusion process satisfies the equation

$$\text{div}(k \text{ grad } u) = 0, \tag{3}$$

where k is the diffusion coefficient.

Elliptic equations are also satisfied by the fields describing steady-state oscillatory processes. For instance, if the solution to the wave equation

$$\frac{\partial^2 u}{\partial t^2} = \Delta u$$

is sought in the form

$$u = e^{i\lambda t} v(x, y, z),$$

then v must satisfy the Helmholtz equation

$$\Delta v = \lambda^2 v = 0. \tag{4}$$

As noted earlier, the potential of the velocity field of an ideal incompressible liquid satisfied Laplace's equation with respect to spatial coordinates at any fixed instant of time.

The classical problems for elliptic equations are boundary-value problems in which one seeks within a domain a solution which would satisfy a certain condition on the boundary of this domain. Examples of such boundary-value problems follow.

Dirichlet Problem for Laplace's Equation. Given a bounded region $\mathcal{D}$ with a smooth boundary surface S in a three-dimensional space and a continuous function f specified on S. The problem is to determine the solution to Laplace's equation (1) within $\mathcal{D}$, satisfying the condition

$$u|_S = f.$$

4.3. Elliptic Equations

Neumann Problem. The problem is to determine the solution to Laplace's equation within $\mathcal{D}$, satisfying the condition

$$\left.\frac{\partial u}{\partial n}\right|_S = f.$$

Here $\partial u/\partial n|_S$ is the directional derivative of u along the normal to the boundary S.

General Boundary-Value Problem. Given are a vector l and a function a on S. The problem is to determine the solution of Laplace's equation (1) within $\mathcal{D}$, subject to the condition (on S)

$$l \cdot \operatorname{grad} u + au = f.$$

In Dirichlet's problem u is assumed to be continuous in a closed region $\overline{\mathcal{D}}$, in the Neumann and general problems u is assumed to be continuously differentiable in the closure $\overline{\mathcal{D}}$. These boundary-value formulations are considered for other elliptic equations.

The theory of boundary-value problems is concerned with the uniqueness and existence of the solution and with its continuous dependence upon data.

That the solution to the Dirichlet problem for Laplace's equation is unique follows from the fundamental properties of solutions to Laplace's equation stated by the principle of the maximum.

Principle of the Maximum. A solution to Laplace's equation defined in a bounded domain $\mathcal{D}$ and continuous in the closure $\overline{\mathcal{D}}$ attains a maximum (and minimum) value on the boundary of $\mathcal{D}$. If a solution to Laplace's equation attains a maximum (or minimum) value at a point interior to $\mathcal{D}$, then this solution is constant throughout $\mathcal{D}$.

By this maximum modulus principle, if a solution of equation (1) vanishes on the boundary of $\mathcal{D}$,

$$u|_S = 0,$$

then it is identically zero. Since the problem is linear, this means that the solution is unique.

The maximum principle in a somewhat modified formulation holds also for solutions of elliptic equations in a general form.

Consider the equation

$$\sum_{j,k=1}^{n} a_{jk} \frac{\partial^2 u}{\partial x_j \partial x_k} + \sum_{j=1}^{n} b_j \frac{\partial u}{\partial x_j} + cu = 0 \qquad (5)$$

in a bounded region $\mathcal{D}$ of an n-dimensional space. The coefficients a_{jk}, b_j, c are assumed to be continuous. Suppose also that

$$\sum_{j,k=1}^{n} a_{jk} \xi_j \xi_k \geq \delta \sum_{j=1}^{n} \xi_j^2, \quad \delta > 0, \quad c < 0. \tag{6}$$

The first of these conditions means that equation (5) is elliptic. For solutions of equation (5) the following theorem holds

THEOREM (Maximum Principle). *A solution of equation (5) cannot attain its positive maximum and negative minimum values in the region $\mathcal{D}$.*

This principle implies that the solution to the Dirichlet problem for equation (5) is unique.

In Neumann's problem, the uniqueness follows from Green's formulas.

THEOREM (Green's Formulas). *Let u, v be twice continuously differentiable in a bounded domain $\mathcal{D}$ of a three-dimensional space and continuously differentiable in the closure $\overline{\mathcal{D}}$. Then*

$$\int_{\mathcal{D}} (u'_x v'_x + u'_y v'_y + u'_z v'_z) \, dx\, dy\, dz + \int_{\mathcal{D}} v \cdot \Delta u\, dx\, dy\, dz = \int_{S} v \frac{\partial u}{\partial n} d\sigma, \tag{7}$$

$$\int_{\mathcal{D}} (u \cdot \Delta v - v \cdot \Delta u)\, dx\, dy\, dz = \int \left(u \frac{\partial v}{\partial n} - v \frac{\partial u}{\partial n} \right) d\sigma.$$

Let u now be a solution of Laplace's equation. Setting $v = u$ in the first line of equation (7) yields

$$\int_{\mathcal{D}} (u'^2_x + u'^2_y + u'^2_z)\, dx\, dy\, dz = \int_{S} u \frac{\partial u}{\partial n} d\sigma. \tag{8}$$

Given

$$\left. \frac{\partial u}{\partial n} \right|_S = 0,$$

this implies

$$\operatorname{grad} u = 0,$$

so

$$u = \operatorname{const}.$$

Thus, for the Neumann problem, the following assertion holds.

4.3. Elliptic Equations

THEOREM (Uniqueness, Neumann problem). *If u_1, u_2 are the solutions of Laplace's equation and*

$$\left.\frac{\partial u_1}{\partial n}\right|_S = \left.\frac{\partial u_2}{\partial n}\right|_S,$$

then

$$u_1 - u_2 = \text{const.}$$

Now we proceed to the existence of solutions to boundary-value problems.

THEOREM (Existence, Dirichlet problem). *Let f be a continuous function given on a smooth surface[1] S bounding a domain $\mathcal{D}$. Then there exists a function U twice continuously differentiable in $\mathcal{D}$ and continuous in the closure $\overline{\mathcal{D}}$, which satisfies the conditions*

$$\Delta U = 0, \quad u|_S = f.$$

It should be noted that the existence theorem for the solution of the Dirichlet problem holds under weaker constraints on the surface S, specifically for the so-called Lyapunov surfaces.

THEOREM (Existence, Neumann Problem). *Let f be a continuously differentiable function, given on S and satisfying the condition*

$$\int_S f\,d\sigma = 0.$$

Then there exists a function U twice continuously differentiable in $\mathcal{D}$ and continuously differentiable in the closure $\overline{\mathcal{D}}$, satisfying the conditions

$$\Delta U = 0, \quad \left.\frac{\partial u}{\partial n}\right|_S = f.$$

For some geometrically simple domains, the solutions of the Dirichlet and Neumann problems can be deduced in explicit form. These expressions for a ball and a half space are known as Poisson's integral formulas. In such cases, to prove the existence theorems it suffices to verify that the functions defined by these formulas do satisfy Laplace's equation and the boundary condition.

For domains of a general form, other methods are known to approach

---

[1] By a smooth surface we imply a surface which is defined by a twice continuously differentiable function in a neighborhood of any of its points.

the proof of the existence theorems. One of the classical methods consists in the reduction of the boundary-value problems to integral equations.

Solutions of boundary-value problems are sought in the form of potentials. The solution to the Dirichlet problem is sought in the form of a "double-layer" potential:

$$u = \int_S \frac{\partial}{\partial n}\left(\frac{1}{r}\right)\varphi\, d\sigma,$$

and the solution of the Neumann problem is sought in the form of the "single-layer" potential:

$$u = \int_S \frac{1}{r}\varphi\, d\sigma.$$

Solving a boundary-value problem is now equivalent to solving the Fredholm equation of the second kind for the density φ

$$\varphi + \int_S G\varphi\, d\sigma = \frac{1}{2\pi} f, \qquad (9)$$

which is approached via the solution of the *associated homogeneous integral equation*

$$\psi + \int_S G^* \psi\, d\sigma = 0. \qquad (10)$$

There are two situations.

(1) Equation (10) has only the trivial solution $\psi = 0$. Then, the solution of equation (9) is unique and exists for any right-hand side.

(2) Equation (10) has a finite number of linearly independent solutions

$$\psi_k + \int_S G^* \psi_k\, d\sigma = 0, \quad k = 1, \ldots, n.$$

Then the homogeneous equation (9) (with $f = 0$) also has n linearly independent solutions. The necessary and sufficient condition for a solution of equation (9) to exist is

$$\int_S f \psi_k\, d\sigma = 0, \quad k = 1, \ldots, n.$$

These statements are known as the Fredholm theorems.

An analysis of boundary-value problems consists in proving that the associated integral equation is equivalent to some boundary-value problem which is referred to as dual to the initial problem. From the uniqueness of the solution of the dual boundary-value problem follows the uniqueness

4.4. Hyperbolic and Parabolic Equations

of the solution of the associated integral equation, and, by the Fredholm theorems, the existence of the solution of the integral equation corresponding to the initial boundary-value problem.

Boundary-value problems for the Helmholtz equation (4) are reduced to integral equations with the aid of the following potentials

$$\int_S \frac{\cos(\lambda r)}{r} \varphi \, d\sigma,$$

$$\int_S \frac{\partial}{\partial n}\left(\frac{\cos(\lambda r)}{r}\right) \varphi \, d\sigma.$$

Boundary-value problems for elliptic equations of general form with variable coefficients can also be reduced to Fredholm integral equations of the second kind, but the proof of the existence of the corresponding potentials is equivalent to proving the existence of a solution to the initial boundary-value problem.

The classical method of proving the existence theorems concerning the boundary-value problems for general elliptic equations is to reduce these problems to calculus of variations problems of finding functional extrema.

4.4. Hyperbolic and Parabolic Equations

The simplest equation of hyperbolic type is the wave equation, introduced by d'Alembert for the displacement of a string fixed at its ends and with a given initial displacement

$$\frac{\partial^2 u}{\partial t^2} - \frac{\partial^2 u}{\partial x^2} = 0. \tag{1}$$

A solution to this equation $u(x, t)$ describes small vibrations of the string, i.e., deviations of the string about the equilibrium position.

This equation is handled by solving the Cauchy problem, namely, to find a solution of equation (1) subject to the conditions

$$u(x, 0) = f_0(x),$$
$$u'_t(x, 0) = f_1(x). \tag{2}$$

The theory of the Cauchy problem (2) for equation (1) evolves rather easily from the general representation of solutions to equation (1). It demonstrates that any solution of equation (1) can be represented in the form

$$u(x, t) = \varphi(x - t) + \psi(x + t), \tag{3}$$

where φ, ψ are some twice continuously differentiable functions.

The representation (3) yields the formula due to d'Alembert which supplies the solution to the Cauchy problem

$$u(x, t) = \frac{1}{2}[f_0(x - t) + f_0(x + t)] + \frac{1}{2}\int_{x-t}^{x+t} f_1(\xi)d\xi, \quad (4)$$

and also proves its uniqueness.

The Cauchy problem for equation (1) occurs very often in practical situations and also in intermediate or mixed problems, in which the Cauchy data are augmented by some boundary conditions. Examples of simplest mixed problems are:

(a) vibrations of a string with fixed ends

$$u(0, t) = u(l, t) = 0;$$

(b) vibrations of a string with free ends

$$u'_x(x, t) = u'_x(l, t) = 0.$$

Solutions to mixed problems can be obtained by resorting to the representation (3) and by separation of variables (Fourier's method).

The general hyperbolic equation in two independent variables can be reduced by a change of variables to the form

$$\frac{\partial^2 u}{\partial t^2} - \frac{\partial^2 u}{\partial x^2} + a\frac{\partial u}{\partial t} + b\frac{\partial u}{\partial x} + cu = 0. \quad (5)$$

The Cauchy problem for this equation with variable coefficients reduces to a special case of the Fredholm integral equations of the second kind, known as Volterra's equations. The theory of integral equations tells us that the unique solution to a Volterra equation always exists and can be obtained by the method of sequential approximations.

Solutions to Cauchy problems in two, three or more spatial variables are given explicitly by the Kirchhoff formulas.

The Cauchy and mixed problems for hyperbolic equations can be reduced to Volterra's integral equations of the second kind. Uniqueness is established with the aid of the energy inequalities.

The simplest parabolic equation is the heat equation, in one dimension,

$$\frac{\partial u}{\partial t} - \frac{\partial^2 u}{\partial x^2} = 0. \quad (6)$$

The Cauchy problem for this equation seeks to determine a solution

4.5. Concept of Well-Posedness

satisfying the condition

$$u(x, 0) = f(x). \tag{7}$$

To guarantee the uniqueness of the solution, certain restrictions are imposed on the behavior of $u(x, t)$ as $|x| \to \infty$.

In the simplest case of

$$\lim_{|x| \to \infty} u(x, t) = 0,$$

the uniqueness follows from the maximum principle.

The solution to the problem (7) is given by Poisson's formula

$$u(x, t) = \frac{1}{2\sqrt{\pi t}} \int_{-\infty}^{\infty} \exp[-(x - \xi)^2 / 4t] f(\xi) d\xi. \tag{8}$$

In mixed problems for equation (6), in addition to the initial condition (7), boundary conditions such as

$$u(0, t) = u(l, t) = 0,$$

or

$$u_x(0, t) = u_x(l, t) = 0$$

are satisfied.

The theorems on uniqueness, existence and continuous dependence on the data (stability) have been established also for the Cauchy and mixed problems of general parabolic equations with variable coefficients.

4.5. Concept of Well-Posedness

The concept of well-posed initial-value problems was formulated earlier this century by the prominent French mathematician Hadamard.

A problem is said to be well-posed (in the sense of Hadamard) if it has the following properties:

(i) *the solution to the problem exists*;
(ii) *the solution is unique*;
(iii) *the solution depends continuously on the data.*

The first requirement means that the problem must not have redundant data which render it overspecified.

The requirement (ii) means that the data must be sufficient to adequating the problem.

The requirement (iii) is associated with the following circumstance. If the problem is related to a physical phenomenon, its data cannot be considered known exactly. We can only assume that the exact values are approximated arbitrarily closely. Consequently, if the solution does not depend continuously on the data, it is not actually determined.

These well-posedness requirements need a refinement. The theory of boundary-value problems treats both the solution and the data as elements of some functional spaces, and the well-posedness conditions are formulated as follows.

(i) *The solution of the problem exists for any data from a closed subspace in a linear normed space* $C^{(K)}$, L_p, $W_p^{(l)}$, *etc., and it belongs to one of these spaces.*

(ii) *The solution is unique in one of these spaces.*

(iii) *Infinitesimal variations of data* (*in the respective space*) *produce infinitesimal variations in the solution* (*in the space where the solution is considered*).

Establishing the theorems on the uniqueness of a solution, its existence, and continuous dependence on the data is equivalent to proving that the problem is well-posed.

For elliptic equations, well-posed problems are those in which the boundary conditions are given on the whole boundary of the domain where the solution needs be defined.

For hyperbolic and parabolic equations, well-posed problems are Cauchy and mixed problems in which the data are given on a part of the domain's boundary. From the standpoint of a physical problem described by the mathematical model concerned, the Cauchy data mean that we know the state of the field at the initial time instant. The problem is to determine the field at later time.

Having formulated the concept of well-posedness, Hadamard gave an example of an ill-posed problem for differential equations which, he believed, did not correspond to any real physical setting. It was the Cauchy problem for Laplace's equations.

This problem, along with some other ill-posed problems in the sense of Hadamard, will be discussed in the forthcoming chapter. Here we confine ourselves to noting that the Cauchy problem for Laplace's equation, and also a number of other ill-posed problems, satisfy the conditions of the classical Cauchy-Kowalewski theorem. This theorem states that the unique solution to the Cauchy problem for Laplace's equation exists in the space of analytic functions.

Chapter 5. ILL-POSED PROBLEMS

5.1. Ill-Posed Cauchy Problems

Any problem that is not well-posed, that is, does not possess at least one of the properties of the well-posed problem, is said to be ill-posed. However, the theory of ill-posed problems pays most attention to the third requirement.

In this section we examine examples of ill-posed Cauchy problems for second-order equations.

Cauchy Problem for Laplace's Equation. The simplest example of ill-posed problems for Laplace's equation is a mixed problem in two dimensions.

The problem is to determine in a rectangle $\{0 \leq x \leq \pi, 0 \leq y \leq y_0\}$ a function of two variables $u(x, y)$ satisfying the following conditions

$$\Delta u = 0,$$
$$u(0, y) = u(\pi, y) = 0,$$
$$u(x, 0) = f_0(x),$$
$$u_y(x, 0) = f_1(x).$$
(1)

The solution of Laplace's equation satisfying the homogeneous conditions on the edges of the semi-indefinite stripe is representable in the form

$$u(x, y) = \sum_{k=1}^{\infty} \sin kx (a_k e^{ky} + b_k e^{-ky}).$$

From the initial conditions

$$a_k = \frac{1}{\pi} \left(\int_0^\pi f_0(x) \sin kx\, dx + \frac{1}{k} \int_0^\pi f_1(x) \sin kx\, dx \right),$$

$$b_k = \frac{1}{\pi} \left(\int_0^\pi f_0(x) \sin kx\, dx - \frac{1}{k} \int_0^\pi f_1(x) \sin kx\, dx \right).$$

Thus, the solution to the problem (1) is unique.

These formulas suggest that for a solution to the problem (1) to exist it is no longer sufficient to have the data $f_0(x)$, $f_1(x)$ belong to a functional space whose norm is defined with a finite number of derivatives.

Now we present Hadamard's example for the problem (1).

Consider the solutions to Laplace's equation

$$u_n(x, y) = a_n e^{ny} \sin nx.$$

The Cauchy data for $u_n(x)$ are

$$f_{0n}(x) = a_n \sin nx,$$
$$f_{1n}(x) = na_n \sin x.$$

Obviously, an appropriate choice of n and a_n may render the Cauchy data for u_n arbitrarily small in the norm, while the function will be arbitrarily large for any fixed y.

Let us formulate the Cauchy problem for Laplace's equation in the general form.

Let $\mathcal{D}$ be a bounded domain in an n-dimensional space, S be a smooth boundary of $\mathcal{D}$, S_1 be part of S, and f_0, f_1 be functions specified on S_1.

The problem is to determine a function u obeying Laplace's equation $\mathcal{D}$ and satisfying the boundary conditions

$$\Delta u = 0,$$
$$u|_{S_1} = f_0, \qquad (2)$$
$$\left.\frac{\partial u}{\partial n}\right|_S = f_1.$$

The problem (2) possesses the same properties as the problem (1). The solution is unique, its existence is not guaranteed by the fact that the data belong to either of functional spaces indicated above, and there is no continuous dependence of the solution on the data.

The Cauchy problem for general elliptic equations with variable coefficients has similar properties.

Cauchy Problem for Backward Heat Equation. The problem is to determine within the rectangle

$$0 \leq x \leq \pi,$$
$$0 \leq t \leq t_0,$$

the function of two variables $u(x, t)$ satisfying the conditions

$$\frac{\partial u}{\partial t} = -\frac{\partial^2 u}{\partial x^2},$$
$$u(0, t) = u(\pi, t) = 0, \qquad (3)$$
$$u(x, 0) = f(x).$$

5.1. Ill-Posed Cauchy Problems

The solution to the problem (3) is of the form

$$u(x, t) = \sum_{k=1}^{\infty} a_k e^{k^2 t} \sin kx,$$

$$a_k = \frac{2}{\pi} \int_0^\pi f(x) \sin kx\, dx.$$

From this formula follows the uniqueness of the solution and Hadamard's example

$$u_n(x, t) = a_n \exp(n^2 t) \sin nx.$$

Cauchy Problem for Heat Equation with Data on Time-Like Manifold. The problem is to determine within the rectangle

$$0 \leq x \leq l, \qquad 0 \leq t \leq t_0,$$

a function $u(x, t)$ subject to the conditions

$$\frac{\partial u}{\partial t} = \frac{\partial^2 u}{\partial x^2},$$

$$u(0, t) = f_0(t), \qquad (4)$$

$$u_x(0, t) = f_1(t).$$

This problem satisfies the conditions of the Cauchy-Kowalewski theorem. The solution is unique, but the proof is not so simple as it was in problem (3). Hadamard's example for this case is

$$u_n(x, t) = a_n e^{nx} \cdot \sin(2n^2 t + nx).$$

Cauchy Problem for Wave Equation with Data on Time-Like Manifold. Determine the function of three variables $u(x, y, t)$ in the region

$$0 \leq x, \qquad 0 \leq y \leq \pi, \qquad 0 \leq t \leq \pi,$$

subject to the conditions

$$\frac{\partial^2 u}{\partial t^2} = \Delta u,$$

$$u(x, y, 0) = u(x, y, \pi) = 0,$$

$$u(0, y, t) = f_0(y, t), \qquad (5)$$

$$u_x(0, y, t) = f_1(y, t).$$

The solution to the problem (5) is unique. The corresponding Hadamard's example is

$$u_x(x, y, t) = a_n e^{nx} \sin(\sqrt{5}\,ny) \sin(2nt).$$

5.2. Analytic Continuation and Interior Problems

There are several settings of the analytic continuation problem. We give two of them for functions of a complex variable.

Formulation 1. Let $f(z)$ be an analytic function of complex variable regular within a bounded region $\mathcal{D}$ on the complex plane and continuous in the closure $\overline{\mathcal{D}}$,

$$|f(z)| \leq c, \quad z \subset \mathcal{D}, \tag{1}$$

Γ be the boundary of $\mathcal{D}$, Γ_1, Γ_2 be parts of $\Gamma: \Gamma_1 \cup \Gamma_2 = \Gamma$, $\Gamma_1 \cap \Gamma_2 = \emptyset$. Suppose also that $f(z)$ is specified on Γ_1, and the problem is to determine $f(z)$ within the interior part of $\mathcal{D}$. If $\Gamma = \Gamma_1$, the solution is provided by the Cauchy integral,

$$f(z) = \frac{1}{2\pi i} \int_\Gamma \frac{f(\xi)}{\xi - z}\, d\xi.$$

We intend to show that if $\Gamma_1 \neq \Gamma$, then the problem of analytic continuation of $f(z)$ is equivalent to the Cauchy problem for Laplace's equation.

1. Suppose that a harmonic function u and its normal derivative u_n are prescribed over Γ_1. Denote by $f(z)$ the function

$$f(z) = u(z) + iv(z), \quad (z = z + iy),$$

where v is conjugated to u. It is known that on Γ_1

$$v(z) = \int_{z_0}^{z} \frac{\partial}{\partial n} u(z)\, ds + C_1,$$

where z_0 is one of the ends of Γ_1. Thus, if u, u_n are prescribed on Γ_1, then the analytic function $f(z)$ may be deemed given on Γ_1.

2. Suppose that a function $f(z) = u(z) + iv(z)$ is prescribed on Γ_1 with u, v being conjugate harmonic functions. From the Cauchy-Riemann conditions it follows that

$$\left.\frac{\partial u}{\partial n}\right|_{\Gamma_1} = \left.\frac{\partial v}{\partial s}\right|_{\Gamma_1}.$$

5.2. Analytic Continuation and Interior Problems

where v_s is the derivative of v along Γ_1. Taking the derivative of $f(z)$ and $\bar{f}(z)$ along Γ_1 yields

$$\left.\frac{\partial u}{\partial n}\right|_{\Gamma_1} = \frac{1}{2}\frac{\partial}{\partial s}(f - \bar{f})\bigg|_{\Gamma_1},$$

and we arrive at the Cauchy data for $u(x, y)$.

Thus, if $\Gamma_1 \neq \Gamma$, the problem of analytic continuation is equivalent to the Cauchy problem for Laplace's equation and so is ill-posed in the classical sense.

The fact that the solution to this problem of analytic continuation is unique is proved in a textbook on functions of a complex variable.

Formulation 2. Suppose that the values of an analytic function $f(z)$ are given not on Γ_1, but rather on some set $\mathcal{M}$ lying within the domain $\mathcal{D}$. If $\mathcal{M}$ has at least one limit point in $\mathcal{D}$, then the solution of the analytic continuation problem is unique. This assertion is a classical theorem of the theory of analytic functions. If $\mathcal{M}$ is a sequence having one (or a finite number of) limit points $\mathcal{D}$, then the analytic continuation problem is more unstable in a certain sense than even the one in the Cauchy problem for Laplace's equation.

The second formulation of the problem of analytic continuation is a special case of a class of problems for differential equations, which we will call *interior problems*. By these we mean the following type: determine a solution to some equation (or a set of equations), regular within a domain $\mathcal{D}$, by the values of solution (possibly its derivatives) specified on a set $\mathcal{M}$ interior to $\mathcal{D}$.

We illustrate this point by a sample interior problem for the diffusion equation in two dimensions.

Let $u(x, y, t)$ be the solution to the equation

$$\frac{\partial u}{\partial t} = \Delta u$$

regular within the cylinder

$$x^2 + y^2 < 1,$$
$$0 < t < t_0$$

and let $\{x_k, y_k\}$ be a sequence of points within the unit circle such that

$$\lim_{k \to \infty} x_k = \lim_{k \to \infty} y_k = 0.$$

The problem is to determine the function u, given that

$$u(x_k, y_k, t) = f_{0k}(t),$$
$$u'_x(x_k, y_k, t) = f_{1k}(t),$$
$$u'_y(x_k, y_k, t) = f_{2k}(t).$$

5.3. Physical Problems Leading to Ill-Posed Formulations

Many physical fields have potentials obeying Laplace's equation. Suppose that we possess the measurement data of a two-dimensional field along some curve, or of a three-dimensional field on a surface and we need to determine the field in the region bounded by this curve or this surface. This physical setting leads to the Cauchy problem for Laplace's equation. Indeed, the gradient of a potential determines the potential itself up to a constant.

A similar situation takes place also for fields whose potentials satisfy the Helmholtz equation or other elliptic equations. Given a data on a curve or a surface over a time interval for a field obeying the wave equation we eventually arrive at the Cauchy problem for the wave equation with the data on a time-like equation manifold. Similar physical settings lead to Cauchy problems for parabolic equations with the data on a time-like manifold.

The theory of differential equations tells us that the solutions to elliptic equations with constant coefficients are analytic functions. The solutions to parabolic equations within the domain of regularity are also analytic in spatial variables. Thus, the evaluation of physical fields within a domain by measurement data on some sets leads to problems of analytical continuation.

An example may be the following physical problem which reduces to an ill-posed Cauchy problem for backward heat equation.

Consider a rod of material with a difference of temperature or concentration along its axis. The processes of temperature variation and diffusion are described by the heat equation. The role of Cauchy data can be played by the readings of instruments measuring the profile of temperature or concentration at a given time. If we are interested in the future distribution of temperature or concentration, we formulate the classical Cauchy problem for heat equation. However, if we are interested in the history of the process, i.e., in what the distribution was like in the past, proceeding only along these lines we arrive at the aforementioned classical ill-posed problem.

5.4. Reduction of Ill-Posed Problems to Integral Equations

Classical well-posed boundary-value problems for differential equations can be reduced to Fredholm integral equations of the second kind. Some boundary-value problems are reducible to singular integral equations which obey Noether's theorem.

Ill-posed problems, exemplified earlier, can be reduced to Fredholm integral equations of the first kind whose properties differ essentially from those of Fredholm's equations of the second kind, and those of the classical singular equations. In this section we present examples of such a reduction.

Cauchy Problem for Backward Heat Equation. The problem is to determine within a stripe

$$-\infty < x < \infty, \qquad 0 \leq t \leq t_0$$

a bounded function $u(x, t)$ satisfying the conditions

$$\frac{\partial u}{\partial t} = -\frac{\partial^2 u}{\partial x^2}, \qquad (1)$$

$$u(x, 0) = f(x),$$

where $f(x)$ is assumed to be bounded and continuous.

Suppose that the solution to the problem (1) exists, and that it is continuous in the bounded stripe.
Denote

$$\varphi(x) = u(x, t_0).$$

Then, by Poisson's formula,

$$u(x, t) = \frac{1}{2\sqrt{\pi(t_0 - t)}} \int_{-\infty}^{\infty} \exp\left(-\frac{(x - \xi)^2}{4(t_0 - t)}\right) \varphi(\xi) d\xi,$$

hence, the solution of (1) is equivalent to the solution of the following integral equation for $\varphi(x)$:

$$\frac{1}{2\sqrt{\pi t_0}} \int_{-\infty}^{\infty} \exp\left(-\frac{(x - \xi)^2}{4t_0}\right) \varphi(\xi) d\xi = f(x).$$

Interior Problem for Laplace's Equation on Half Plane. We are to determine on the half plane $y \geq 0$, a bounded function $u(x, y)$ subject to

the conditions
$$\Delta u = 0,$$
$$u(x, y_0) = f(x), \quad y_0 > 0.$$

Suppose that a solution exists. Denote
$$\varphi(x) = u(x, 0).$$

Then, by Poisson's formula,
$$u(x, y) = \frac{1}{\pi} \int_{-\infty}^{\infty} \frac{y}{y^2 + (x - \xi)^2} \varphi(\xi) d\xi$$

and the solution of the problem (2) is equivalent to the solution of the integral equation
$$\frac{1}{\pi} \int_{-\infty}^{\infty} \frac{y_0}{y_0^2 + (x - \xi)^2} \varphi(\xi) d\xi = f(x).$$

Cauchy Problem for Laplace's Equation. Let $\mathcal{D}$ be a domain in three dimensions bounded by a smooth surface S; S_1, S_2 be parts of S such that
$$S = S_1 \cup S_2, \quad S_1 \cap S_2 = \emptyset,$$
and f_0, f_2 be continuous functions prescribed on S_1.

We wish to determine a function $u(x)$ twice continuously differentiable in $\mathcal{D}$ and continuously differentiable in $\overline{\mathcal{D}}$ subject to the conditions
$$\Delta u = 0,$$
$$u\big|_{S_1} = f_0, \quad (3)$$
$$\frac{\partial u}{\partial n}\bigg|_{S_1} = f_1.$$

Suppose that the solution to the problem (3) exists. We seek this solution in the form of "single-layer" potentials
$$u = \int_{S_1} \frac{1}{r} \varphi_1 d\sigma + \int_{S_2} \frac{1}{r} \varphi_2 d\sigma.$$

The Cauchy conditions imply that φ_1, φ_2 satisfy the following system

5.4. Reduction of Ill-Posed Problems to Integral Equations

of integral equations

$$\int_{S_1} \frac{1}{r} \varphi_1 \, d\sigma + \int_{S_2} \frac{1}{r} \varphi_2 \, d\sigma = f_0, \quad x \in S_1, \tag{4}$$

$$2\pi\varphi_1 + \int_{S_1} \frac{\partial}{\partial n}\left(\frac{1}{r}\right) \varphi_1 \, d\sigma + \int_{S_2} \frac{\partial}{\partial n}\left(\frac{1}{r}\right) \varphi_2 \, d\sigma = f_1, \quad x \in S_1.$$

Denoting

$$f_1 - \int_{S_2} \frac{\partial}{\partial n}\left(\frac{1}{r}\right) \varphi_2 \, d\sigma = 2\pi f_2,$$

recasts the last equation to the form

$$\varphi_1 + \frac{1}{2\pi} \int_{S_1} \frac{\partial}{\partial n}\left(\frac{1}{r}\right) \varphi_1 \, d\sigma = f_2. \tag{5}$$

Consider (5) as an integral equation for φ_1. It can be demonstrated that a solution of equation (5) exists and can be represented as

$$\varphi_1 = f_2 + \int_{S_1} G f_2 \, d\sigma,$$

where G is a function of a pair of points on S_1.

Substituting the expression for φ_1 into the first line of equations (4) we obtain the integral equation of the first kind for φ_2

$$\int_{S_2} K \varphi_2 \, d\sigma = f_0.$$

Mixed Problem for Heat Equation with Cauchy Data on Time-Like Manifold. The problem is to determine within the rectangle

$$0 \leq x \leq 1, \quad 0 \leq t \leq t_0$$

a function $u(x, t)$ satisfying the conditions

$$\frac{\partial u}{\partial t} = \frac{\partial^2 u}{\partial x^2},$$

$$u(0, t) = f(t), \quad u_x(0, t) = 0, \quad u(x, 0) = 0. \tag{6}$$

We shall seek a solution of this problem in the form of a heat potential

$$u(x, t) = \frac{1}{2} \int_0^t \left[\exp\left(-\frac{(x-1)^2}{4(t-\tau)}\right) + \exp\left(-\frac{(x+1)^2}{4(t-\tau)}\right) \right] \varphi(\tau) \, d\tau.$$

A function admitting such a representation satisfies the heat equation, and also the second and third boundary conditions in (6).

The first boundary condition in (6) leads to the Volterra-type integral equation in function φ

$$\int_0^t \exp\left(-\frac{l^2}{4(t-\tau)}\right)\varphi(\tau)d\tau = f(t).$$

5.5. Well-Posedness of Integral Equations

Integral equations of the first kind are a particular case of a more general class of equations – linear operator equations of the first kind.

By a linear operator equation of the first kind we mean the equation

$$Au = f, \qquad (1)$$

where u, f are elements of some linear normed spaces U and $\mathscr{F}$, and A is a linear completely continuous operator.

In an infinite dimensional space U equation (1) is not well-posed, since the complete continuity of A implies that A^{-1}, if exists, is unbounded. Therefore, arbitrarily small changes of f in the right-hand side may produce arbitrarily large variations of the solution u.

The property of well-posedness obviously depends on spaces on which the problem is set. A problem can be well-posed for one pair of spaces, and ill-posed for another pair. In principle, for any operator equation a pair of spaces can be chosen such that it renders the problem on hand well-posed.

We illustrate this statement by the example of equation (1). Suppose for simplicity that the solution to this equation is unique, i.e., the operator A^{-1} exists. Denote by $\mathscr{F}_A$ the subspaces of elements from $\mathscr{F}$ which admit representation in the form

$$f = Au, \quad u \in U,$$

under the norm

$$\|f\|_{\mathscr{F}_A} = \|A^{-1}f\|_u.$$

Clearly, $\|A^{-1}\| = 1$, and for the pair of spaces U, $\mathscr{F}_A$ the problem is well-posed.

Such a trivial resolution of the well-posedness problem is not satisfactory for the following reasons. If an operator equation of the first kind relates to an actual physical problem, its right-hand side is specified by the recording of physical instruments. Experimental measurements are always subject to error, and the errors are small not in arbitrary spaces, but rather in spaces defined by specific features of the measurement process. More often the experimenter specifies the maximum error (norm in C) or the r.m.s. error (space l_2). The measurement process can be set so that the measurement error be small along with its derivative (in space C^1). However, it is hard to imagine a measurement process with error being small in the norm of the space $\mathscr{F}_A$, for the integral operators corresponding to the integral equations of the first kind discussed in the preceding section.

5.6. Two Ill-Posed Problems

Differentiation. Differentiation is a classical problem of mathematical analysis. Solving this problem is equivalent to solving the integral equation of the first kind,

$$\int_0^x u(\xi)\,d\xi = f(x). \tag{1}$$

The problem of differentiation is well-posed if equation is considered for pairs of spaces C, C^1; $\mathscr{L}_2$, W_2^1; etc., while it is ill-posed for the pairs of spaces C, C; $\mathscr{L}_2$, $\mathscr{L}_2$. Since, as noted above, an estimate of the norm of measurement errors is given most frequently in spaces C and $\mathscr{L}_2$, the differentiation of a function constructed by the readings of some instruments (experimental curve) is a typical ill-posed problem.

Interpretation of Instrumental Readings. The operation of many instruments recording time-varying physical fields is schematically presented in Fig. 1. A signal $\varphi(t)$ enters the instrument, and a function $f(t)$ is produced at its output. In the simplest case, the functions $\varphi(t)$, $f(t)$ are related by

$$\int_0^t g(t-\tau)\varphi(\tau)\,d\tau = f(x),$$

where $g(t)$ is known as the unit impulse response of the instrument. Theoretically, $g(t)$ is the output recorded by the instrument with Dirac's delta function $\delta(t)$ applied to its input. In practice, to obtain the function $g(t)$ characterizing the instrument, a sufficiently short input impulse is used.

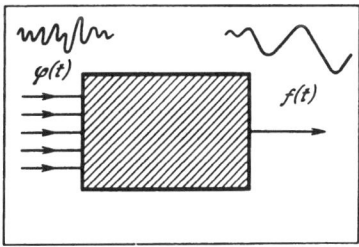

Fig. 1

Thus, the problem of interpretation of physical measurements, i.e., one of evaluations of the form of an input signal, reduces to solving the integral equation of the first kind.

The relation between an input signal $\varphi(t)$ and the output function $f(t)$ may be more involved. For a "linear" instrument the relation is of the form

$$\int_0^t K(t, \tau)\varphi(\tau)d\tau = f(t).$$

A nonlinear relation between $\varphi(t)$ and $f(t)$ is also possible, for instance,

$$\int_0^t K(t, \tau, \varphi(\tau))d\tau = f(t).$$

This way of operation is characteristic of the instruments recording time-varying electromagnetic fields, stress and strains in continuous media, and seismographic records of the Earth's crust vibrations.

The simplest equations may be solved with the use of the Fourier and Laplace transformations. Let $\tilde{g}(\lambda)$, $\tilde{\varphi}(\lambda)$, $\tilde{f}(\lambda)$ be the Fourier transforms of the functions $g(t)$, $\varphi(t)$, $f(t)$

$$\tilde{g}(\lambda) = \int_0^\infty e^{i\lambda t} g(t)dt, \quad \tilde{\varphi}(\lambda) = \int_0^\infty e^{i\lambda t} \varphi(t)dt, \quad \tilde{f}(\lambda) = \int_0^\infty e^{i\lambda t} f(t)dt.$$

Then, by the convolution theorem

$$\tilde{g}(\lambda)\tilde{\varphi}(\lambda) = \tilde{f}(\lambda),$$

whence, taking the inverse Fourier transform, we obtain the solution in the form

$$\varphi(t) = \frac{1}{2\pi}\int_{-\infty}^\infty e^{-i\lambda t}\tilde{\varphi}(\lambda)d\lambda = \frac{1}{2\pi}\int_{-\infty}^\infty e^{-i\lambda t}\frac{\tilde{f}(\lambda)}{\tilde{g}(\lambda)}d\lambda.$$

This formula is unstable, since $\tilde{g}(\lambda)$, the Fourier transform of the impulse response of a real instrument, tends to zero as $\lambda \to \infty$. For large λ, arbitrarily small perturbations in the evaluation of $\tilde{f}(\lambda)$ may produce large variations of the solution $\varphi(t)$.

5.7. Tomography

The word tomography appeared in the 20s of this century with the advance of X-ray tomographs. An intensive development of

5.7. Tomography

tomography began in the 70s when powerful computers became commercially available. A modern tomograph is a combination of a hardware responsible for X-raying (or nuclear magnetic resonance analysis) of the studied object and a software performing a transformation of the radiographic data into the image of a slice of the object. The latter is based on a mathematical theory.

Although until recently the term tomography has been related only to medicine, the same principles had long been used in geophysics for the construction of images of sections of the Earth's crust by geophysical measurements performed on the terrain.

At the present time tomographs and tomography are widely used in different branches of science and technology.

Mathematical problems arising in the interpretation of tomographic data (imagery of a slice of the object) relate to inverse problems of mathematical physics.

The mathematical formalism underlying modern X-ray tomography is based on the inverse Radon transformation. The application of Radon transformation is based on a certain model of interaction of X-rays with the object. This model ignores a number of factors which can be of significance in different tomographic applications. The model leading to the Radon transformation assumes that X-rays propagate along straight lines, overlooking in particular the divergence and scattering of these rays.

Let us consider a flux ψ of X-rays irradiating an object characterized by an absorption coefficient a. It is assumed that the exposure is performed within a thin layer, so the absorption coefficient is a function of two variables, which is other than zero within a bounded region $\mathcal{D}$.

In tomographs, a studied layer of the body is irradiated by a number of very narrow (plane) beams incident at different angles. Thus, the flux ψ is a function of variables x, y and angle α:

$$\psi = \psi(x, y, \alpha).$$

Under the above assumptions, the flux ψ obeys the simplest transport equation

$$\psi_x \cos \alpha + \psi_y \sin \alpha + a\psi = 0. \qquad (1)$$

The substitution $u = \ln \psi$ carries equation (1) to

$$u_x \cos \alpha + u_y \sin \alpha + a = 0. \qquad (2)$$

Suppose that the initial flux is constant, $\psi = c$. Then after passage through the object, the function u will be given by

$$u(x, y, \alpha) = \int_{\Gamma(x, y, \alpha)} a(\xi, \eta) ds, \qquad (3)$$

where Γ is a straight line across the domain $\mathcal{D}$.

Now, the initial flux ψ is known, the beam intensity after its passage through the object is measured. Thus, we may deem the quantity $u(x, y, \alpha)$ known and consider equation (3) as an integral equation for the distribution of absorption intensity $a(x, y)$.

In a more detailed formulation, the problem of solving equation (3) is as follows. Given the function

$$\int_{\Gamma(x, y, \alpha)} a(\xi, \eta) ds = u(x, y, \alpha), \tag{4}$$

we seek to determine the function $a(x, y)$ equal to zero outside the bounded domain $\mathcal{D}$.

The transformation relating $a(x, y)$ to $u(x, y, \alpha)$ is known as the Radon transformation. Thus, solving equation (4) is essentially the inversion of the Radon transform.

There are several formulas for inversion of a Radon transform. If the quantities u, a in equation (4) are elements of the space C, then the problem of solving (4) is ill-posed. Equation (4) is not an operator equation of the first kind since the Radon operator is not completely continuous.

The Radon problem is the one of integral geometry. We formulate the general problem of integral geometry on a plane below.

Let $\Gamma(p, q)$ be a family of curves on a plane depending on two parameters and having the following properties: each point $\mathcal{D}$ is crossed by a bunch of curves from this family. Further, let $g(x, y, p, q)$ be a prescribed function of four variables. Given the function

$$\int_{\Gamma(p, q)} g(x, y, p, q) u(x, y) ds = f(p, q), \tag{5}$$

it is required to determine $u(x, y)$ identical to zero outside $\mathcal{D}$.

Now, consider the geophysical problem of interpretation of seismic data, which leads to the problem of integral geometry, that is, of solving an equation of type (5). This problem is called the problem of seismic tomography.

Say $v(x, y)$ is the propagation velocity of a wave signal in a two-dimensional medium, $\tau(x_1, y_1, x_2, y_2)$ is the time when the wave from a point (x_1, y_1) arrives at (x_2, y_2).

We have

$$\tau = \int_{\Gamma(x_1, y_1, x_2, y_2)} n(x, y) ds, \quad n = \frac{1}{v}, \tag{6}$$

where Γ is the ray path connecting (x_1, y_1) and (x_2, y_2). The function

5.7. Tomography

τ satisfies the eikonal equation

$$\left(\frac{\partial \tau}{\partial x_1}\right)^2 + \left(\frac{\partial \tau}{\partial y_1}\right)^2 = n^2(x_1, y_1),$$

$$\left(\frac{\partial \tau}{\partial x_2}\right)^2 + \left(\frac{\partial \tau}{\partial y_2}\right)^2 = n^2(x_2, y_2).$$
(7)

The direct kinematic problem is to determine the function τ, given v or n. Mathematically, this is the classical Cauchy problem for the first-order equations (7), or the boundary-value problem for the corresponding system of differential equations.

The inverse problem is the one of solving the operator equation (6). Since the rays Γ depend on the sought function n, equation (6) is nonlinear.

Suppose that

$$n = n_0 + n_1,$$

where n_0 is prescribed, and n_1 is sufficiently small so that in our equations the terms of order n_1^2 are negligible.

In geophysics, this corresponds to the case when the distance between the pairs of points is not large, then n_0 is simply constant (seismic sounding between two wells), or when n_0 depends on one variable and must be determined from the solution of the one-dimensional inverse problem (seismic sounding of sedimentary layer). This hypothesis can be used also in the problem of acoustic stratification of the ocean, where the velocity of sound departs but slightly from its average value.

According to our assumptions, equations (7) take on the form

$$\frac{\partial \tau_0}{\partial x_j}\frac{\partial \tau_1}{\partial x_j} + \frac{\partial \tau_0}{\partial y_j}\frac{\partial \tau_1}{\partial y_j} = n_0(x_j, y_j)n_1(x_j, y_j) \qquad (8)$$

whence

$$\tau_1(x_1, y_1, x_2, y_2) = \int_{\Gamma(x_1, y_1, x_2, y_2)} n_1(x, y)\,ds, \qquad (9)$$

where Γ_0 is the ray path connecting points (x_1, y_1) and (x_2, y_2) in the medium corresponding to the function $n_0(x, y)$. Since n_0 is assumed to be prescribed, one can consider known the family of rays Γ_0 as well.

Chapter 6. OPERATOR AND INTEGRAL EQUATIONS

6.1. Fredholm Equations of the Second Kind

Real Hilbert Space. *We say Fredholm's equation of the second kind of the equation for the element u of a Hilbert space U*

$$u + Au = f, \qquad (1)$$

where A is a linear compact operator, and f is a given element of U.

This equation is considered along with the adjoint equation,

$$v + A^*v = g, \qquad (2)$$

and the associated homogeneous equations

$$u + Au = 0, \qquad (1')$$
$$v + A^*v = 0. \qquad (2')$$

In these equations, A^* is the adjoint of A, and v, g are elements of U.

If $\|A\| < 1$, then for any right-hand side f, a solution of equation (1) exists. This solution is unique and can be represented by the convergent series

$$u = f + \sum_{k=1}^{\infty} (-1)^k A^k f. \qquad (3)$$

Since

$$\|A^*\| = \|A\|,$$

the same assertion holds for the adjoint equation (2). Rewriting equation (1) in the form

$$(I + A)u = f,$$

where I is the identity operator, from equation (3) it follows that the operator

$$\sum_{k=0}^{\infty} (-1)^k A^k = (I + A)^{-1}$$

is continuous.

In the general case, for equations (1) and (2), the following theorem holds.

6.1. Fredholm Equations of the Second Kind

THEOREM (Fredholm). *Two possibilities occur for equations* (1) *and* (2).

(1) *The solution of equation* (1) *is unique, then it exists for any right-hand side f. The solution of equation* (2) *is also unique and exists for any g.*

(2) *The homogeneous equation* (1') *has a finite number of linearly independent solutions* $\{u_k\}$, $k = 1, ..., n$. *Then the homogeneous adjoint equation* (2') *has the same number of linearly independent solutions* $\{v_k\}$, $k = 1, ..., n$.

The necessary and sufficient conditions for a solution of equation (1) *to exist is that the right-hand side f be orthogonal to all the solutions of the adjoint homogeneous equation* (2'), *i.e., that*

$$(f, v_k) = 0, \quad k = 1, ..., n. \tag{4}$$

We consider first the case of A being a finite-rank operator. Then

$$Au = \sum_{k=1}^{m} (\psi_k, u)\varphi_k,$$

where $\{\varphi_k\}$ and $\{\varphi_k\}$ ($k = 1, ..., m$) are linear independent systems of elements.

Denoting

$$(\psi_k, u) = x_k,$$

we have from equation (1)

$$u = f - \sum_{k=1}^{m} x_k \varphi_k,$$

which suggests that solving equation (1) is equivalent to determining the numbers $\{x_k\}$, $k = 1, ..., m$.

The numbers $\{x_k\}$ are determined from the following set of linear equations

$$x_k + \sum_{j=1}^{m} a_{kj} x_j = b_k \tag{5}$$

with $a_{kj} = (\psi_k, \varphi_j)$ and $b_k = (\psi_k, f)$.

The system adjoint to (5) is

$$y_k + \sum_{j=1}^{m} a_{jk} y_j = c_k = (\varphi_k, g). \tag{6}$$

The problems of solving the adjoint operator equation (2) and the system (6) are equivalent.

By the rules of linear algebra, if the determinant of the system (5) is nonzero, then equations (5) and (6) are uniquely solvable and their solutions exist for any $\{b_k\}$ and $\{c_k\}$. Thus, the first case of the Fredholm theorem takes place.

Now suppose that the determinant of equation (5) is zero, and the rank of the matrix in equation (5) is $r < m$. Then the homogenous systems (5), (6) have $m - r$ linearly independent solutions, like the homogenous systems of operator equations (1'), (2').

The necessary and sufficient condition for a solution of the set (5) to exist is that the vector $\{b_k\}$ be orthogonal to all the nontrivial solutions of the homogeneous ($g = 0$) system (6).

Any nontrivial solution to the operator equation (2) admits the representation

$$v = \sum_{k=1}^{m} y_k \psi,$$

where $\{y_k\}$ is a nontrivial solution of the set (6) with $g = 0$. Hence, the necessary and sufficient condition that the solution to equation (1) exists is that

$$\sum_{k=1}^{m} b_k y_k = (\sum_{k=1}^{m} y_k \psi_k, f) = (v, f) = 0.$$

Now we will consider the general case. In Section 3.6 we have shown that a compact operator A admits the representation

$$A = A_1 + A_2,$$

where $\|A_1\| < 1$, and A_2 has a finite rank.

Equation (1) will be equivalent to

$$u + (I + A_1)^{-1} A_2 u = f_1 = (I + A_1)^{-1} f. \tag{7}$$

The equation adjoint to (7) will take the form

$$W + A_2^* (I + A_1^*)^{-1} W = g_1. \tag{8}$$

Equation (7) is the Fredholm equation of the second kind with finite-rank operator, for which the Fredholm theorem has already been established. Equation (8) is equivalent to equation (2). Indeed, letting in equation (2)

$$v = (I + A_1^*)^{-1} W, \quad g = g_1,$$

we obtain equation (8).

6.1. Fredholm Equations of the Second Kind

Thus, for the general Fredholm equation of the second kind the following assertions may be deemed established:

THEOREM (1). *If the solution to equation* (1) *is unique, then so is the solution to equation* (2), *and the solutions to equations* (1) *and* (2) *exist for any right-hand sides f and g.*

(2) *The homogeneous equation* (1') *can have only a finite number of linear independent solutions. In this case the homogeneous equation* (2') *has the same number of linearly independent solutions.*

PROOF. By the Fredholm theorem about operators of finite rank, for a solution to equation (7) to exist it is necessary and sufficient that f_1 be orthogonal to all the nontrivial solutions to the homogeneous adjoint equation (8)

$$(W, f_1) = 0,$$

$$W + A_2^*(I + A_1^*)^{-1}W = 0. \tag{9}$$

Letting in (9)

$$W = (I + A_1^*)v,$$

$$f_1 = (I + A_1)^{-1}f,$$

we obtain

$$(v, f) = 0, \quad v + A_1^*v + A_2^*v = 0. \tag{10}$$

Thus, Fredholm's theorem is established completely. ∎

Banach Space. When in equation (1) u and f are elements of a Banach space $\mathscr{F}$, then equation (1) is a Fredholm equation of the second kind in the Banach space. The quantities v, g in the adjoint equation (2) will be elements of the dual Banach space $\mathscr{F}^*$.

The formulation and proof of Fredholm's theorem for the equations in a Banach space closely resemble those for a Hilbert space, the difference being that in a Banach space we consider values of functionals instead of inner products.

Now we consider the posedness of problems for Fredholm equations of the second kind. In the first formulation of Fredholm's theorem, the equation is uniquely solvable. The continuous dependence of the solution upon the right side follows from the properties of operators discussed in Section 3.6. Consequently, in this case the problem for equation (1) is well-posed.

In the second case, a solution to equation (1) not always exists and is not unique.

We give a wider solvability criterion which ensures that the problem for Fredholm equations of the second kind is always well-posed.

Let U, $\mathcal{F}$ be some Banach spaces, A be a bounded linear operator with domain U and range in $\mathcal{F}$. Consider the equation

$$Au = f, \quad u \in U, \quad f \in \mathcal{F}. \tag{11}$$

DEFINITION. *The problem* (11) *is well-posed in the sense of Ficherat if the operator* A *is normally solvable.*

6.2. Equations of the First Kind

Let U, $\mathcal{F}$ be Banach spaces, and A be a linear compact operator with domain U and range from $\mathcal{F}$.

We say an equation of the first kind of

$$Au = f, \quad u \in U, \quad f \in \mathcal{F}. \tag{1}$$

If the solution of equation (1) is unique or, equivalently, the kernel of A is zero,

$$N(A) = 0,$$

then, as we noted in Section 3.6, the operator A^{-1} exists, but is not bounded. Hence, the problem of solving equation (1) is ill-posed.

In the first half of this century equations of the first kind were given comparatively little attention. The forthcoming Picard theorem may be the only classical result on the general equation of the first kind.

Let in equation (1), U, $\mathcal{F}$ be Hilbert spaces and $U = \mathcal{F}$.

We begin with the case when A is self-adjoint. The operator A has then a complete orthogonal system of eigenvectors (see Section 3.8).

Denote by $\{\varphi_k^0\}$, $k = 1, 2, \ldots$, the set of eigenvectors associated with the zero eigenvalue, and by $\{\varphi_k\}$, $k = 1, 2, \ldots$, the set of eigenvectors associated with the nonzero eigenvalues $\{\lambda_k\}$. The vectors $\{\varphi_k^0\}$ constitute a basis in the kernel $N(A)$. We now have

THEOREM (Picard). *The necessary and sufficient condition for a solution of equation* (1) *to exist is that the right side f satisfies the two conditions*

$$(1) \ (f, \varphi_k^0) = 0, \quad k = 1, 2, \ldots,$$

$$(2) \ \sum_k \frac{1}{\lambda_k^2} f_k^2 < \infty, \quad f_k = (f, \varphi_k).$$

6.2. Equations of the First Kind

PROOF. Suppose that a solution of equation (1) exists. We represent the solution as a series in $\{\varphi_k^0\}$, $\{\varphi_k\}$

$$u = \sum_k a_k^0 \varphi_k^0 + \sum_k a_k \varphi_k,$$

$$a_k^0 = (u, \varphi_k^0), \quad a_k = (u, \varphi_k).$$

Then

$$Au = \sum_k \lambda_k a_k \varphi_k = \sum_k f_k \varphi_k. \tag{2}$$

From (2) it follows that f is orthogonal to the vectors $\{\varphi_k^0\}$, and the series

$$\sum_k \frac{1}{\lambda_k^2} f_k^2 = \sum_k a_k^2$$

converges.

If

$$(f, \varphi_k^0) = 0, \quad \sum_k \frac{1}{\lambda_k^2} f_k^2 < \infty,$$

then any vector of the form

$$\sum_k a_k^0 \varphi_k^0 + \sum_k \frac{1}{\lambda_k} f_k \varphi_k$$

is a solution of equation (1). ∎

Now we proceed to the general case. Denote by $\{\psi_k^0\}$, $k = 1, 2, \ldots$, the orthogonal set of eigenvectors of the operator A^*A, which are associated with the zero eigenvalue, and by $\{\psi_k\}$, $k = 1, 2, \ldots$, the eigenvectors associated with the nonzero eigenvalues $\{\lambda_k^0\}$. The respective sets of eigenvectors for AA^* will be denoted by $\{\varphi_k^0\}$ and $\{\varphi_k\}$.

THEOREM (Picard, general case). *The necessary and sufficient condition that the solution to equation* (1) *exists is that the right side f satisfies the conditions*

(1) $(f, \{\varphi_k^0\}) = 0.$

(2) $\sum_{k=1}^{\infty} \frac{1}{\lambda_k^2} f_k^2 < \infty, \quad f_k = (f, \psi_k).$

PROOF. Suppose that the solution (1) exists. We will seek it in the form of a series

$$u = \sum_k a_k^0 \psi_k^0 + \sum_k a_k \psi_k,$$

$$a_k^0 = (u, \psi_k^0), \quad a_k = (u, \psi_k).$$

Then

$$Au = \sum_k \lambda_k a_k \varphi_k = \sum_k f_k \varphi_k. \tag{3}$$

From equation (3) it follows that f is orthogonal to the vectors $\{\varphi_k^0\}$, and the series

$$\sum_k \frac{1}{\lambda_k^2} f_k^2 = \sum_k a_k^2$$

converges. If

$$(f, \varphi_k^0) = 0 \text{ and } \sum_k^\infty \frac{1}{\lambda_k^2} f_k^2 < \infty,$$

then any vector of the form

$$\sum_k a_k^0 \psi_k^0 + \sum_k \frac{1}{\lambda_k} f_k \psi_k$$

is a solution of equation (1). ∎

It is instructive to note that if ψ_k^0 are absent then the solution of equation (1) is unique. If φ_k^0 are absent, the set of f for which a solution to equation (1) exists is dense in U.

6.3. Well-Posedness in the Sense of Tychonoff

Consider operator equations of the first kind.

$$Au = f, \quad u \in U, \quad f \in \mathcal{F}. \tag{1}$$

Here $U, \mathcal{F}$ are Banach spaces, and A is a compact operator. Define a set in U, $\mathcal{M} \in U$.

The problem of solving equation (1) is said to be well-posed in the sense of Tychonoff [13, 14] *if the following conditions are fulfilled*:

(i) *it is known beforehand that a solution of equation (1) exists and belongs to $\mathcal{M}$;*

6.3. Well-Posedness in the Sense of Tychonoff

(ii) *the solution of* (1) *is unique*;

(iii) *the solution u depends continuously on the right side f, if perturbations of f do not lead to the solution outside the set $\mathcal{M}$.*

Denote by $\mathcal{M}_A$ the image of $\mathcal{M}$ in $\mathcal{F}$ mapped there by the operator A,

$$\mathcal{M}_A = A\mathcal{M}.$$

Then the last condition can be reformulated as

(iii) the operator A^{-1} is continuous on $\mathcal{M}_A$.

The set $\mathcal{M}$ is said to be one of well-posedness. Since generally $\mathcal{M}$ is not a linear space, the problem of solving equation (1) in this setting becomes nonlinear.

The concept of well-posedness in the sense of Tychonoff differs from the classical notion. In analysis of well-posedness in the sense of Tychonoff, no existence theorem is proved. That the solution exists and belongs to the set of well-posedness is postulated in the problem statement. If the equation is derived as a mathematical model of a physical phenomenon, additional physical considerations are normally invoked to this end.

Studies of uniqueness in problems well-posed in the sense of Tychonoff in essence do not differ from those in classically well-posed problems. Uniqueness is established via the uniqueness theorems. Although for problems well-posed in the sense of Tychonoff it will suffice to prove the uniqueness in $\mathcal{M}$, the uniqueness theorems are normally proved for the whole space U.

As a well-posedness set $\mathcal{M}$ one normally considers a compact set. In this case the condition (iii) of the inverse operator being continuous follows from the condition (ii).

THEOREM 1 (Tychonoff). *Let the solution to equation* (1) *be unique, and $\mathcal{M}$ compact. Then A^{-1} is continuous on $\mathcal{M}_A$.*

PROOF. Suppose that the assertion of the theorem is not valid. Then there exist $u_0 \in \mathcal{M}$ and $\varepsilon_0 > 0$ such that for any $\delta > 0$ there is $u_1 \in \mathcal{M}$ such that

$$\|Au_0 - Au_1\| < \delta, \quad \|u_0 - u_1\| > \varepsilon_0.$$

Let $\{\delta_k\}$ be a sequence tending to zero as $k \to \infty$, and let $\{u_k\}$ be a sequence of vectors such that

$$\|u_k - u_0\| > \varepsilon_0, \quad \|Au_k - Au_0\| < \delta_k, \quad u_k \in \mathcal{M}.$$

Since $\mathcal{M}$ is compact, $\{u_k\}$ has a convergent subsequence. Clearly, we can assume that this subsequence coincides with the sequence itself.

Let

$$\lim_{k \to \infty} u_k = u^0.$$

Then, since A is continuous,

$$\|Au^0 - Au_0\| = 0,$$
$$Au^0 = Au_0, \quad \|u^0 - u_0\| \geq \varepsilon_0,$$

which is in contradiction with the uniqueness of the solution. ∎

This theorem admits the following generalization.

THEOREM 2. *Let the solution to equation* (1) *be unique, and the well-posedness set* $\mathcal{M}$ *be the algebraic sum*

$$\mathcal{M} = \mathcal{M}_1 + U_1,$$

where $\mathcal{M}_1$ *is a compact set, and* U_1 *a finite-dimensional subspace of* U. *Then* A^{-1} *is uniformly continuous on* $\mathcal{M}_A$.

PROOF. Suppose that the theorem is not true. Then there are sequences

$$u_{1k}, u_{2k} \in \mathcal{M}_1, \quad v_{1k}, v_{2k} \in U_1,$$

and an $\varepsilon > 0$, such that

$$\|A(u_{1k} + v_{1k} - u_{2k} - v_{2k})\| \leq \delta_k,$$
$$\|u_{1k} + v_{1k} - u_{2k} - v_{2k}\| \geq \varepsilon,$$
$$\lim_{k \to \infty} \delta_k = 0.$$

Denote by $\mathcal{M}_2$ the set of elements admitting the representation

$$u = u_1 - u_2, \quad u_1, u_2 \in \mathcal{M}_1.$$

Clearly, $\mathcal{M}_2$ is compact. Then there exist sequences

$$\{u_k\} \in \mathcal{M}_2, \quad \{v_k\} \in U_1,$$

such that

$$\|A(u_k + v_k)\| \leq \delta_k, \quad \|u_k + v_k\| \geq \varepsilon.$$

If the sequence v_k is bounded, then from $\{v_k + u_k\}$ one can pick a convergent subsequence. The limit of this subsequence satisfies the relations

$$\|Au_0\| = 0, \quad \|u_0\| \geq \varepsilon,$$

which contradicts the uniqueness of the solution to equation (1).

Now, suppose that the sequence $\{v_k\}$ is unbounded. In the subspace U_1 equation (1) is equivalent to a finite set of linear algebraic

6.4. Regularization

equations, whence for $u \in U_1$ we have

$$\|Au\| \geq q\|u\|,$$

where $q > 0$ is a constant. Thus the sequence of $\|A v_k\|$ is unbounded, while the sequence of $\|A u_k\|$ is bounded as is the operator A. This contradicts the inequality

$$\|A(u_k + v_k)\| \leq \delta_k,$$

and establishes the theorem. ∎

Let us turn to the physical settings leading to ill-posed formulations discussed in Part One.

In evaluations of time-independent or time-varying physical fields in a space by measurements derived on the data in a part of the space, the unknown fields are real and obey relevant equations. Furthermore, the fields are induced by certain types of sources and cannot be too complicated or have intense high-frequency components in time or space domain. These considerations make it natural to assume *a priori* that the solution exists and belongs to the given compact set.

6.4. Regularization

Like in the preceding section, we will deal with the operator equation of the first kind

$$Au = f, \quad u \in U, \quad f \in \mathscr{F}. \tag{1}$$

We assume that this equation is uniquely solvable, and define V to be a subset of U.

DEFINITION. *A family $\{R_\alpha\}$ of linear operators $0 < \alpha < \alpha_0$, from $\mathscr{F}$ into U, is said to be regularizing* [15, 18] *for equation* (1) *on V if*

(i) *for any α, $0 < \alpha < \alpha_0$, the operator R is defined over all $\mathscr{F}$ and bounded;*

(ii) *for any $u \subset V$,*

$$\lim_{\alpha \to 0} R_\alpha Au = u.$$

If in (ii) the convergence is uniform on V, then R_α is said to be *uniformly regularizing* [17]. Since the operator A^{-1} is unbounded, for $V = U$, there is no uniformly regularizing family.

In some circumstances, regularizing families depending on an integer-valued parameter n prove to be more convenient.

DEFINITION. *A linear operator family $\{R_n\}$, $n = 1, ..., $ from $\mathscr{F}$ into U, is said to be regularizing for equation* (1) *on a set V if*

(i) *for any n, R_n is defined over all $\mathcal{F}$ and bounded*;
(ii) *for any $u \in V$,*

$$\lim_{n \to \infty} R_n A u = u.$$

Suppose that a regularizing family $\{R_\alpha\}$ for equation (1) is known, and consider the problem of evaluating an approximate solution to equation (1) on the approximate data [14, 16, 17].

Suppose that we know a vector f_ε,

$$\|f - f_\varepsilon\| \leq \varepsilon.$$

Denote by $u_{\alpha\varepsilon}$ the result of applying R_α to the approximate right side,

$$u_{\alpha\varepsilon} = R_\alpha f_\varepsilon,$$

and estimate the difference $u - u_{\alpha\varepsilon}$:

$$\|u - u_{\alpha\varepsilon}\| \leq \|u_{\alpha\varepsilon} - R_\alpha A u\| + \|u - R_\alpha A u\|$$
$$\leq \|R_\alpha\|\varepsilon + \|u - R_\alpha A u\|. \quad (2)$$

If $u \in V$, then the right-hand side tends to zero as $\alpha \to 0$. If R_α is a uniformly regularizing family, then

$$\|u - R_\alpha A u\| \leq \beta(\alpha),$$
$$\beta(\alpha) = \sup \|R_\alpha A v - v\|, \quad \lim_{\alpha \to 0} \beta(\alpha) = 0,$$

hence

$$\|u - u_{\alpha\varepsilon}\| \leq \|R_\alpha\|\varepsilon + \beta(\alpha).$$

With an approximate choice of α we will have a guaranteed estimate of accuracy of the approximate solution $u_{\alpha\varepsilon}$. In particular, we can set

$$\alpha = \alpha_0(\varepsilon),$$

where α_0 is given by

$$\|R_{\alpha_0}\|\varepsilon + \beta(\alpha_0) = \inf_\alpha [\|R_\alpha\|\varepsilon + \beta(\alpha)].$$

Some examples of regularizing families for equations in Hilbert space $U = F$ follow.

Reduction to the Equation of the Second Kind. Let the operator A

6.4. Regularization

be self-adjoint and positive. In addition to equation (1) we consider the family of equations

$$\alpha u_\alpha + A u_\alpha = (\alpha I + A) u_\alpha = f. \tag{3}$$

Denote

$$R_\alpha = (\alpha I + A)^{-1}.$$

We will show that $\{R_\alpha\}$ is a regularizing family for equation (1) on the whole U.

Since A is positive and compact,

$$\|R_\alpha\| = \frac{1}{\alpha}.$$

Consider the difference

$$u - R_\alpha A u = \alpha (\alpha E + A)^{-1} u.$$

Denote by $\{\varphi_k\}$, $\{\lambda_k\}$, $\lambda_k \geq \lambda_{k+1} > 0$ the complete orthonormed set of eigenvectors and eigenvalues of A. Then

$$u = \sum_{k=1}^\infty a_k \varphi_k, \qquad a_k = (u, \varphi_k),$$

$$u - R_\alpha A u = \alpha \sum_{k=1}^\infty \frac{a_k}{\alpha + \lambda_k} \varphi_k,$$

$$\|u - R_\alpha A u\|^2 = \alpha^2 \sum_{k=1}^\infty \frac{a_k^2}{(\alpha + \lambda_k)^2} = \alpha^2 \sum_{k=1}^n \frac{a_k^2}{(\alpha + \lambda_k)^2}$$

$$+ \alpha^2 \sum_{k=n+1}^\infty \frac{a_k^2}{(\alpha + \lambda_k)^2} \leq \frac{\alpha^2}{\lambda_n^2} \|u\|^2 + \sum_{k=n+1}^\infty a_k^2. \tag{4}$$

Since the series $\sum_{k=1}^\infty a_k^2$ converges, the second addend on the right-hand side of equation (4) is arbitrarily small for a sufficiently large n. The first addend will be arbitrarily small for a sufficiently small α and fixed n, whence

$$\lim_{\alpha \to 0} \|u - R_\alpha A u\| = 0$$

as desired.

If A is not positive and self-adjoint, equation (1) can be reduced to an equation with a positive and self-adjoint operator by applying the operator A^*, to its both sides

$$A_1 u = f_1, \quad A_1 = A^* A, \quad f_1 = A^* f.$$

The regularizing family for the initial equation (1) will be of the form

$$R_\alpha = (\alpha I + A^* A)^{-1} A^*.$$

Expansion in Eigenvectors. Let A be a self-adjoint operator, $\{\varphi_k\}$ be its eigenvectors, and $\{\lambda_k\}$ be the corresponding eigenvalues, $|\lambda_{k+1}| \leq |\lambda_k|$.

Denote by R_n the operator

$$R_n = \sum_{k=1}^{n} \frac{1}{\lambda_k} f_k \varphi_k, \quad f_k = (f, \varphi_k).$$

Clearly, the operators R_n are continuous, and

$$\|R_n\| = \frac{1}{|\lambda_n|},$$

$$\lim_{n \to \infty} R_n A u = \lim_{n \to \infty} \sum_{k=1}^{n} a_k \varphi_k = u, \quad a_k = (u, \varphi_k).$$

Thus, $\{R_n\}$ is a regularizing family depending on an integer-valued parameter.

Method of Sequential Approximations. Let A be positive and self-adjoint, $\{\varphi_k\}$, $\{\lambda_k\}$ be the sequences of eigenvectors and eigenvalues of A, respectively, and $\|A\| = \lambda_1 \leq 1$. Consider the sequence

$$u_{k+1} = u_k - A u_k + f \quad u_0 = f, \quad k = 0, \ldots .$$

We show that

$$\lim_{n \to \infty} u_n = u.$$

Indeed,

$$u_n = \sum_{k=0}^{n} (I - A)^k A u, \quad u - u_n = (I - A)^{n+1} u.$$

6.4. Regularization

Expanding u, u_n in $\{\varphi_k\}$ we obtain

$$u - u_n = \sum_{k=1}^{\infty} (I - \lambda_k)^{n+1} a_k \varphi_k, \quad a_k = (u, \varphi_k)$$

with

$$\|u - u_n\|^2 = \sum_{k=1}^{m} (1 - \lambda_k)^{2(n+1)} a_k^2 + \sum_{k=m+1}^{\infty} (1 - \lambda_k)^{2(n+1)} a_k^2$$

$$\leq (1 - \lambda_m)^{2(n+1)} \|u\|^2 + \sum_{k=m+1}^{\infty} a_k^2. \quad (5)$$

The second addend on the right-hand side is arbitrarily small for a sufficiently large m, and the first one is arbitrarily small for a sufficiently large n at a fixed m.

Now, the regularizing family corresponding to the above method of successive approximations is of the form

$$R_n = \sum_{k=0}^{n} (I - A)^k, \quad \|R_n\| = n + 1.$$

Below we will show that regularizing families yield criteria for existence of solutions.

THEOREM. *Let $\{R_\alpha\}$ be a regularizing family for equation* (1) *on the whole space $U = \mathscr{F}$, and let the operators R_α and A commute*:

$$R_\alpha A = A R_\alpha.$$

Then, if for some f_0 there exists a limit

$$\lim_{\alpha \to 0} R_\alpha f_0 = u_0,$$

then this limit is a solution of equation (1) *with the right side f_0*:

$$A u_0 = f_0.$$

PROOF. Since A is continuous,

$$\lim_{\alpha \to 0} A R_\alpha f_0 = A u_0,$$

we have

$$A u_0 = \lim_{\alpha \to 0} R_\alpha A f_0 = f_0,$$

as claimed. ■

6.5. Equations with Bounded Operators

The above notions and results for operator equations of the first kind can be extended to a wider class of operator equations, namely the linear operator equations with bounded operator.

Let U, $\mathscr{F}$ be Banach spaces, and A be a continuous operator with domain U and range from $\mathscr{F}$.

Consider the equation

$$Au = f, \quad u \in U, \quad f \in \mathscr{F}. \tag{1}$$

Note in passing that some authors refer to these equations and those defined in Section 4.2, as equations of the first kind. It is clear, however, that the class of equations defined above also includes the Fredholm equation of the second kind.

If A has a bounded inverse A^{-1}, then the problem (1) is well-posed.

We will be concerned with the case when this problem is, generally, ill-posed. Moreover, the operator A is not normally resolvable; that is, the problem (1) is not well-posed in the sense of Ficherat. We will confine ourselves to the case of A being a self-adjoint operator in Hilbert space.

THEOREM (Picard). *Let $\{E_\lambda\}$ be a family of projection-valued operators such that*

$$A = \int \lambda \, dE_\lambda.$$

Then a necessary and sufficient condition for a solution of equation (1) *to exist is that the right side f satisfies the conditions*:

$$(f, \varphi^0) = 0, \quad \varphi^0 \in N(A),$$

$$\int \frac{1}{\lambda^2} (dE_\lambda f, f) < \infty.$$

PROOF. We suppose that the solution to equation (1) exists, and represent it in the form of an integral

$$u = \int dE_\lambda u.$$

Then

$$f = Au = \int \lambda \, dE_\lambda u \tag{2}$$

and the conditions of the theorem are satisfied.

6.5. Equations with Bounded Operators

If f satisfies the conditions of the theorem, then any vector of the form

$$\varphi^0 = \int \frac{1}{\lambda} dE_\lambda f, \quad \varphi^0 \in N(A),$$

will be a solution of equation (1). ∎

The concept of well-posedness in the sense of Tychonoff, and the formulation and proof of Tychonoff's theorem for equation (1) with a bounded operator A, differs nowhere from those for the case of a compact operator. The same relates to the definition of the regularizing family, and the general argument regarding the applicability of regularizing family to the evaluation of an approximate solution to equation (1) by the approximate right-hand side.

Let us look at how the first example of regularizing families for equations in Hilbert spaces discussed in the preceding section translates to the bounded operator case.

Let

$$U = \mathscr{F},$$

and A be self-adjoint and positive. Denote

$$R_\alpha = (\alpha I + A)^{-1}.$$

We intend to show that $\{R_\alpha\}$ is a regularizing family for equation (1). Since A is positive,

$$\|R_\alpha\| \leq 1/\alpha.$$

Consider the difference

$$u - R_\alpha A u = \alpha(\alpha I + A)^{-1} u.$$

From the fact that A is positive and self-adjoint it follows that the range of A is everywhere dense. Hence, for any $\delta > 0$ there exists a vector v such that

$$u = Av + u_1, \quad \|u_1\| \leq \delta.$$

We have

$$\|\alpha(\alpha I + A)^{-1} u\| \leq \|\alpha(\alpha I + A)^{-1} Av\| + \alpha\|(\alpha I + A)^{-1} u_1\|$$
$$\leq \alpha\|v\| + \delta. \quad (3)$$

The first form on the right-hand side is arbitrarily small for any fixed δ and sufficiently small α, hence

$$\lim_{\alpha \to 0} \| u - R_\alpha A u \| = 0.$$

6.6. Equations with Weak Singularities

Let D be a finite domain in R^n, $x = (x_1, ..., x_n)$, $\xi = (\xi_1, ..., \xi_n)$ be elements (points) of R^n, and $u(x)$ be a function defined in D.

By integral equations of the first and second kind, respectively for $u(x)$ we mean equations of the form

$$\int_D K(x, \xi) u(\xi) d\xi = f(x), \tag{1}$$

$$u(x) + \int_D K(x, \xi) u(\xi) d\xi = f(x). \tag{2}$$

Here the function $K(x, \xi)$ is such that the integral operator in equations (1), (2) is the integral operator in the narrow sense defined in Section 3.6.1. In equation (1), $f(x)$ can be defined within a domain D_1 different from D. Correspondingly, the kernel $K(\cdot)$ will be defined on $D \times D_1$. The functions $u(x)$ and $f(x)$ can be treated as elements of different functional spaces.

In equations of the second kind, the domains of $u(x)$ and $f(x)$ and the respective functional spaces coincide.

Among the integral equations of the first kind we distinguish a class of equations which can be reduced to equations of the second kind by applying differential or pseudodifferential operators. In this section we will be concerned with integral equations of the first kind that belong to this class. We restrict our attention to the case when the operator converting an equation of the first kind into an equation of the second kind is a power of the Laplacian. The solution $u(x)$ will be deemed an element of the Hilbert space $L_2(D)$.

Let in equation (1)

$$K(x, \xi) = \frac{1}{|x - \xi|^{n-\alpha}} + K_0(x, \xi), \tag{3}$$

where

$$0 < \alpha < n \tag{4}$$

is an even integer, $n \geq 3$, and $K_0(x, \xi) \in c^\alpha \; \overline{(D \times D)}$. It is known that

$$\Delta^{\frac{\alpha}{2}} |x - \xi|^{-n+\alpha} = c\delta(x - \xi),$$

6.7. Volterra Equations for Function of One Variable

where $c = c(n, \alpha)$ is a nonzero constant. Consequently, acting on equation (1) by the operator $c^{-1} \Delta^{\alpha/2}$ yields

$$u(x) + \int_D K_1(x, \xi) u(\xi) d\xi = f_1(x), \tag{5}$$

$$K_1(x, \xi) = c^{-1} \Delta^{\alpha/2} K_0(x, \xi),$$

$$f_1(x) = c^{-1} \Delta^{\alpha/2} f(x).$$

Thus, the solution of the first-kind equation (1) reduces to the solution of the second-kind equation (5). If the solution to equation (5) is unique, then so is the solution to equation (1). Furthermore, the problem of solving equation (1) will be well-posed in the sense of Ficherat whenever

$$u(x) \in L_2(D), \quad f(x) \in W_2^\alpha.$$

Reduction of a first-kind equation with the kernel (3) to a second kind equation is possible also under the condition (4) alone, without requiring that α be even. To this end one should determine the fractional powers of the Laplacian first. If in equation (1) $f(x)$ is given within the domain $D_1 = R^n$, this evaluation is easily accomplished by Fourier thansformation. This can be done also in a more general case of $D_1 \supset \overline{D}$.

6.7. Volterra Equations for Functions of One Variable

By integral Volterra equations of the first and second kind for a function $u(x)$ we mean equations of the form

$$\int_0^x K(x, \xi) u(\xi) d\xi = f(x), \tag{1}$$

$$u(x) + \int_0^x K(x, \xi) u(\xi) d\xi = f(x). \tag{2}$$

We shall consider these equations on a finite interval $x \in [0, l]$. Since the integral operator in Volterra equations is quasinilpotent, a solution of the second-kind equation always exists, is unique, and admits representation by a convergent Neumann series.

Generally speaking, Volterra equations of the first kind enjoy almost the same variety of cases as general integral equations of the first kind. This section will be concerned with the Volterra equations of the first kind belonging to the class of weakly singular integral equations of the first kind reducible to equations of the second kind. The sought solution $u(x)$ will be deemed an element of the Banach space $C[0, l]$.

THEOREM. *Let in equation* (1) $K(x, \xi)$ *admit the representation*

$$K(x, \xi) = \frac{(x - \xi)^{\alpha - 1}}{\Gamma(\alpha)} + K_0(x, \xi),$$

where $\alpha \in (0, 1)$, $\Gamma(\alpha)$ *is the gamma function, and*

$$K_0(x, \xi) \in C^1(\overline{\Omega}), \quad \Omega = \{0 < \xi < x < l\}.$$

Then
  (i) *the solution to* (1) *is unique*;
  (ii) *for any* $f(x) \in C^1[0, l]$ *with* $f(0) = 0$, *there exists a solution* $u \in C[0, l]$.

PROOF. Define the differentiation operator of order α by the formula

$$(D^\alpha f)(x) = \frac{1}{\Gamma(1 - \alpha)} \int_0^x \frac{f'(\xi)d\xi}{(x - \xi)^\alpha} = \frac{1}{\Gamma(1 - \alpha)} \frac{d}{dx} \int_0^x \frac{f(\xi)d\xi}{(x - \xi)^\alpha}. \quad (3)$$

Since

$$\int_S^x (x - \xi)^{-\alpha}(\xi - S)^{\alpha - 1} d\xi = \Gamma(\alpha)\Gamma(1 - \alpha),$$

applying the operator D^α to the both sides of equation yields

$$u(x) + \int_0^x K_1(x, \xi)u(\xi)d\xi = f_1(x), \quad (4)$$

where $f_1 = D^\alpha f$, and the continuous kernel $K_1(x, \xi)$ is expressed in terms of the function K_0. It is not hard to show that equations (1) and (4) are equivalent.

The assertion of the theorem follows from the properties of solutions of the integral Volterra equations of the second kind. ∎

Equation (1) is a special case of the general integral equation of the first kind and so amenable to the general methods of regularization discussed in Section 6.4. However, the integral operator in equation (1) is not self-adjoint. Application of a self-adjoint operator to equation (1) reduces the stability of its solution. In view of this fact and because Volterra's operators offer certain computational advantages, it is reasonable to regularize equation (1) with the aid of these operators.

We illustrate the Volterra regularization equation (1) for the case of $K(x, x) = 1$. Assume in addition that $u(x)$ satisfies the condition $u(0) = 0$.

6.7. Volterra Equations for Function of One Variable

Denote a solution to the equation

$$\delta u_\delta(x) + \int_0^x K(x, \xi) u_\delta(\xi) d\xi = f(x) \tag{5}$$

by $u_\delta(x)$ ($\delta > 0$).
A solution to this equation exists, is unique and admits the representation

$$u_\delta(x) = \frac{1}{\delta} f(x) + \frac{1}{\delta} \int_0^x R(x, \xi, \delta) f(\xi) d\xi, \tag{6}$$

where $R(\cdot)$ is a continuous function.

For a fixed δ, the right-hand side of (6) is the result of applying a continuous operator to $f(x)$. Let us show that the family of these operators depending on a parameter δ is regularizing in respect to equation (1). To do so, it suffices to demonstrate that the family of functions $u_\delta(x)$ converges to the solution $u(x)$ as $\delta \to 0$.

Differentiating equation (5) we get

$$\delta u'_\delta(x) + u_\delta(x) + \int_0^x K_1(x, \xi) u_\delta(\xi) d\xi = f_1(x). \tag{7}$$

From this equation it follows that $u_\delta(x)$ satisfies the integral equation

$$u_\delta(x) + \int_0^x K_\delta(x, \xi) u_\delta(\xi) d\xi = f_\delta(x), \tag{8}$$

$$K_\delta(x, \xi) = \frac{1}{\delta} \int_\xi^x \exp\left(-\frac{x-y}{\delta}\right) K_1(y, \xi) dy,$$

$$f_\delta(x) = \frac{1}{\delta} \int_0^x \exp\left(-\frac{x-\xi}{\delta}\right) f_1(\xi) d\xi.$$

Consider the difference

$$u(x) - u_\delta(x) = f_1(x) - f_\delta(x) - \int_0^x [K_1(x, \xi) - K_\delta(x, \xi)] u(\xi) d\xi$$

$$- \int_0^x K_\delta(x, \xi) [u(\xi) - u_\delta(\xi)] d\xi. \tag{9}$$

It is not hard to show that the differences $f_1(\cdot) - f_\delta(\cdot)$ and

$K_1(\cdot) - K_\delta(\cdot)$ obey the inequalities

$$|f_1(x) - f_\delta(x)| \leq \exp\left(-\frac{x}{\delta}\right) \mathcal{M}(x),$$

$$|K_1(x, \xi) - K_\delta(x, \xi)| \leq \exp\left(-\frac{x-\xi}{\delta}\right) \mathcal{M}_1(x, \xi), \quad (10)$$

where

$$\mathcal{M}(x) = \max |f(\xi)|, \quad \xi \in [0, x],$$
$$\mathcal{M}_1(x, \xi) = \max |K_1(x, y)|, \quad y \in [x, \xi].$$

From the inequalities (10) it follows that the first two terms on the right-hand side of equation (9) tend to zero as $\delta \to 0$. Observing that $K_\delta(x, \xi)$ is uniformly bounded in δ we get

$$\lim_{\delta \to 0} \|u(x) - u_\delta(x)\|_c = 0.$$

6.8. Operator Volterra Equations

Let $u(t)$ be a continuous function of scalar argument t with range in a Hilbert space U, and $A(t, \tau)$ be a family of continuous operators with domain U and range from U. The family $A(t, \tau)$ is assumed to depend continuously on variables (t, τ). By Volterra equations of the first and second kind for $u(t)$ we mean

$$\int_0^t A(t, \tau) u(\tau) d\tau = f(t), \quad (1)$$

and

$$Bu(t) + \int_0^t A(t, \tau) u(\tau) d\tau = f(t), \quad (2)$$

respectively. Here B is a continuous operator, and

$$N(B) = 0. \quad (3)$$

We will consider equations (1) and (2) on a finite interval $t \in [0, T]$. If B^{-1} is bounded, the second kind equation (2) is equivalent to

$$u(t) + \int_0^t A(t, \tau) u(\tau) d\tau = f(t).$$

6.7. Volterra Equations for Function of One Variable

Since the integral operator in (2) is quasinilpotent in the Banach space

$$U_T = C([0, T]; U)$$

under the norm

$$\|u(t)\|_{U_T} = \max \|u(t)\|_U,$$

equation (2) is always uniquely solvable and the solution depends continuously on the right-hand side f.

If the Volterra operator equation of the first kind (1) is reducible to the second-kind equation (2) with bounded operator B^{-1} by the differentiation with respect to t, then the results of the preceding section for scalar Volterra operations carry over to the operator equations as they stand.

Now let us look at the case of B^{-1} unbounded.

THEOREM 1. *Let the operator B be self-adjoint, and let it commute with all $A(t, \tau)$. Then the solution to equation (2) is unique.*

PROOF. Denote by B_T the extension of B to the space U_T, and by A_T the integral operator in equation (2) as one from U_T into U_T. Since B is self-adjoint and $N(B) = 0$, for any $u(t) \in U_T$, there exists a $c > 0$ such that (see Section 3.9)

$$\|B_T^n u(t)\|_{U_T} \geq c^n \|u(t)\|_{U_T}.$$

On the other hand, since the integral operator in equation (2) is quasinilpotent, we conclude that for any $\delta > 0$ there exists an n_0 such that for all $n \geq n_0$

$$\|A_T^n u(t)\|_{U_T} \leq \delta^n \|u(t)\|_{U_T}. \blacksquare$$

Note in passing that for the norm of A_T^n it is not hard to obtain the following inequality

$$\|A_T^n\| \leq \frac{a^n T}{n!},$$

where

$$a = \max \|A(t, \tau)\|.$$

Consider an example of Volterra operator equations of the first kind that can be reduced to equations of the second kind with unbounded B^{-1}. Let all operators $A(t, \tau)$ be self-adjoint and commutative. Then there exist a family of projection-valued operators $\{E_\lambda\}$ and a scalar-valued function $a(t, \tau, \lambda)$ such that

$$A(t, \tau) = \int a(t, \tau, \lambda) dE_\lambda.$$

In this case, equation (1) is equivalent to a family of scalar Volterra equations depending on a parameter λ

$$\int_0^t a(t, \tau, \lambda) u(\tau, \lambda) d\tau = f(\tau, \lambda). \tag{4}$$

If for any λ the function $a(t, \tau, \lambda)$ is continuously differentiable with respect to t, and $a(t, t, \lambda) > 0$, then for a fixed λ, equation (4) is equivalent to the following Volterra equation of the second kind

$$u(t, \lambda) + \int_0^t a_1(t, t, \lambda) u(\tau, \lambda) d\tau = f_1(t, \lambda), \tag{5}$$

$$a_1(t, \tau, \lambda) = \frac{a_t(t, \tau, \lambda)}{a(t, t, \lambda)},$$

$$f_1(t, \lambda) = f_t(t, \lambda)/a(t, t, \lambda).$$

For a fixed λ, the solution to (5) is unique, and continuously depends on the right-hand side $f_1(\cdot)$. This implies the uniqueness of solutions to equations (4) and (1). However, the function $a_1(\cdot)$ may tend to infinity as $\lambda \to \infty$, and so the solution to equation (2) may not depend continuously on the right-hand side.

Introducing the designation

$$\mu(\lambda) = \max |a_1(t, \tau, \lambda)|,$$

it is an easy matter to see that in this case equation (1) reduces to an equation (2) with

$$B = \int \mu(\lambda)^{-1} dE_\lambda.$$

Chapter 7. EVOLUTION EQUATIONS

7.1. Cauchy Problems and Operator Semigroups

Let $u(t)$ be a function of scalar argument t with values in a Banach space U, and A be a linear operator with domain $D(A)$ dense everywhere in U and with values from U. Consider the equation

$$\frac{du}{dt} = Au. \tag{1}$$

By a solution to equation (1) we mean a function continuously differentiable in the strong sense, belonging to $D(A)$ for all $t \geqslant 0$ and obeying equation (1) [20].

The Cauchy problem for equation (1) is to determine a solution to (1) satisfying the condition

$$u(0) = u_0, \quad u_0 \in D(A).$$

In agreement with the earlier given definitions, *the Cauchy problem for equation (1) is said to be well-posed provided that the following conditions are fulfilled*:
 (i) *for any $u_0 \in D(A)$ the solution exists*;
 (ii) *the solution is unique*;
 (iii) *the solution depends continuously on the initial values, i.e., from*

$$\lim_{n \to \infty} u_n(0) = 0, \quad u_n(0) \in D(A),$$

it follows that for any t

$$\lim_{n \to \infty} u_n(t) = 0.$$

Generally we shall consider the Cauchy problem on a finite interval, $t \in [0, T]$. However, for a well-posed Cauchy problem, the well-posedness on an interval $[0, T]$ implies the well-posedness on any other interval $[0, T_1]$, hence on the entire semiaxis $[0, \infty]$.

Indeed, consider an interval $[0, 2T]$, and let $u(t)$ be the solution on $[0, T]$. Let us construct a solution $v(t)$ of equation (1) such that

$$v(0) = u(T),$$

and define the function $W(t)$ as

$$W(t) = u(t), \quad t \in [0, T]$$
$$W(t) = v(t - T), \quad t \in [T, 2T].$$

Clearly, $W(t)$ will be a solution of equation (1) on the interval $[0, 2T]$ with the data

$$W(0) = u_0.$$

Denote by $B(t)$ the operator associating with an element $u_0 \in D(A)$ the value of a solution $u(t)$ to the Cauchy problem for equation (1) for a fixed $t > 0$. If the Cauchy problem is well-posed, then the operator $B(t)$ is defined on $D(A)$ and continuous. Consequently, $B(t)$ can be extended to a bounded linear operator defined on the whole U. It is easy to see that the family of operators $B(t)$ is a semigroup, that is,

$$B(t_1 + t_2) = B(t_1)B(t_2), \quad t_1, t_2 > 0.$$

Indeed, let $u_0 \in D(A)$. Then the function

$$W(t) = u(t + \tau) = B(t + \tau)u_0$$

satisfies equation (1) in t and the initial condition

$$W(0) = u(\tau) = B(\tau)u_0.$$

The function

$$v(t) = B(t)B(\tau)u_0$$

is also a solution to equation (1) with the initial value $B(\tau)u_0$. Since the solution is unique, we obtain $v(t) = W(t)$.

Thus, the operators $B(t + \tau)$ and $B(t)B(\tau)$ coincide on the set $D(A)$ everywhere dense in U, hence on the whole U.

If u_0 does not belong to $D(A)$, then the function $B(t)u_0$ is called a *generalized solution* to equation (1).

Consider now the complete equation,

$$\frac{du}{dt} = Au + f, \qquad (2)$$

where f is a given continuous function with values in U. If the Cauchy problem for the corresponding homogeneous equation (1) is well-posed,

7.2. Equations in Hilbert Spaces

then the solution of equation (2) is given by the formula

$$u(t) = B(t)u(0) + \int_0^t B(t-\tau)f(\tau)d\tau, \qquad (3)$$

where $B(t)$ is the operator defining the solution to the Cauchy problem for the homogeneous equation.

Indeed, differentiating equation (1) yields

$$\frac{du}{dt} = B'(t)u(0) + B(0)f(t) + \int_0^t B'(t-\tau)f(\tau)d\tau$$

$$= AB(t)u(0) + f(t) + A\int_0^t B(t-\tau)f(\tau)d\tau = Au(t) + f(t).$$

7.2. Equations in Hilbert Spaces

Let $u(t)$ be a function with values in a real Hilbert space over the field of real numbers R, and A be a self-adjoint operator. Consider the equation

$$\frac{\partial u}{\partial t} = Au. \qquad (1)$$

THEOREM 1. *Let $u(t)$ be a solution of equation (1) on the interval $[0, T]$. Then*

$$\|u(t)\| \leq \|u(T)\|^{t/T} \|u(0)\|^{(T-t)/T}. \qquad (2)$$

COROLLARY. *The solution to the Cauchy problem for equation (1) is unique.*

PROOF. First, suppose that $u(t) \neq 0$, $t \in [0, T]$, and $u(t)$ is twice continuously differentiable.

Consider the functions

$$\varphi(t) = \|u(t)\|^2 = (u, u),$$

$$\psi(t) = \ln \varphi(t).$$

The derivatives of φ are

$$\varphi'(t) = 2(u, u_t) = 2(u, Au),$$

$$\varphi''(t) = 2(Au, Au) + 2(u, A^2u) = 4(Au, Au),$$

whence by the Cauchy-Bunyakovski inequality

$$\psi''(t) = \frac{1}{\varphi^2(t)} [\varphi(t)\varphi''(t) - \varphi'^2(t)] \geq 0. \tag{3}$$

This inequality implies

$$\psi(t) \leq \frac{t}{T} \psi(t) + \frac{T-t}{T} \psi(0),$$

hence also (2).

Proceeding to the case when $u(t)$ does not have the second derivative, we consider the function

$$u_h(t) = \int_{t-h}^{t+h} g_h(\tau) u(\tau) d\tau$$

where $g_h(t)$ is a smooth function satisfying the conditions

$$g_h(-h) = g_h(h) = 0,$$

$$\int_{-h}^{h} g_h(t) dt = 1.$$

For instance, we can let

$$g_h(t) = \frac{1}{2h} \left(1 + \cos \frac{\pi}{h} t \right).$$

Clearly, $u_h(t)$ will be a twice continuously differentiable solution to equation (1) on $[h, T - h]$.

From the inequality

$$\psi_h''(t) \geq 0$$

with

$$\psi_h(t) = \ln \|u_h(t)\|$$

it follows that for any t_1, t_2 and t such that

$$h \leq t_1 \leq t \leq t_2 \leq T - h, \quad \text{and} \quad t_1 < t_2,$$

we have

$$\|u_h(t)\| \leq \|u_h(t_2)\|^{\frac{t-t_1}{t_2-t_1}} \|u_h(t_1)\|^{\frac{t_2-t}{t_2-t_1}}. \tag{4}$$

7.2. Equations in Hilbert Spaces

Passing to the limit as $h \to 0$, we find that (4) and consequently (2) hold for $u(t)$.

From the inequality (4) it follows that if for some $t_0 \in [0, T]$, $u(t_0) = 0$, then $u(t) = 0$ for any $t \in [0, T]$.

Thus, the inequality (2) is valid in this case as well. ∎

THEOREM 2. *The necessary and sufficient condition for the Cauchy problem (1) to be well-posed is that the operator A be semibounded from above.*

PROOF. Let $\{B_\lambda\}$ be a family of projection operators induced by A (see Section 3.9):

$$A = \int_{-\infty}^{\infty} \lambda \, dE_\lambda.$$

If A is semibounded from above then

$$A = \int_{-\infty}^{a} \lambda \, dE_\lambda$$

and the solution to the Cauchy problem for equation (1) is of the form

$$u(t) = \left(\int_{-\infty}^{a} e^\lambda \, dE_\lambda \right) u_0, \quad u(0) = u_0. \tag{5}$$

The function defined by (5) is indeed a solution to the problem, and its uniqueness has been established above. The continuous dependence on the data is obvious.

Now, let A be not semibounded from above. Then for any $a > 0$ there exists $b > a$ such that the subspace

$$U_{ab} = (E_b - E_a) U$$

is not empty.

Let $u_0 \in U_{ab}$. Then the norm of the solution to the Cauchy problem with the initial condition u_0 satisfies the inequality

$$\|u(t)\| = \left\| \left(\int_{a}^{b} e^\lambda \, dE_\lambda \right) \right\| \geq e^a \|u_0\|. \tag{6}$$

This inequality implies that arbitrarily small Cauchy data can produce solutions arbitrarily large in the norm, that is, the solution does not depend continuously on the data any longer.

The theorems established above admit a generalization to the evolution equations with a normal operator in a Hilbert space over the field of complex numbers. Let in equation (1) U be a Hilbert space over the field of complex numbers, and A be a normal operator.

THEOREM 3. *Let $u(t)$ be the solution of equation (1) on the interval $[0, T]$. Then the inequality (2) holds.*

PROOF. We assume as before that $u(t) \neq 0$, $t \in [0, T]$, $u(t)$ is twice continuously differentiable and $\varphi = (u, u)$, $\psi = \ln \varphi(t)$.
Differentiating φ and ψ we get

$$\varphi'(t) = (u, (A + A^*)u),$$

$$\varphi''(t) = (Au, (A + A^*)u) + (u, (A + A^*)Au)$$
$$= ((A + A^*)u, (A + A^*)u),$$

$$\varphi''(t) = \frac{1}{\varphi^2(t)}[\varphi(t)\varphi''(t) - (\varphi')^2(t)] \geq 0.$$

The assertion now follows exactly as above. ∎

Note that a normal operator A can be recast as

$$A = X + iY,$$

where X and Y are self-adjoint and commutative.

THEOREM 4. *The necessary and sufficient condition that the Cauchy problem for equation (1) be well-posed is that the operator X be semibounded from above.*

The proof of this theorem is entirely analogous to that of Theorem 2.

7.3. Equations with Variable Operator

Let $u(t)$ be a function with values in a Hilbert space U over the field of real numbers R, and $A(t)$ be a family of operators depending on a parameter $t \in [0, T]$. Consider the equation

$$\frac{du}{dt} = Au. \qquad (1)$$

Suppose

$$A = A_0 + A_1,$$

where the operator A_0 is constant and self-adjoint, $A_1(t)$ has a continuous derivative $A_1'(t)$, and the operators A_1 and A_1' are bounded.

7.3. Equations with Variable Operator

THEOREM. *Let there exist constants* a, $b_0 \geq 0$ *such that for any* u,

$$(A_0 u, A_1^* u) - (A_0 u, A_1 u) \geq -a(u, Au) - b_0(u, u). \tag{2}$$

Then any solution of equation (1) *satisfies the inequality*

$$\|u(t)\| \leq \|u(T)\|^{w(t)} \|u(0)\|^{1-w(t)} c(t), \tag{3}$$

where

$$w(t) = \frac{1 - e^{-at}}{1 - e^{-aT}},$$

$$c(t) = \exp\left\{ \frac{b}{a} \frac{(1 - e^{-at})T - (e^{-aT} - 1)t}{1 - e^{-aT}} \right\},$$

$$b = 2b_0 + 4a_1^2 + 2a_2,$$

$$a_1 = \max \|A_1(t)\|, \quad a_2 = \max \|A_1'(t)\|.$$

COROLLARY. *The solution to the Cauchy problem for equation* (1) *is unique.*

PROOF. Differentiating $\varphi(t) = \|u(t)\|^2 = (u, u)$ yields

$$\varphi'(t) = 2(u, Au) = 2(u, A_0 u) + 2(u, A_1 u),$$

$$\varphi''(t) = 2(Au, Au) + 2(u, A^2 u) + 2(u, A_1' u)$$

$$= 4(A_0 u, A_0 u) + 2(A_0 u, A_1^* u) - 2(A_0 u, A_1 u)$$

$$+ 2(A_1^* u, A_1 u) - 2(A_1 u, A_1 u) + 2(u, A_1' u)$$

$$\geq 4(A_0 u, A_0 u) - 2a(u, Au) - b(u, u). \tag{4}$$

These relations imply that if $\varphi(t) \neq 0$, then the second derivative of $\psi(t) = \ln \varphi(t)$ satisfies the inequality

$$\psi''(t) + a\psi'(t) + b \geq 0. \tag{5}$$

From the theory of ordinary differential equations it is known that a function $\psi(t)$ satisfying (5) also obeys

$$\psi(t) \leq \psi_0(t), \tag{6}$$

where $\psi_0(t)$ is the solution of the differential equation

$$\psi_0''(t) + a\psi_0'(t) + b = 0$$

with the boundary conditions
$$\psi_0(0) = \psi(0), \quad \psi_0(T) = \psi(T).$$
It is not hard to verify that
$$\psi_0(t) = c_1 + c_2 e^{-at} - \frac{b}{a} t \qquad (7)$$
with
$$c_1 = \frac{\psi(T) - \psi(0) e^{-at} + \frac{b}{a} T}{1 - e^{-aT}}$$
and
$$c_2 = \frac{\psi(0) \psi(T) - \frac{b}{a} T}{1 - e^{-aT}}.$$

Now, the inequality (3) we wish to prove readily follows from (6) and (7). ∎

7.4. Equations of the Second Order

Let $u(t)$ be a function with values in a Hilbert space over the field of real numbers, and A be a self-adjoint operator. Consider the equation
$$\frac{d^2 u}{dt^2} = Au. \qquad (1)$$

THEOREM 1. *Any solution of equation* (1) *satisfies the inequality*
$$\|u(t)\|^2 \leqslant c(t)(\|u(T)\|^2 + |a|)^{\frac{t}{T}} (\|u(0)\|^2 + |a|)^{\frac{T-t}{T}} - |a| \qquad (2)$$
with $c(t) = e^{2t(T-t)}$ and $a = \frac{1}{2}[(Au(0), u(0)) - (u'(0), u'(0))].$

PROOF. Consider the function
$$\varphi(t) = (u, u).$$
Its derivatives read
$$\varphi'(t) = 2(u, u'), \quad \varphi''(t) = 2(u', u') + 2(u, Au).$$
If we differentiate the second term in the expression for $\varphi''(t)$, then
$$\frac{d}{dt}(u, Au) = 2(u', Au) = 2(u', u'') = \frac{d}{dt}(u', u').$$

7.4. Equations of the Second Order

Consequently, $\varphi''(t) = 4(u', u') + 4a$.

Now, consider the function $\psi(t) = \ln(\varphi(t) + |a|)$.
Differentiating $\psi(t)$ yields

$$\psi''(t) = \frac{\varphi''(t)(\varphi(t) + |a|) - (\varphi'(t))^2}{(\varphi(t) + |a|)^2}$$

$$= 4\frac{[(u', u') + a][(u, u) + |a|] - (u, u')^2}{[\varphi(t) + |a|]^2} \geq -4,$$

$$\psi(t) \leq \frac{t}{T}\psi(T) + \frac{T-t}{T}\psi(0) + 2t(T-t),$$

whence the assertion of the theorem immediately follows. ∎

THEOREM 2. *The necessary and sufficient condition that the Cauchy problem for equation* (1) *be well-posed is that the operator A be semibounded from above.*

PROOF. Let $\{E_\lambda\}$ be the family of projection operators induced by the operator

$$A = \int_{-\infty}^{\infty} \lambda dE_\lambda.$$

If A is semibounded from above, then

$$A = \int_{-\infty}^{a} \lambda dE_\lambda.$$

For certainty let $a > 0$. In this case, the solution of the Cauchy problem has the form

$$u(t) = \left(\int_{-\infty}^{0} \cos(\sqrt{-\lambda}\, t)\, dE_\lambda\right)u_0 + \left(\int_{0}^{a} \operatorname{ch}(\sqrt{\lambda}\, t)\, dE_\lambda\right)u_0$$

$$+ \left(\int_{-\infty}^{0} \frac{\sin(\sqrt{-\lambda}\, t)}{\sqrt{-\lambda}}\, dE_\lambda\right)u_1 + \left(\int_{0}^{a} \frac{\operatorname{sh}(\sqrt{\lambda}\, t)}{\sqrt{\lambda}}\, dE_\lambda\right)u_1,$$

$$u_0 = u(0), \quad u_1 = u'(0). \quad ∎$$

If A is not semibounded from above, then at sufficiently large $a > 0$ and $b > a$ the solution of the Cauchy problem with the data

$$u_0 \in E_b - E_a, \quad u_1 = 0,$$

can be arbitrarily large in the norm for any Cauchy data, however small.

7.5. Well-Posed and Ill-Posed Cauchy Problems

Let D be a bounded region of the n-dimensional space R^n with a smooth boundary S. Consider a self-adjoint elliptic differential operator on D subject to a Dirichlet boundary condition:

$$Lu = \sum_{j,k=1}^{n} \frac{\partial}{\partial x_j} a_{jk} \frac{\partial u}{\partial x_k} + cu,$$

$$u|_S = 0,$$

$$\sum_{j,k=1}^{n} a_{jk} \xi_j \xi_k \geq \delta |\xi|^2, \quad \delta > 0.$$

Here a_{jk} are continuously differentiable functions, and c is a continuous function.

As is known, this operator has a complete orthogonal set of eigenfunctions. The sequence of eigenvalues of the operator is bounded from above and tends to $-\infty$.

Consider mixed problems for differential equations in cylindrical domains $D_T = D \times [0, T]$

$$\frac{\partial u}{\partial t} = Lu, \quad u|_{t=0} = u_0, \tag{1}$$

$$\frac{\partial u}{\partial t} = -Lu, \quad u|_{t=0} = u_0, \tag{2}$$

$$\frac{\partial^2 u}{\partial t^2} = Lu, \quad u|_{t=0} = u_0, \quad u'|_{t=0} = u_1, \tag{3}$$

$$\frac{\partial^2 u}{\partial t^2} = -Lu, \quad u|_{t=0} = u_0, \quad u'|_{t=0} = u_1, \tag{4}$$

$$u|_S = 0.$$

According to the results of Sections 7.2 and 7.5, the problems (1) and (3) are well-posed, while the problems (2) and (4) are ill-posed. All these problems are uniquely solvable. The theorems established in Sections 7.2 and 7.5 furnish estimators of the conventional stability for the ill-posed problems (2), (4).

Chapter 8. PROBLEMS OF INTEGRAL GEOMETRY

8.1. Determining a Function from Its Spherical Averages

Inverting the Radon transform, a classical problem of integral geometry, was considered in Section 3.4. In what follows is another classical problem of integral geometry associated with the ill-posed Cauchy problem for the wave equation. The presentation is based on the results outlined in the books [19, 21].

Consider the problem of determining a function $u(x_1, ..., x_n)$ in an n-dimensional space $(x_1, ..., x_n)$, $n \geq 2$, given its average values over spheres (circles for $n = 2$) of arbitrary radius r ($0 < r < \infty$) whose centers trace out a set of points in a fixed plane. For convenience we embed this plane into the coordinate plane $x_n = 0$, and use the designation (x, y), where $x = (x_1, ..., x_{n-1})$, for an arbitrary point $(x_1, ..., x_n)$. Thus, $u = u(x, y)$. We denote a sphere of radius r about $(x, 0)$ by $S(x, r)$. In this notation, the problem under consideration consists in evaluating a function $u(x, y)$ by its integrals

$$\frac{1}{w_n} \int_{S(x,r)} u(\xi, \eta) d\omega = v(x, r). \tag{1}$$

Here (ξ, η) is a variable point on the sphere $S(x, r)$, $\xi = (\xi_1, ..., \xi_{n-1})$, w_n is the area of a unit sphere in the n-dimensional space:

$$w_n = \frac{2\pi^{n/2}}{\Gamma(n/2)},$$

and $d\omega$ is an elementary solid angle subtended by the elementary surface area, $dS = r^{n-1} d\omega$.

Clearly, any function $u(x, y)$ odd in the variable y is a solution to the homogeneous equation (1), where $v(x, r) = 0$. It is reasonable therefore to state the problem of determining only the even component of $u(x, y)$, that is the function $u_1(x, y) = \frac{1}{2}[u(x, y) + u(x_1 - y)]$ from the function $v(x, r)$. This is equivalent to considering the set of functions $u(x, y)$ even in y. Clearly, the problem in this formulation is equivalent to one of determining a function $u(x, y)$ satisfying the condition $u(x, y) = 0$, $y \leq 0$. The following uniqueness theorem is valid.

THEOREM 1. *Any function $u(x, y)$, continuous in the domain $D(x_0, r_0) = \{(x, y) : |x - x_0|^2 + y^2 < r_0^2\}$ and even in y is uniquely determined within this domain by fixing a function $v(x, r)$ in the region $G_\varepsilon(x_0, r) = \{(x, r) : |x - x_0| < \varepsilon, 0 < r < r_0 - |x - x_0|\}$, where ε is any fixed positive number such that $\varepsilon \leq r_0$.*

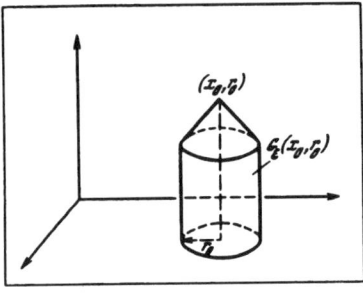
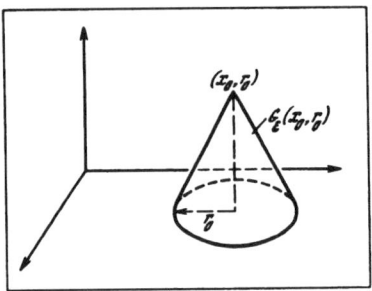

Fig. 2   Fig. 3

PROOF. This proof is based on computing all the moments of $u(x, y)$ with respect to x on each fixed sphere $S(x, r)$ such that $(x, r) \in G_\varepsilon(x_0, r_0)$. The region $G_\varepsilon(x_0, r_0)$ resembles a pointed pencil (Fig. 2). At $\varepsilon = r_0$ this "pencil" degenerates into a piece of cone (Fig. 3). For each point $(x, r) \in G_\varepsilon(x_0, r_0)$, $0 < \varepsilon < r_0$ the sphere $S(x, r)$ is entirely contained in the domain $D(x_0, r_0)$.

Consider the result of applying the operator

$$A_i v = \frac{\partial}{\partial x_i} \int_0^r \tau^{n-1} v(x, \tau) d\tau, \quad i = 1, \ldots, n-1,$$

to equation (1). After some calculations we find

$$A_i v = \frac{1}{w_n} \frac{\partial}{\partial x_1} \int_0^r \tau^{n-1} \left[\int_{S(x,\tau)} u(\xi, \eta) d\omega \right] d\tau$$

$$= \frac{1}{w_n} \frac{\partial}{\partial x_i} \int_{|\xi - x|_i^2 + \eta^r \le r^2} \int_{x_i - \sqrt{r^2 - \eta^2 - |\xi - x|_i^2}}^{x_i - \sqrt{r^2 - \eta^2 - |\xi - x|_i^2}}$$

$$d\eta d\xi_1 \ldots d\xi_{i-1} \ldots d\xi_{i+1} \ldots d\xi_{n-1} = \frac{1}{w_n} \int_{S(x,r)} u(\xi, \eta) \cos(n, \xi_i) dS$$

$$= \frac{r^{n-2}}{w_n} \int_{S(x,r)} u(\xi, \eta)(\xi_i - x_i) d\omega.$$

Here $|\xi - \prime x|_i$ stands for the expression $\sqrt{|\xi - x|^2 - |\xi_i - x_i|^2}$ which numerically coincides with the projection of the straight line joining the points $(\xi, 0)$ and $(x, 0)$ on the plane $x_i = 0$, $y = 0$, and n is the unit vector of the outward normal to $S(x, r)$.

8.1. Determining a Function from Its Spherical Averages

Denoting by L_i the operator defined by the identity

$$L_i v = \frac{1}{r^{n-2}} A_i v + x_i v(x, r), \quad i = 1, \ldots, n-1$$

we find from the above formula that the result of applying L_i to equation (1) is equivalent to calculating the spherical average for the function $u_i(x, y) = x_i u(x, y)$. Indeed,

$$L_i v = \frac{1}{w_n} \int_{S(x,r)} u(\xi, \eta) \xi_i \, d\omega.$$

It is clear that the application of the superposition of operators $L_j \cdot L_i$ to the function $v(x, y)$ is equivalent to calculating the spherical average of $u(x, y) x_i x_j$. Generally, if $P_m(x)$ is a polynomial of degree m with constant coefficients, then applying the operator $P_m(L)$, $L = (L_1, \ldots, L_m)$, to equation (1) leads to the equality

$$P_m(L) v = \frac{1}{w_n} \int_{S(x,r)} u(\xi, \eta) P_m(\xi) \, d\omega.$$

Now we fix a point (x, r). Let $D_0(x, r) = \{ \xi : |\xi - x| \leq r \}$. On the sphere $S(x, r)$ this equality can be represented in the form

$$P_m(L) v = \int_{D_0(x,r)} \varphi(\xi, x, r) P_m(\xi) \, d\xi, \tag{2}$$

where the function $\varphi(\xi, x, r)$ (x and r are fixed) is related to $u(x, y)$ by the formula

$$\varphi(\xi, x, r) = \frac{2}{w_n r^{n-2}} \frac{u(\xi, \sqrt{r^2 - |x - \xi|^2})}{\sqrt{r^2 - |x - \xi|^2}}.$$

Here we took into account that the function $u(x, y)$ is even in y.

The set of equations (2) corresponding to different polynomials $P_m(\xi)$ uniquely defines the function $\varphi(\xi, x, r)$ and consequently the function $u(x, y)$. ∎

To construct the function $u(x, y)$ by a function $v(x, r)$, one can take as $P_m(\xi)$ a set of polynomials orthogonal in $D_0(x, r)$. Then $\varphi(\xi, x, r)$ can be represented explicitly as a Fourier series. This would require, however, more stringent conditions on the function $u(x, y)$ to guarantee the convergence of the Fourier series.

COROLLARY. *If the function $u(x, y)$ is continuous within a domain $D(x_0, r_0)$ along with its partial derivative $u_y(x, y)$, then it is uniquely determined in $D(x_0, r_0)$ by the spherical averages of both $u(x, y)$*

and its partial derivative $u_y(x, y)$ over all the spheres $S(x, r)$ such that $(x, r) \in G_\varepsilon(x_0, r_0)$, $0 < \varepsilon \leq r_0$.

Indeed, in this case the part of $u(x, y)$ even in y, and the even part of the partial derivative $u_y(x, y)$ are determined uniquely. Here we note that any function $u(x, y)$ of the class under consideration can be represented as

$$u(x, y) = \frac{1}{2}[u(x, y) + u(x, -y)] + \frac{1}{2}\int_0^y [u_y(x, y) + u_y(x, y)]dy,$$

and, consequently, can be uniquely determined.

The proved theorem tells us that giving a function $v(x, r)$ within a region $G_\varepsilon(x_0, r_0)$ for any positive ε however small defines the function $u(x, y)$ in $D(x_0, \tau_0)$. If $u(x, y)$ is given within $D(x_0, r_0)$ one can compute the integrals over all spheres interior to the domain $D(x_0, r_0)$, and so determine $v(x, r)$ for $(x, r) \in G_\varepsilon(x_0, r_0)$ for $\varepsilon = r_0$. Hence, giving a function $v(x, r)$ in $G_\varepsilon(x_0, \tau_0)$, uniquely determines it in a wider region $G_{r_0}(x_0, r) \supset G_\varepsilon(x_0, r_0)$.

This result indicates that the function $v(x, r)$ cannot be chosen arbitrarily. Moreover, any $v(x, r)$ which admits the representation (1) possesses a property analogous to that of analytic functions: it is uniquely defined in $G_{r_0}(x_0, r_0)$ by its values within an arbitrarily thin domain $G_\varepsilon(x_0, r_0)$. Also it is clear that to nonanalytic functions $u(x, y)$ there correspond nonanalytic $v(x, r)$. To summarize, a constructive description of the class of functions $v(x, r)$ representable in the form (1) is a rather hard task.

We demonstrate now that the problem of solving equation (1) is classically ill-posed, namely, that it is strongly unstable with respect to small perturbations of $v(x, r)$. This demonstration can be done most conveniently in three dimensions, so in what follows we presume $n = 3$. Furthermore, we suppose here that $u(x, y)$ is twice continuously differentiable in the whole space (x, y) and is even in y. Let us introduce the function

$$w(x, y, r) = \frac{r}{w_\xi} \int_{|\xi-x|^2+(\eta-y)^2 = r^2} u(\xi, \eta)d\omega, \qquad (3)$$

which, up to a factor r, is the spherical average over the sphere of radius r about (x, y). From a course of mathematical physics (see, e.g., 21) it is known that in the half-space $r > 0$ the function $w(x, y, r)$ satisfies the wave equation

$$\frac{\partial^2 w}{\partial r^2} = \Delta_{xy} w, \qquad (4)$$

8.1. Determining a Function from Its Spherical Averages

where Δ_{xy} is the Laplacian with respect to x, y. From the formula (3) we have

$$w(x, y, 0) = 0, \tag{5}$$

and from equation (1)

$$w(x, y, r)\big|_{y=0} = rv(x, r). \tag{6}$$

Finally, the condition that $u(x, y)$ is even in y leads to the identity

$$w_y(x, y, r)\big|_{y=0} = 0. \tag{7}$$

Evaluation of a function $w(x, y, r)$ satisfying equation (4) and the conditions (5)–(7) is a boundary-value problem, which obviously is equivalent to the Cauchy problem (4), (6), (7) in the whole space (x, y, r) if $v(x, r)$ is evenly extended to the region $r < 0$.

THEOREM 2. *Let problem (4)–(7) be equivalent to solving equation (1) in the class of functions even in y.*

PROOF. If $u(x, y)$ is a solution of equation (1) such that $u(x, y) = u(x_1 - y)$, then the function $w(x, y, r)$ computed by the formula (3) is a solution of the problem (4)–(7). This solution is unique because the homogeneous problem (4)–(7) has the zero solution alone. Thus, to any function $u(x, y)$ there corresponds exactly one solution of the problem (4)–(7).

Now, we show that the converse statement is also true, namely, that every solution of the problem (4)–(7) induces exactly one solution of equation (1) such that $u(x_1 - y) = u(x, y)$. Let $w(x, y, r)$ be a solution of the problem (4)–(7). Denote

$$w_r(x, y, 0) = u(x, y). \tag{8}$$

Then for $r \geq 0$, $w(x, y, r)$ can be represented uniquely as a solution in the form (3) of the Cauchy problem with the data (5), (8). Comparing the formula (3) for $y = 0$ with the formula (6) we see that $u(x, y)$ is a solution of equation (1). Taking the derivative with respect to y of the both sides of the identity (3), setting $y = 0$, and allowing for (7), we find

$$\int_{S(x,r)} u_r(\xi, \eta)\,d\omega = 0. \tag{9}$$

From this identity and Theorem 1 it follows that the even component of $u_y(x, y)$ is zero:

$$u_y(x, -y) + u_y(x, y) = 0,$$

that is, $u(x, -y) = u(x, y)$.

Thus, the function $u(x, y)$ computed from a solution of the problem (4)–(7) by the formula (8) is endowed with the property of evenness in the variable y. By Theorem (1), to any solution of the problem (4)–(7) there corresponds only one such function.

By this reasoning we have proved that there exists a one-to-one correspondence between the solutions of the problem (4)–(7) and those of equation (1) provided that $u(x, -y) = u(x, y)$. It is this statement that constitutes the equivalence theorem. ∎

Now, by way of example, we illustrate that the problem (4)–(7) is unstable. Let

$$v(x, r) = \frac{1}{k^s} \frac{\sin kr}{r} \sin(kx_1) \cdot \sin(kx_2).$$

For sufficiently large k and s this function is small together with any finite number of its derivatives. One can readily verify that in this case the solution to the problem (4)–(7) is given by

$$w(x, y, r) = \frac{1}{k^s} \sin(kr) \sin(kx_1) \sin(kx_2) \cosh(ky).$$

However, for any fixed y, this solution tends to infinity as $k \to \infty$. The same property is exhibited by the function $u(x, y)$ computed by the formula (8) and yielding a solution to equation (1):

$$u(x, y) = \frac{1}{k^{s-1}} \sin(kx_1) \sin(kx_2) \cosh(ky).$$

This means that the solutions to the problem (4)–(7), and the equivalent problem of solving equation (1) are strongly unstable.

8.2. General Problems of Plane Integral Geometry

Let D be an open, simply connected region with boundary Γ in the (x, y) plane. Let the boundary be given by $x = g(s)$, $y = p(s)$, where s is the length measured from a fixed point of Γ in the positive direction consistent with the choice of orientation on Γ; $g(s), p(s)$ are functions of the class $C^1[0, l]$, $g(0) = g(l)$, $p(0) = p(l)$; and l is the length of Γ. Let also $L(t_1, t_2)$ be a two-parameter family of curves in $\overline{D}$ endowed with the following properties:

(i) *Each pair of points from $\overline{D}$ is joined by one curve of the family. Any curve $L(t_1, t_2)$ crosses Γ at the points (x_1, y_1), (x_2, y_2), where*

$$x_1 = g(s_1), \quad x_2 = g(s_2),$$
$$y_1 = p(s_1), \quad y_2 = p(s_2),$$

8.2. General Problems of Plane Integral Geometry

while the other points of L (t_1, t_2) belong to D. The lengths of the curves L (t_1, t_2) are bounded as a totality.

(ii) *The equation of the curve passing through (x_0, y_0) in the direction $v^0 = (\cos\theta_0, \sin\theta_0)$ is given by*

$$x = f_1(s, \theta_0, x_0, y_0) = x_0 + s\cos\theta_0 + s^2 \tilde{f}_1(x, \theta_0, x_0, y_0), \quad (1)$$
$$y = f_2(s, \theta_0, x_0, y_0) = y_0 + s\sin\theta_0 + s^2 \tilde{f}_2(x, \theta_0, x_0, y_0),$$

where $\tilde{f}_j(\cdot)$ are functions of the arc length S measured from $(x_0, y_0) \in \overline{D}$ and the parameters $\theta_0 \in [0, 2\pi]$ and (x_0, y_0); they are continuously differentiable and bounded along with the derivatives, $f_j(s, 0, x_0, y_0) = f_j(s, 2\pi, x_0, y_0)$

$$\frac{1}{s} \frac{D(f_1, f_2)}{D(s, \theta_0)} \geq c_0 > 0, \quad (2)$$

over the whole domain of the variables (s, θ, x_0, y_0).

We note that for s close to zero the inequality (2) readily follows from equations (1), and so is valid for finite s. In this case, it is equivalent to the positiveness of the Jacobian

$$\frac{D(f_1, f_2)}{D(s_1, \theta_0)}.$$

The following lemma holds for the family of curves L.

LEMMA 1. *Let the family of curves satisfy the conditions stated above and the functions $\tilde{f}_j(\cdot)$ have continuous and bounded derivatives of the order up to $m > 1$. Then equations (1) define s and v^0 as single-valued and m-times continuously differentiable functions of (x_0, y_0), (x, y) for all (x_0, y_0), $(x, y) \in \overline{D}$ and $(x_0, y_0) \neq (x, y)$, moreover, in a neighborhood of the set $(x, y) = (x_0, y_0)$ the following estimates hold true for these functions:*

$$\left| D^\alpha s(x_0, y_0, x, y) \right| \leq \frac{c}{[(x-x_0)^2 (y-y_0)^2]^{\frac{|\alpha|-1}{2}}}, \quad (3)$$

$$\left| D^\alpha v^0(x_0, y_0, x, y) \right| \leq \frac{c}{[(x-x_0)^2 + (y-y_0)^2]^{\frac{|\alpha|}{2}}},$$

with $|\alpha| \leq m$.

PROOF. From equations (1) we have

$$s = \frac{[(x-x_0)^2 - (y-y_0)^2]^{1/2}}{[(\cos\theta_0 + s\tilde{f}_1(\cdot))^2 + (\sin\theta_0 + s\tilde{f}_2(\cdot))^2]^{1/2}},$$

$$\cos\theta_0 = \frac{x-x_0}{[(x-x_0)^2 + (y-y_0)^2]^{1/2}} [(\cos\theta_0 + s\tilde{f}_1)^2 + (\sin\theta_0 + s\tilde{f}_2)^2]^{1/2} + [(x-x_0)^2 + (y-y_0)^2]^{1/2}$$

$$\times \frac{\tilde{f}_1(\cdot)}{[(\cos\theta_0 + s\tilde{f}_1)^2 + (\sin\theta_0 + s\tilde{f}_2)^2]^{1/2}}.$$

By the theorem on implicit functions we find that within a sufficiently small domain $(x-x_0)^2 + (y-y_0)^2 < \delta^2$ the functions s and v have the form

$$s = [(x-x_0)^2 + (y-y_0)^2]^{1/2}$$
$$\times \{1 + [(x-x_0)^2 + (y-y_0)^2]^{1/2}\varphi(\cdot)\},$$

$$\cos\theta = \frac{x-x_0}{[(x-x_0)^2 + (y-y_0)^2]^{1/2}}$$
$$\times \{1 + [(x-x_0)^2 + (y-y_0)^2]^{1/2}\varphi(\cdot)\}$$
$$+ [(x-x_0)^2 + (y-y_0)^2]^{1/2}\varphi(\cdot),$$

$$\sin\theta = \frac{y-y_0}{[(x-x_0)^2 + (y-y_0)^2]^{1/2}}$$
$$\times \{1 + [(x-x_0)^2 + (y-y_0)^2]^{1/2}\varphi(\cdot)\}$$
$$+ [(x-x_0)^2 + (y-y_0)^2]^{1/2}\varphi(\cdot),$$

$$\varphi = \varphi(x_0, y_0, x, y),$$
$$\psi = \psi(x_0, y_0, x, y),$$

where φ, ψ are m-times continuously differentiable functions. This representation leads in particular, to the estimate (3). For $(x-x_0)^2 + (y-y_0)^2 > \delta^2$, the immediate implication is $s > \delta$, and, hence $\frac{D(f_1, f_2)}{D(s, \theta_0)} \geq c_0 \delta > 0$. Then, the smoothness of s and v^0 postulated in the lemma follows from the smoothness of $\tilde{f}(\cdot)$, while the existence and uniqueness of these functions follow from the condition (i) imposed on the family of curves. ∎

Let us denote by $L(x, y, x_0, y_0)$ the segment of a curve from the family L, which passes through the points (x, y), (x_0, y_0) and is

8.2. General Problems of Plane Integral Geometry

confined between these points, and by $v = (\cos\theta, \sin\theta)$ the unit vector tangential to $L(x, y, x_0, y_0)$ at (x, y). This vector also can be treated as function of (x, y), (x_0, y_0), having the same properties of smoothness as does the vector v^0. Indeed, if $v^0 = h(x, y, x_0, y_0)$, then, because the points (x, y) and (x_0, y_0) are interchangeable, we have $v = -h(x_0, y_0, x, y)$.

LEMMA 2. *The following inequality holds*

$$\frac{\partial}{\partial s_1}\theta(g(s_1), p(s_1), x, y) \geq 0, \quad (x, y) \in D, \quad s_1 \in [0, l]. \quad (4)$$

PROOF. We will make use of the fact that the function $\theta(x_0, y_0, x, y)$ considered as a function of (x_0, y_0) at a fixed $(x, y) \in D$, has as its level lines the segments of curves from L crossing (x, y) and confined between (x, y) and the boundary Γ. Consequently, grad $\theta(x_0, y_0, x, y)$ at (x_0, y_0) is directed along the normal to $L(x_0, y_0, x, y)$ toward the growing θ. Since for a fixed (x, y) the lines from L intersect only at (x, y), then on any closed curve containing $(x, y) \in D$, in particular on Γ, the positive direction corresponds to higher values of θ. Letting $x_0 = g(s_1)$, $y_0 = p(s_1)$ we see that higher s_1 are associated with higher values of $\theta(g(s_1), p(s_1), x, y)$. This fact leads us to the inequality (4). ∎

Now, for $u(x, y) \in C^1(\overline{D})$ and $\rho(x_0, y_0, x, y) \in C^1(\overline{D} \times \overline{D})$, we consider the function $w(x_0, y_0, x, y)$, defined by

$$w(x_0, y_0, x, y) = \int_{L(x_0, y_0, x, y)} \rho(x_0, y_0, x_1, y_1) u(x_1, y_1) ds, \quad (5)$$

and state the following problem: find $u(x, y)$ within $\overline{D}$ given $\rho(x_0, y_0, x, y)$ and

$$v(t_1, t_2) = w(g(t_1), p(t_1), g(t_2), p(t_2)), \quad t_1, t_2 \in [0, l]. \quad (6)$$

THEOREM 1. *If the family of curves satisfies the conditions (i) and (ii), and the weighting function* $\rho(x_0, y_0, x, y) \in C^1(\overline{D} \times \overline{D})$ *satisfies the constraints*

$$\rho(x_0, y_0, x, y) \in \rho_0 > 0, \quad (7)$$

$$\left|\frac{\partial}{\partial t}\ln\rho(g(t_1), p(t_1), x, y)\right| \leq q\frac{\partial}{\partial t_1}\theta(g(t_1), p(t_1), x, y), \quad (8)$$

$$0 \leq q < 1,$$

then this problem is uniquely solvable for $u \in C^1(\overline{D})$, and the stability of the solution can be estimated by

$$\|u\|_{L_2(D)} \leq \frac{1}{\rho_0\sqrt{1-q^2}} \frac{1}{2\sqrt{\pi}} \|\operatorname{grad}_{t_1,t_2} v(t_1, t_2)\|_{L_2(Q)},$$

$$Q = [0, l] \times [0, l]. \qquad (9)$$

PROOF. First we show that this estimate (9) is valid when the smoothness of $u(\cdot)$, $\rho(\cdot)$, and the function $\tilde{f}(\cdot)$ appearing in equation (1) is one order higher than that required by the theorem. In this case, for $w(\cdot)$ the following lemma holds.

LEMMA 3. *Let $\tilde{f}, \rho,$ and u be twice continuously differentiable with respect to their arguments, and bounded along with their partial derivatives. Then the function $w(x_0, y_0, x, y)$ is twice continuously differentiable with respect to x_0, y_0, x, y everywhere except the points $(x_0, y_0) = (x, y)$ in the neighborhood of which the following estimate holds*

$$|D^\alpha w(\cdot)| \leq c[(x-x_0)^2 + (y-y_0)^2]^{\frac{1-|\alpha|}{2}}, \quad |\alpha| \leq 2. \qquad (10)$$

The proof of this lemma follows from the representation
$w(x_0, y_0, x, y)$
$$= \int_0^{s(x_0, y_0, x, y)} \rho[x_0, y_0, f_1(s, \theta_0(x_0, y_0, x, y) f_2(\cdot)] u(f_1, f_2) ds$$

and Lemma 1.

Now, we derive a differential equation for $w(x_0, y_0, x, y)$. To do so, we note that

$$w(x_0, y_0, f_1(s, \theta_0, x_0, y_0), f_2(\cdot)) = \int_0^s \rho[x_0, y_0, f_1, f_2] u(f_1, f_2) ds_1.$$

Differentiating this equality with respect to s and using

$$(f_1(s, \theta_0, x_0, y_0), f_2(\cdot)) = (x, y),$$
$$(f_{1s}(\cdot), f_{2s}(\cdot)) = \nu(x_0, y_0, x, y),$$

we find

$$(\operatorname*{grad}_{(x,y)} w(x_0, y_0, x, y), \nu(x_0, y_0, x, y)) = \rho(x_0, y_0, x, y) u(x, y). \qquad (11)$$

Dividing both sides by $\rho(x_0, y_0, x, y)$, setting $x_0 = g(t_1)$ and $y_0 = p(t_1)$,

8.2. General Problems of Plane Integral Geometry

and differentiating the resulting identity with respect to t_1 yields the following equation for the function $w(g(t_1), p(t_1), x, y) = \tilde{w}(t, x, y)$

$$\frac{\partial}{\partial t_1}\left\{\frac{1}{\tilde{\rho}(t_1, x, y)}(\nabla_{xy}\tilde{w}, \tilde{v})\right\} = 0, \tag{12}$$

where

$$\tilde{\rho}(\cdot) = \rho(g(t_1), p(t_1), x, y) \text{ and } \tilde{v}(\cdot) = v(g(t_1), p(t_1), x, y).$$

This equation relates to the mixed, hyperbolic type. It has a cylindrical domain $[0, l] \times \overline{D}$. On the boundary of this domain we have the periodicity condition

$$\tilde{w}(0, x, y) = \tilde{w}(l, x, y), \quad (x, y) \in \overline{D}, \tag{13}$$

and a condition which follows from the formula (6):

$$\tilde{w}(t_1, g(t_2) \, p(t_2)) = v(t_1, t_2). \tag{14}$$

Thus, the function $\tilde{w}(\cdot)$ is a solution of the nonclassical problem (12)–(14). We note that from the formulas (5) and (6) it follows that $v(t_2, t_1) = 0$, $t_1 \in [0, l]$. Under this condition, the problem (12)–(14) and the above problem of integral geometry are equivalent. Therefore, it will suffice to find $\tilde{w}(t_1, x, y)$ to determine then $u(x, y)$ from the formula (11). To investigate the uniqueness and stability for solutions of the problem (12)–(14), we apply the method of energy estimates.

For $x \neq g(t_1)$, $y \neq p(t_1)$, and $\beta = (-\sin\tilde{\theta}, \cos\tilde{\theta})$, $\tilde{\theta} = \tilde{\theta}(t_1, x, y)$ we have two obvious identities:

$$\tilde{\rho}(\nabla_{xy}\tilde{w} \cdot \beta)\frac{\partial}{\partial t_1}\left[\frac{1}{\tilde{\rho}}\nabla_{xy}\tilde{w} \cdot \tilde{v}\right] = (\nabla_{xy}\tilde{w} \cdot \beta)(\nabla_{xy}\tilde{w} \cdot \tilde{v}_{t_1})$$

$$+ (\nabla_{xy}\tilde{w} \cdot \beta)(\nabla_{xy}w_{t_1} \cdot v) - (\nabla_{xy}\tilde{w} \cdot \beta)(\nabla_{xy}\tilde{w} \cdot \tilde{v})\frac{\partial}{\partial t_1}\ln\tilde{\rho},$$

$$\tilde{\rho}(\nabla_{xy}\tilde{w} \cdot \beta)\frac{\partial}{\partial t_1}\left[\frac{1}{\tilde{\rho}}(\nabla_{xy}\tilde{w} \cdot \tilde{v})\right] = \frac{\partial}{\partial t_1}[(\nabla_{xy}\tilde{w} \cdot \beta)(\nabla_{xy}\tilde{w} \cdot \tilde{v})]$$

$$- (\nabla_{xy}\tilde{w} \cdot \tilde{v})(\nabla_{xy}\tilde{w} \cdot \beta_{t_1}) - (\nabla_{xy}\tilde{w} \cdot \tilde{v})(\nabla_{xy}\tilde{w} \cdot \beta)$$

$$- (\nabla_{xy}\tilde{w} \cdot \beta)(\nabla_{xy}\tilde{w} \cdot \tilde{v})\frac{\partial}{\partial t_1}\ln\tilde{\rho}.$$

Adding these identities and observing that

$$v_{t_1} = \beta\frac{\partial\tilde{\theta}}{\partial t_1}, \quad \beta_{t_1} = -\tilde{v}\frac{\partial\tilde{\theta}}{\partial t_1},$$

$$(\nabla_{xy}\tilde{w}\cdot\beta)(\nabla_{xy}\tilde{w}_{t_1}\cdot\tilde{v}) - (\nabla_{xy}\tilde{w}\cdot\tilde{v})(\nabla_{xy}\tilde{w}_{t_1}\cdot\beta)$$

$$= \frac{\partial}{\partial x}(\tilde{w}_{t_1}\tilde{w}_y) - \frac{\partial}{\partial y}(\tilde{w}_{t_1}\tilde{w}_x),$$

we obtain

$$2\rho(\nabla_{xy}\tilde{w}\cdot\beta)\frac{\partial}{\partial t_1}\left(\frac{1}{\tilde{\rho}}\nabla_{xy}\tilde{w}\cdot\tilde{v}\right) = \left\{[(\nabla_{xy}\tilde{w}\cdot\tilde{v})^2 + (\nabla_{xy}\tilde{w}\cdot\beta)^2]\frac{\partial\tilde{\theta}}{\partial t_1}\right.$$

$$\left. - 2(\nabla_{xy}\tilde{w}\cdot\tilde{v})(\nabla_{xy}\tilde{w}\cdot\beta)\frac{\partial}{\partial t_1}\ln\tilde{\rho}\right\} + \frac{\partial}{\partial t_1}[(\nabla_{xy}\tilde{w}\cdot\tilde{v})(\nabla_{xy}\tilde{w}\cdot\beta)]$$

$$+ \frac{\partial}{\partial x}(\tilde{w}_{t_1}\tilde{w}) - \frac{\partial}{\partial y}(\tilde{w}_{t_1}\tilde{w}_x). \quad (15)$$

The solutions of equation (12) cause the left-hand side of this identity to vanish. Assume that $\tilde{w}(t_1, x, y)$ is a solution of equation (12) and integrate the identity (15) over the region G_ε obtained by excluding the set $\{(t_1, x, y) : t_1 \in [0, l], (x, y) \in \overline{D}, (x - g(t_1))^2 + (y - p(t_1))^2 \leq \varepsilon^2\}$ for sufficiently small $\varepsilon > 0$ from $G = [0, l] \times \overline{D}$. By the Gauss theorem we obtain

$$\int_{G_\varepsilon}\left\{[(\nabla_{xy}\tilde{w}\cdot\tilde{v})^2(\nabla_{xy}\tilde{w}\cdot\beta)^2]\frac{\partial\tilde{\theta}}{\partial t_1}\right.$$

$$\left. - 2(\nabla_{xy}\tilde{w}\cdot\tilde{v})(\nabla_{xy}\tilde{w}\cdot\beta)\frac{\partial}{\partial t_1}\ln\tilde{\rho}\right\}dx\,dy\,dt_1$$

$$+ \int_{S_\varepsilon}\{[(\nabla_{xy}\tilde{w}\cdot\tilde{v})(\nabla_{xy}\tilde{w}\cdot\beta)]\cos(\angle n, t_1)$$

$$+ \tilde{w}_{t_1}[\tilde{w}_y\cos(\angle n, x) - \tilde{w}_x\cos(\angle n, y)]\}dS = 0.$$

Here S_ε is the boundary of D_ε, n is the outward normal to S_ε, and dS is the element of area.

We pass to the limit by letting $\varepsilon \to 0$ in this identity. In view of the estimates of Lemma 3 the integrals over G_ε and S_ε converge as improper ones to the integrals over G and S, respectively (S is the boundary of G). The integral over S can be partitioned into integrals over the upper and lower bases of the cylinder G, and its side surface $[0, l] \times \Gamma$. On both bases of the cylinder $\cos(\angle n, x) = \cos(\angle n, y) = 0$, while $\cos(\angle n, t_1)$

8.2. General Problems of Plane Integral Geometry

has opposite signs at corresponding points. Because of the periodicity condition (13) the sum of the integrals over the upper and lower bases of the cylinder vanishes. On the side surface $\cos(\angle n, t_1) = 0$, and for

$$x = g(t_2), \quad y = p(t_2),$$

$$\tilde{w}_{t_1} = v_{t_1}(t_1, t_2), \quad \tilde{w}_y \cos(\angle n, x) - \tilde{w}_x \cos(\angle n, y)$$

$$= \frac{\partial}{\partial t_2} \tilde{w}(t_1, g(t_2), p(t_2)) = v_{t_2}(t_1, t_2)$$

and $dS = dt_1\, dt_2$. Finally,

$$\int_G \left\{ [(\nabla_{xy}\tilde{w}\cdot\tilde{v})^2 + (\nabla_{xy}\tilde{w}\cdot\beta)^2]\frac{\partial\tilde{\theta}}{\partial t_1} \right.$$

$$\left. - 2(\nabla_{xy}\tilde{w}\cdot\tilde{v})(\nabla_{xy}\tilde{w}\cdot\beta)\frac{\partial}{\partial t_1}\ln\tilde{\rho} \right\} dx\,dy\,dt_2 = -\int_0^l\int_0^l v_{t_1}v_{t_2}\,dt_1\,dt_2. \quad (16)$$

By Lemma 2, $\partial\tilde{\theta}/\partial t_1 \geq 0$. At all points where $\partial\tilde{\theta}/\partial t_1 = 0$ the expression in brackets vanishes because of the condition (8). For points where $\partial\tilde{\theta}/\partial t_1 > 0$ we have

$$[(\nabla_{xy}\tilde{w}\cdot\tilde{v})^2 + (\nabla_{xy}\tilde{w}\cdot\beta)^2]\frac{\partial\tilde{\theta}}{\partial t_1} - 2(\nabla_{xy}\tilde{w}\cdot\tilde{v})(\nabla_{xy}\tilde{w}\cdot\beta)\frac{\partial}{\partial t_1}\ln\tilde{\rho}$$

$$= (\nabla_{xy}\tilde{w}\cdot\tilde{v})^2\left[\frac{\partial\tilde{\theta}}{\partial t_1} - \left(\frac{\partial}{\partial t_1}\ln\tilde{\rho}\right)^2\left(\frac{\partial\tilde{\theta}}{\partial t_2}\right)^{-1}\right]$$

$$+ \left[(\nabla_{xy}\tilde{w}\cdot\tilde{v})\frac{\partial}{\partial t_2}\ln\tilde{\rho}\left(\frac{\partial\tilde{\theta}}{\partial t_2}\right)^{-1/2} - (\nabla_{xy}\tilde{w}\cdot\beta)\left(\frac{\partial\tilde{\theta}}{\partial t_2}\right)^{1/2}\right]$$

$$\geq (\nabla_{xy}\tilde{w}\cdot\tilde{v})^2(1-q^2)\frac{\partial\tilde{\theta}}{\partial t_1} \geq \rho_0^2(1-q^2)u^2(x,y)\frac{\partial\tilde{\theta}}{\partial t_1}.$$

Strengthening the inequality (16) we obtain

$$\rho_0^2(1-q^2)\int_D u^2(x,y)\,dx\,dy \int_0^{2\pi}\frac{\partial\tilde{\theta}}{\partial t_1}dt_1 \leq \frac{1}{2}\int_0^l\int_0^l |\nabla v|^2\,dt_1\,dt_2.$$

This inequality leads to the estimate (9).

To complete the proof we notice that the expression on the right-hand side of the estimate (9) makes sense for $u \in C^1(\overline{D})$, $\rho(x_0, y_0, x, y) \in C^1(\Gamma \times \overline{D})$ and a family of curves L satisfying the conditions (i) and (ii). Therefore, approximating the functions u, ρ, and $\tilde{f}$ by the functions u_n, ρ_n, and $\tilde{f}_n$ for which the estimate (9) has already been established, and passing in it to the limit by letting $n \to \infty$, we arrive at the estimate (9) under the conditions of the theorem. This estimate implies in particular that the solution to the problem of integral geometry in the space $C^1(\overline{D})$ is unique. ∎

8.3. General Problems of Integral Geometry in Space

Consider now a bounded region D in a multidimensional space R_n, $x = (x_1, ..., x_n)$. Let Ω be an open domain belonging to D. For an even-dimensional space, we shall assume that the boundaries of D and Ω are uniformly separated from one another by a positive distance h, thus Ω is strictly interior to D. In an odd-dimensional space, Ω may coincide with D. Assume also that, for any point $x \in D$ and any unit vector $\nu = (\nu_1, ..., \nu_n)$, there is a unique smooth hypersurface $S(x, \nu)$ which passes through x and has the normal ν at this point.

We denote by U the class of functions $u(x)$ whose support is contained in Ω and $u(x) \in L_2(\Omega)$. For an $u(x) \in U$, we consider the problem of determining $u(x)$ from the equation

$$v(x, \nu) = \int_{S(x, \nu)} p(\xi, x, \nu) u(\xi) \, dS, \quad x \in D, \; |\nu| = 1, \qquad (1)$$

where $\rho(\xi, x, \nu)$ is a given smooth function of its arguments, and dS is an element of surface.

Generally speaking, this problem is overdetermined, because $u(x)$ depends on n variables and v depends on $2n - 1$ variables. We note, however, an important case of u and v having the same number of variables.

Consider the surface $S(x, y)$ and an arbitrary point $x^0 \in S(x, \nu)$. Let ν^0 be a normal to $S(x, \nu)$ at x^0. Then $S(x^0, \nu^0) = S(x, \nu)$. However, if $\rho(\xi, x^0, \nu^0) = \rho(\xi, x, \nu)$ for $\xi \in S(x, \nu)$, then $v(x^0, \nu^0) = v(x, \nu)$. Since each point x^0 of the hypersurface $S(x, \nu)$ is defined by $n - 1$ parameters, this identity indicates that in this case there are only n essential parameters on which $v(x, \nu)$ depends. If the weighting function depends only on the point $\xi \in D$ and the surface $S(x, \nu)$, then u and v have an equal number of essential variables. This fact suggests that there exists a better statement of the integral geometry problem for the same family of surfaces, namely,

8.3. General Problems of Integral Geometry in Space

one that is associated with another parametrization of this family, which eliminates overdetermination.

In this section we tackle the problem of solving equation (1) presuming that the diameter of D is sufficiently small. In what follows we shall assume that the equation of the surface $S(x, v)$ can be characterized by a sufficiently smooth function $\varphi(\xi, x, v)$:

$$S(x, v) = \{\xi : \varphi(\xi, x, v) = 0\}, \qquad (2)$$

with $|\nabla_\xi \varphi| \neq 0$ at points of $S(x, v)$.

Allowances on the method of parametrization lead to the following necessary conditions on φ:

$$\varphi(x, x, v) = 0, \quad \Delta_\xi \varphi|_{\xi = x} = v|\Delta_\xi \varphi|_{\xi = x}. \qquad (3)$$

Clearly, we can always assume

$$|\Delta_\xi \varphi|_{\xi = x} = 1, \qquad (4)$$

dividing, when necessary, φ by $|\nabla_\xi \varphi|_{\xi = x}$, therefore henceforth we shall consider equation (4) to take place. Then φ admits the representation

$$\varphi(\xi, x, v) = (v, \xi - x) + \sum_{i,j=1}^{n} a_{ij}(\xi, x, v)(\xi_i - x_i)(\xi_j - x_j), \qquad (5)$$

where

$$a_{ij}(\cdot) = \int_0^1 \varphi_{\xi_i \xi_j}(x + t(\xi - x), x, v)(1 - t)dt.$$

We shall use this representation in our development.

We will formulate the result and prove it first for odd-dimensional spaces, then for even-dimensional spaces [19].

THEOREM 1. *Let n be odd, $s = (n - 1)/2 \geq 1$, $\varphi(\xi, x, v)$ be continuous in $G = \{(\xi, x, v) : \xi \in \overline{D}, x \in \overline{D}, |v| = 1\}$ along with its partial derivatives up to order $n + 2$, and the weight function $\rho(\xi, x, v)$ be continuous within G along with the partial derivatives up to order n and satisfy the condition*

$$\rho(x, x, v) \geq \rho_0 > 0 \qquad (6)$$

in G. Then there exists a number $d^* > 0$, $d^* = d^*(\varphi, \rho)$, such that if diam $D < d^*$ the solution of equation (1) is unique in the class of functions U, and the following stability estimate holds:

$$\|u\|_{L_2(\Omega)} \leq C\|\Delta^s v_1\|_{L_2(\Omega)},$$

$$v_1 = \int_{|v|=1} \frac{v(x, v)}{\rho(x, x, v)} d\omega_v. \tag{7}$$

Here $d\omega_v$ is an elementary area of the unit sphere and Δ^s is the sth power of the Laplacian.

PROOF. Making use of the δ-function and the finiteness of $u(x)$, we rewrite equation (1) as

$$\int_\Omega \rho(\xi, x, v) u(\xi) |\nabla_\xi \varphi(\xi, x, v)| \delta(\varphi(\xi, x, v)) d\xi = v(x, v), \tag{8}$$

divide it by $\rho(\xi, x, v)$, and average it at a fixed $x \in D$ over all v. Observing the notations (7) recasts the identity (8) to the form

$$\int_\Omega K(x, \xi) u(\xi) d\xi = v_1(x), \tag{9}$$

where

$$K(x, \xi) = \int_{|v|=1} \tilde{\rho}(\xi, x, v) \delta(\varphi(\xi, x, v)) d\omega_v,$$

$$\tilde{\rho}(\xi, x, v) = \frac{\rho(\xi, x, v)}{\rho(x, x, v)} |\nabla_\xi \varphi(\xi, x, v)|. \tag{10}$$

It is an easy matter to examine the kernel of equation (10) for close x and ξ. This will be sufficient, since in the proof of the theorem we can assume that the diameter of D is small. We make use of the invariance of the measure $d\omega_v$ with respect to orthogonal transformations of the space $(v_1, \ldots, v_n)$ and take a transform with an orthogonal matrix Q. Then the formula for the kernel takes the form

$$K(x, \xi) = \int_{|v|=1} \tilde{\rho}(\xi, x, Qv) \delta(\varphi(\xi, x, Qv)) d\omega_v. \tag{11}$$

Introduce the unit vector $v^0 = \dfrac{\xi - x}{|\xi - x|}$, $v^0 = (v_1^0, \ldots, v_n^0)$. Depending on the position of v^0 on the unit sphere we shall resort to two different

8.3. General Problems of Integral Geometry in Space

orthogonal transformations in computing the integral (11). Let $e^1 = (1, 0, ..., 0)$, then for $v^0 \cdot e^1 \geq 0$, we take the transformation defined by

$$Qv = v + 2v \cdot e^1 \cdot v^0 - \frac{v \cdot v^0 + v \cdot e^1}{1 + (v^0, e^1)} (v^0 + e^1). \tag{12}$$

It is straightforward to verify directly that for any unit vector v, $(Qv \cdot Qv) = 1$, and

$$Qe^1 = v^0.$$

For $v^0 \cdot e^1 < 0$, we invoke the transformation (12) with e^1 replaced by $-e^1$, and so $Qe^1 = -v^0$.

Clearly, it will suffice to investigate the configuration of x and ξ such that $(v^0 \cdot e^1) \geq 0$. In what follows we restrict consideration to this case. For fixed x, ξ the equation $\varphi(\xi, x, Qv) = 0$ determines a surface of dimension $n - 2$ on the sphere. Indeed, by virtue of equation (5), this identity divided by $|x - \xi|$ rewrites as

$$v^0 \cdot Qv + |x - \xi| \sum_{i,j=1}^{n} a_{ij}(x + |x - \xi| v^0 \cdot x \cdot Qv) v_i^0 v_j^0 = 0. \tag{14}$$

Represent the vector v in the form

$$v = qe^1 + \sqrt{1 - q^2}\, \hat{v}, \tag{15}$$

where $\hat{v}$ is a unit vector orthogonal to e^1. In view of equation (13),

$$v^0 Q v = Q^* v v^0 = e^1 \cdot v = q,$$

and so equation (14) takes the form

$$q + |x - \xi| \sum_{i,j=1}^{n} a_{ij}(x + |x - \xi| v^0 \cdot x \cdot q v^0$$

$$+ \sqrt{1 - q^2}\, Q\hat{v}) v_i^0 v_j^0 = 0. \tag{16}$$

The principle of contraction mapping readily leads to the following

LEMMA 1. *If a function φ satisfies the conditions of the theorem, then for any $q_0 \in (0, 1)$ there is $d_0 = d_0(\varphi, q_0)$ such that for regions Ω with diam $\Omega < d_0$ the equation $\varphi(x, \xi, Qv) = 0$ set for $x \in \Omega$, $\xi \in \Omega$ defines q as a unique, continuous, and bounded function of the arguments $x, |x - \xi|, v^0$, and $\hat{v}$ having continuous and bounded derivatives up to* order n with respect to the argument and satisfying the inequality

$$|q(x, |x - \xi|, v^0, \hat{v})| \leq q_0 < 1. \tag{17}$$

PROOF. Since φ meets the conditions of the theorem, then $a_{ij}(\xi, x, y)$ are uniformly bounded by some constant c_0 for $\xi, x \in D$, and $|v| = 1$. Therefore, writing the equality (16) as $q = A(g)$ we see that whenever $c_0 d_0 \leq q_0$, the operator A maps the set of continuous functions satisfying the inequality (17) into itself. Since the derivatives of a_{ij} with respect to $v_1, \ldots, v_n$ are uniformly bounded provided $|x - \xi|$ is small, the operator A is contractive on this set. Thus, equation (16) defines q as a continuous function of the arguments $x, |x - \xi|, v^0$, and $\hat{v}$ obeying the inequality (17). The existence and boundedness of the derivatives of q up to the order n with respect to its arguments readily follows from the equality (16), since the functions a_{ij}, and the matrix $Q = Q(v^0)$, defined by equation (12), have respective derivatives. From the identity (16) we see also that $q > 0$ as $|x - \xi| \to 0$.

Now we fix $q_0 \in (0, 1)$ and in the remainder of this section we assume that diam $\Omega < d_0(\varphi, q_0)$. The relation (15), in which q is a solution of equation (16) for a fixed arbitrary unit vector $\hat{v}$, $\hat{v}.e^1 = 0$, determines v as a smooth single-valued function of $\hat{v}$ (for fixed x, ξ), and thus defines an $(n-2)$-dimensional surface on the sphere $|v| = 1$. We denote this surface by $\Sigma(x, \xi)$. If $x \to \xi$ in such a way what $(\xi - x)/|\xi - x| \to v^0$, then $\Sigma(x, \xi)$ tends to the section cut on the sphere by a plane passing through its center and orthogonal to e^1.

Let n be a unit normal to $\Sigma(x, \xi)$, lying in a plane tangent to the sphere $|v| = 1$ at v. Invoking the above arguments we can rewrite the formula (11) for the kernel $K(x, \xi)$ of equation (9) in the form

$$K(x, \xi) = \int_{\Sigma(x, \xi)} \tilde{\rho}(\xi, x, Qv) \left| \frac{\partial \varphi}{\partial n} \right|^{-1} d\sigma, \tag{18}$$

where $d\sigma$ is an elementary area of $\Sigma(x, \xi)$. Since n is parallel to the vector

$$\nabla_v \varphi(\xi, x, Qv) - v(v, \nabla_v \varphi(\xi, x, Qv)),$$

on $\Sigma(x, \xi)$, we have

$$\left| \frac{\partial \varphi}{\partial n} \right| = \{ |\nabla_v \varphi(\xi, x, Qv)|^2 - [v.\nabla_v \varphi(\xi, x, Qv)]^2 \}^{1/2}. \tag{19}$$

8.3. General Problems of Integral Geometry in Space

From equation (5) we find

$$\frac{1}{|x-\xi|}\nabla_v \varphi(\xi, x, Qv) = e^1 + |x-\xi| \sum_{i,j=1}^{n} \nabla_v a_{ij}(\xi, x, Qv) v_i^0 v_j^0.$$

Using the formula (19) and the fact that $v = v(x, |\xi - x|, v^0, \hat{v})$ on $\Sigma(x, \xi)$ we have for $\frac{\partial \varphi}{\partial n}$ on this surface the representation

$$\left|\frac{\partial \varphi}{\partial n}\right|^{-1} = \frac{1}{|x-\xi|} + b(x, |\xi - x|, v^0, \hat{v}), \tag{20}$$

where b is a function of its arguments, continuous and bounded along with its partial derivatives up to the order $n - 1$. Similarly, on $\Sigma(x, \xi)$ ρ admits the representation

$$\tilde{\rho}(\xi, x, Qv) = 1 + |x-\xi| c(x, |x - \xi|, v^0, \hat{v}), \tag{21}$$

where the function c has the same properties as b.

The unit-vector $\hat{v}$ is orthogonal to e^1, hence it is completely determined by the angular spherical coordinates $\psi_1, \ldots, \psi_{n-2}$ in the coordinate plane orthogonal to e^1. The surface element d is evaluated as

$$d\sigma = \left[\Gamma\left(\frac{\partial v}{\partial \psi_1}, \ldots, \frac{\partial v}{\partial \psi_{n-2}}\right)\right]^{1/2} d\psi, \tag{22}$$

where $d\psi = d\psi_1, \ldots, d\psi_{n-2}$, and Γ stands for the Gram determinant constructed for the vectors $\frac{\partial v}{\partial \psi_1}, \ldots, \frac{\partial v}{\partial \psi_{n-1}}$:

$$\Gamma(\cdot) = \begin{vmatrix} \left(\frac{\partial v}{\partial \psi_1}, \frac{\partial v}{\partial \psi_1}\right) & \cdots & \left(\frac{\partial v}{\partial \psi_1}, \frac{\partial v}{\partial \psi_{n-2}}\right) \\ \vdots & & \\ \left(\frac{\partial v}{\partial \psi_{n-2}}, \frac{\partial v}{\partial \psi_1}\right) & \cdots & \left(\frac{\partial v}{\partial \psi_{n-2}}, \ldots, \frac{\partial v}{\partial \psi_{n-2}}\right) \end{vmatrix}.$$

From equations (18)–(22) it follows that the kernel $K(x, \xi)$ can be represented in the form

$$K(x, \xi) = \frac{\omega_{n-1}}{|x - \xi|} + K_0(x, |x - \xi|, v^0), \tag{23}$$

where $K_0(\cdot)$ is a function of its arguments, continuous and bounded along with its derivatives up to the order $n - 1$. Acting on equation (9) by the operator Δ^s, $s = (n - 1)/2$, with respect to x, and making use of

$$\Delta^s \frac{1}{|x - \xi|} = (-1)^s (4\pi)^s (s - 1)! \, \delta(x - \xi),$$

$$\omega_{n-1} = \frac{2\pi^s}{\Gamma(s)},$$

we arrive at an equation in $u(x)$,

$$u(x) + \int_\Omega \tilde{K}(x, \xi) u(\xi) d\xi = \frac{(-1)^s}{2^n \pi^{2s}} \Delta^s v_1(x). \tag{24}$$

Here, the kernel

$$\tilde{K}(x, \xi) = \frac{(-1)^s}{2^n \pi^{2s}} \Delta_x^s K_0(x, |x - \xi|, v^0)$$

has an integrable polarity. Indeed, the evaluation of the operator Δ^s necessitates that $K_0(\cdot)$ be differentiated $2s$ times with respect to the variables x, $|x - \xi|$, and v_0, and that the arguments $|x - \xi|$, v^0 be differentiated $2s$ times with respect to x. But $2s = n - 1$, and, hence, the derivatives of $K_0(\cdot)$ with respect to the arguments are continuous and bounded. For the derivatives of $(|x - \xi|, v^0)$ with respect to x, the following estimates hold

$$\left| \frac{\partial^k}{\partial x_i^k} (|x - \xi|) \right| \leq \frac{c_1}{|x - \xi|^{k-1}},$$

$$\left| \frac{\partial^k}{\partial x_i^k} v^0 \right| \leq \frac{c_1}{|x - \xi|^k}, \quad 1 \leq k \leq 2s.$$

From these arguments it follows that the kernel $\tilde{K}(x, \xi)$ is continuous everywhere except the set of points (x, ξ) where $x = \xi$. In a neighborhood

8.3. General Problems of Integral Geometry in Space

of this set, $\tilde{K}(x, \xi)$ may be estimated by

$$|\tilde{K}(x, \xi)| \leq \frac{c_2}{|x - \xi|^{n-1}}.$$

Equation (24) with this type of kernel is known to be uniquely solvable in $L_2(\Omega)$ if the diameter of Ω is sufficiently small. Then the estimate (17) obviously applies to the solution of equation (24). ∎

Let us now examine the case of even-dimensional space. As mentioned above, in this case we assume Ω lying strictly inside D, so the distance between Ω and the boundary of D is larger than some $h > 0$. We suppose in addition that $h > \text{diam } \Omega$. Denote by Ω_h the open set resulting as the union of all open balls of radius h about points $x \in \Omega$.

THEOREM 2. *Let $n = 2s$, $s \geq 1$, and the functions φ, ρ satisfy all conditions of Theorem 1 with n replaced by $n + 1$. Then there exists $d^* = d^*(\varphi, \rho, h) > 0$ such that whenever $\text{diam } \Omega < d^*$, equation (1) is uniquely solvable in the class of functions u with the following stability estimate*

$$\|u(x)\|_{L(\Omega)} \leq c_0 \|\Delta v_2\|_{L_2(\Omega)}, \tag{25}$$

in which

$$v_2(x) = \int_{|x - y| \in h} \frac{v_1(y)}{|x - y|^{n-1}} \, dy, \tag{26}$$

and $v_1(x)$ is computed in terms of $v(x, \nu)$ by the formula (7).

Observe that the estimate (25) is quantitatively different from the estimate (7). The latter involves the values of $v(x, \nu)$ only for $x \in \Omega$, while the estimate (25) includes $v(x, \nu)$ from Ω_h, a domain wider in x.

PROOF. We adopt the following scheme. First, like for n odd, we are going to derive equation (9). For the kernel $K(x, \xi)$ of this equation, the representation (23) holds. By the assumption of the theorem, the smoothness of $K_0(\cdot)$ is higher by one unit, namely, $K_0(\cdot)$ has continuous bounded derivatives up to the order n with respect to its arguments $x, |x - \xi|$, and $\nu^0 = (\xi - x)|\xi - x|^{-1}$. Now we apply to the both sides of this identity the operator of averaging over a ball of radius h, which is defined by the formula (26). The result is the equation

$$\int_\Omega T(x, \xi) u(\xi) d\xi = v_2(x), \quad x \in \Omega, \tag{27}$$

where

$$T(x, \xi) = \int_{|x-y| \leq h} \frac{K(y, \xi)}{|x-y|^{n-1}} dy. \tag{28}$$

Further we show that for $x \in \Omega$, $\xi \in \Omega$,

$$T(x, \xi) = -\omega_{n-1} \omega_n \ln|x - \xi| + T_0(\xi, |\xi - x|, v^0), \tag{29}$$

where $T_0(\xi, \rho, v^0)$ is continuous and bounded along with the derivatives up to the order n everywhere except the set $\rho = 0$ in the vicinity of which $T_0(\xi, \rho, v^0)$ satisfies the inequalities ($|\alpha| + k \leq n$)

$$\left| \frac{\partial^k}{\partial \rho^k} D^\alpha_{\xi, v^0} T_0(\xi, \rho, v^0) \right| \leq c_0 \cdot \begin{cases} \ln \rho, & k = 1 \\ \rho^{1-k}, & k \geq 2. \end{cases} \tag{30}$$

Applying the operator Δ^s to equation (27) and taking into account that

$$\Delta^s_x \ln|x - \xi| = (-1)^{s-1} 2^{n-1} \pi^s (s-1)! \, \delta(x - \xi),$$

we obtain a Fredholm equation similar to equation (24) with a singularity of the form $|x - \xi|^{n-1} \ln|x - \xi|$.

Now it remains to verify that the kernel $T(\cdot)$ does possess the aforementioned properties. To do so, we find the integral (28) separately for each summand of $K(x, \xi)$ in (23). Integrating the first term is equivalent to computing

$$j_1(p) = \int_{|x-y| \leq h} \frac{1}{|x-y|^{n-1}} \frac{1}{|y-\xi|} dy$$

$$= \int_{|v|=1} d\omega_v \int_0^h [r^2 + \rho^2 + 2r\rho(v^0.v)]^{-1/2} dr$$

$$= \int_{|v|=1} \{\ln|h + \rho(v^0.v) + [h^2 + \rho^2 + 2h\rho(v.v^0)]^{1/2}|$$

$$- \ln|1 + (v^0.v)|\} d\omega_v - \omega_n \ln \rho.$$

Here, $y = x + rv$ is a spherical system of coordinates, and $\rho, -v^0$ stand for the spherical coordinates of ξ, i.e., $\rho = |\xi - x|$, and $v^0 = (x - \xi)/|x - \xi|$. This formula indicates that finding the integral (28) of the first summand of $K(x, \xi)$ leads to the smooth major part of the kernel $T(x, \xi)$, namely $\ln \rho$, and a smooth function of ρ alone (which is even analytic for $\rho < h$). Also we have used here that $\rho \leq \text{diam } \Omega < h$ for $x \in \Omega$, $\xi \in \Omega$.

8.3. General Problems of Integral Geometry in Space

We show now that finding the integral (28) of the second summand results in a function having the properties of T_0. It is convenient to compute the integral in a spherical system of coordinates with the origin at $\xi : y = \xi + rv$, $v = (v_1, \ldots, v_n)$, $x = \xi + \rho v^0$, then

$$j_2(\xi, \rho, v^0) = \int_{|x-y| \leq h} \frac{K_0(y, |y-\xi|, v^0)}{|x-y|^{n-1}} dy$$

$$= \int_{|v|=1} d\omega_v \int_0^{r(\rho, v \cdot v^0)} \frac{r^{n-1} K_0(\xi + rv, r, -v)}{[r^2 + \rho^2 - 2r\rho(v \cdot v^0)]^{(n-1)/2}} dr,$$

$$r(\rho, v \cdot v^0) = \rho(v \cdot v^0) + \sqrt{h^2 - \rho^2[1 - (v \cdot v^0)^2]}.$$

Let us introduce now, as we did once before, an orthogonal transformation of the coordinate system, having a matrix Q such that $Q^* v^0 = e^1 = [1, 0, \ldots, 0]$ provided that $v^0 \cdot e^1 \geq 0$. Then, denoting $K_0(\xi + rQv, r, -Qv) = \tilde{K}_0(\xi, r, v, v^0)$, we find

$$j_2(\xi, \rho, v^0) = \int_{|v|=1} d\omega_v \int_0^{r(\rho, v_1)} \frac{r^{n-1} \tilde{K}_0(\xi + r, v, v^0)}{(r^2 + \rho^2 - 2r\rho v_1)^{\frac{n-1}{2}}} dr.$$

This expression shows that the derivatives $D_{\xi, v^0}^\alpha j_2$, $|\alpha| \leq n$, are continuous and bounded for $\xi \in \Omega$, $x \in \Omega$. Indeed,

$$D_{\xi, v^0}^\alpha j_2(\cdot) = \int_{|v|=1} d\omega_v \int_0^{r(\rho, v_1)} \frac{r^{n-1} D_{\xi, v^0}^\alpha \tilde{K}(\cdot)}{(r^2 + \rho^2 - 2r\rho v_1)^{\frac{n-1}{2}}} dr. \quad (31)$$

Now, we are only to understand the derivatives involving differentiation with respect to ρ. The change of variables $r = \rho r_1$ in the inner integral shows at once that, for $\rho > 0$, the integral $j_2(\cdot)$ depends continuously on (ξ, ρ, v^0) and so do its derivatives up to the order n. However, the derivatives involving differentiation with respect to ρ are unbounded for $\rho \to 0$. Indeed, for a sufficiently small ρ we can always choose a $\delta > 0$ such that

$$\rho(1 + \delta) < r(\rho, v_1), \quad v_1 \in [-1, 1].$$

Then, splitting the inner integral into two integrals, one over $[0, \rho(1 + \delta)]$ and the other over $[\rho(1 + \delta), r(\rho, v_1)]$, and changing the variable

316   Chapter 8. Problems of Integral Geometry

$r = \rho r_1$ in the first one, we obtain

$$j_2(\xi, \rho, v^0) = \rho \cdot \int_{|v|=1}^{1+\delta} d\omega_v \int_0^{1+\delta} \frac{r_1^{n-1} \tilde{K}_0(\xi, \rho r_1, v, v^0)}{[r_1^2 + 1 - 2r_1 v_1]^{\frac{n-1}{2}}} dr_1$$

$$+ \int_{|v|=1} d\omega_v \int_{\rho(1+\delta)}^{r(\rho_1, v_1)} \frac{r^{n-1} \tilde{K}_0(\xi, r, v, v^0)}{[r^2 + \rho^2 - 2r\rho v_1]^{\frac{n-1}{2}}} dr. \quad (32)$$

The first integral in this formula is a function which is continuous and bounded along with its derivatives up to the order n. All the singularities in these derivatives are induced by the second integral. We notice that the cofactor of $\tilde{K}_0$ can be represented in the form

$$\frac{r^{n-1}}{(r^2 + \rho^2 - 2r\rho v_1)^{\frac{n-1}{2}}} = \Phi\left(\frac{\rho}{r}, v_1\right),$$

moreover, $\Phi(z, v_1)$ as a function of z, v_1 is bounded within the domain $|z| \leq (1+\delta)^{-1}$, $v_1 \in [-1, 1]$, with any finite number of its derivatives. Consequently, its derivatives with respect to ρ may be estimated as follows

$$\left|\frac{\partial^k}{\partial \rho^k} \Phi\left(\frac{\rho}{r}, v_1\right)\right| \leq \frac{c_0}{r^k}, \quad r \geq \rho(1+\delta). \quad (33)$$

To compute the derivatives $\frac{\partial^k}{\partial \rho^k} D^\alpha_{\xi, v^0}$ of the second term in equation (32), we can introduce D^α_{ξ, v^0} under the inner integral sign. Calculating the derivative $\frac{\partial^k}{\partial \rho^k}$ of the inner integral gives rise to a number of terms due to taking the derivatives of the upper and lower limits of integration, and the integral due to the differentiation of the integrand. The first of the terms are bounded, since $\Phi\left(\frac{\rho}{r}, v_1\right)$ is bounded and independent of ρ at the lower limit and coincides with the analytic function $h^{1-n}[r(\rho, v_1)]^{n-1}$ at the upper limit. The integral arising from the differentiation of the integrand is of the form

$$\int_{\rho(1+\delta)}^{r(\rho, v_1)} \frac{\partial^k}{\partial \rho^k} \Phi\left(\frac{\rho}{r}, v_1\right) D^\alpha_{\xi, v^0} \tilde{K}(\xi, r, v, v^0) dr,$$

8.3. General Problems of Integral Geometry in Space

and in view of the inequality (33) it can be estimated by the integral

$$c_1 \int_{\rho(1+\delta)}^{r(\rho, v_1)} \frac{dr}{r^k}$$

$$= c_1 \cdot \begin{cases} \ln \dfrac{v_1 + \sqrt{\left(\dfrac{h}{\rho}\right)^2 - (1 - v_1^2)}}{1 + \delta}, & \text{for } k = 1 \\[2ex] \dfrac{1}{1-k} \{[r(\rho, v_1)]^{1-k} - [\rho(1+\delta)]^{1-k}\}, & \text{for } k > 1. \end{cases}$$

This expression leads to the estimate (30) and completes the proof of Theorem 2. ∎

NOTE 1. Define $D(x, v)$ to be the set of points in the region D for which $\varphi(\xi, x, v) \geq 0$. A more general problem than that of solving equation (1) is the determining of $u(x)$ from the equation

$$\int_{S(x, v)} \rho(\xi, x, v) u(\xi) dS + \int_{D(x, v)} \rho_1(\xi, x, v) u(\xi) d\xi = v(x, v),$$

for $x \in D$, $|v| = 1$.

When the smoothness of $\rho_1(\xi, x, v)$ is at most one unit worse than that of the function $\rho(\xi, x, v)$, Theorems 1 and 2 hold for this problem.

NOTE 2. If we replace in equation (1) the weighting function $\rho(\xi, x, v)$ with an $m \times m$ matrix weighting function $R(\xi, x, v)$ of an arbitrary finite dimension, and the function $u(\xi)$ with a vector-valued function having components $u_1, \ldots, u_m$, then we arrive at a vector problem of integral geometry. Theorems 1 and 2 remain valid for this problem after the natural substitution of

$$|\det R(x, x, v)| \geq \rho_0 > 0$$

for the condition (6), and $R^{-1}(x, x, v)$ for ρ^{-1} in the estimate (7).

NOTE 3. As a corollary of Theorems 1 and 2, we have the following estimate valid for any $n \geq 2$:

$$\|u\|_{L_2(\Omega)} \leq c_0 \sup_{|v|=1} \|v(x, v)\|_{w_2^l(\overline{D})}, \quad l = 2\left[\frac{n}{2}\right]. \tag{34}$$

So far we have considered the problem of integral geometry for the case when for every point x and any direction v there exists a smooth hypersurface $S(x, v)$ passing through x and having the normal v. A more general statement of this problem is possible, in which point x is crossed by the surfaces $S(x, v)$ only for those v which belong to a set of points $\omega(x)$ on the unit sphere. In general the set $\omega(x)$ does not coincide with the unit sphere. It turns out that the structure of $\omega(x)$ is closely connected with the stability of the problem of integral geometry [19].

Consider for simplicity the case of $\omega(x)$ independent of x, $\omega(x) = \omega_0$. Let $\omega_{\delta\alpha} = \{ v : |v| = 1, |v.\alpha| \leq \delta \}$ be a spherical belt. If for some unit vector α and $\delta > 0$, $\omega_{\delta\alpha} \cap \omega_0 \neq \emptyset$, then estimates of the type (34) (with the set ω_0 substituted for the set $|v| = 1$ under the sup sign) cannot exist for any l for the integral geometry problem. This implies, in particular, that for any finite k, l and a constant $c_0 > 0$, estimates of the form

$$\|u\|_{W_2^k(\Omega)} \leq c_0 \sup_{v \in \omega_0} \|v(x, v)\|_{W_2^l(\bar{D})} \tag{35}$$

cannot exist too. We shall demonstrate this for the special case of $n = 2$ by setting $\varphi(\xi, x, v) = v(\xi - x)$, $\rho = 1$. This example will completely display the idea underlying the proof.

Let $\omega_{\delta\alpha} \cap \omega_0 = \emptyset$ for $\delta > 0$ and $\alpha = (0, 1)$. Pick a function $u_\lambda \in C_0^\infty(\Omega)$ of the form

$$u_\lambda(x) = \sin(\lambda x_1) \psi(x),$$

where $\psi(x)$ is an infinitely differentiable, different from identical zero, finite function with support in Ω, and λ is a sufficiently large numerical-valued parameter. Then

$$\|u_\lambda\|_{L_2(\Omega)}^2 = \frac{1}{2}\|\psi\|_{L_2(\Omega)} - \frac{1}{2}\int_\Omega \cos(2\lambda x_1)\psi^2(x)\,dx.$$

Integrating this expression by parts a sufficient number of times yields

$$\left| \int_\Omega \cos(\lambda x_1)\psi^2(x)\,dx \right| \leq \frac{c_k}{|\lambda|^k}.$$

On the other hand, for $x \in D$, $v \in \omega_0$, we have

$$v_\lambda(x, v) = \frac{1}{|v_2|} \int_{-\infty}^{\infty} u_\lambda\left(\xi_1, x_2 - \frac{v_1}{v_2}(\xi_1 - x_1)\right) d\xi_1$$

$$= \frac{1}{|v_2|} \int_{-\infty}^{\infty} \sin(\lambda \xi_1) \psi\left(\xi_1, \xi x_2 - \frac{v_1}{v_2}(\xi_1 - x_1)\right) d\xi_1.$$

Since for $v \in \omega_0$ the inequality $|v \cdot \alpha| = |v_2| \geq \delta$ holds, then

$|v_2|^{-1} \leq \delta^{-1}$, and consequently

$$|D_{xv}^\alpha v_\lambda(x, v)| \leq \frac{c_{\alpha k}}{|\lambda|^k}, \quad x \in D, \quad v \in w_0.$$

Letting $\lambda \to \infty$ we obtain that $\|v_\lambda\|_{w_2^l(D)} \to 0$ uniformly over all $v \in \omega_0$ and for any finite l, and $\|u_\lambda\|_{L_2(\Omega)} \to 2^{-\frac{1}{2}} \|\psi\|_{L_2(\Omega)} \neq 0$, thus proving the above statement.

8.4. Volterra-Type Problems of Integral Geometry with Motion-Group Invariant Manifolds

In this section we consider the problems of integral geometry on a plane. The results of this section can be carried over to problems of this type in a space of any dimension.

Let $u(x, y)$ be a function of two variables. We will consider $u(x, y)$ in the half plane $y \geq 0$, and assume that it is continuous and finite in x:

$$u(x, y) = 0, \quad |x| \geq l > 0.$$

Further on, consider the functions $\varphi(y, \eta)$, $g(y, \eta)$, $y \geq \eta$, satisfying the following conditions

$$\varphi(y, \eta) = \sqrt{y - \eta}\, \varphi_0(y, \eta),$$

$$g(y, \eta) = \frac{1}{\sqrt{y - \eta}}\, g_0(y, \eta),$$

where $\varphi_0(y, \eta)$, $g_0(y, \eta)$ are continuously differentiable functions

$$\varphi_0(y, \eta) \geq \varphi^0 > 0,$$

$$g_0(y, \eta) \geq g^0 > 0.$$

Consider an equation in $u(x, y)$:

$$\int_0^y g(y, \eta)[u(x - \varphi, \eta) + u(x + \varphi, \eta)]d\eta = f(x, y). \quad (1)$$

Solving equation (1) is a problem of integral geometry, namely, one of determining a function, given the integrals of this function over a specified family of curves. In our case, both the family of curves defined by the function $\varphi(y, \eta)$ and the weighting function $g(y, \eta)$ are invariant with respect to the group of motions parallel to the x axis.

Chapter 8. Problems of Integral Geometry

Analysis of equation (1) may be approached by two virtually equivalent methods.

Method 1

Taking the Fourier transform of equation (1) in x yields

$$\int_0^y g(y, \eta) \cos(\lambda\varphi) v(\lambda, \eta) d\eta = \tilde{f}(\lambda, y) \qquad (2)$$

with $v(\lambda, \eta) = \int_{-\infty}^{\infty} e^{i\lambda x} u(x, \eta) dx$ and $\tilde{f}(\lambda, y) = \int_{-\infty}^{\infty} e^{i\lambda x} f(x, y) dx$.

Thus, equation (1) reduces to a family of Volterra equations of the first kind. The properties of the functions $\varphi(y, \eta)$ and $g(y, \eta)$ ensure that equation (2) satisfies the conditions formulated in Section 6.7. Consequently, equation (2), and hence equation (1), is uniquely solvable.

Applying to equation (2) the operator of fractional differentiation gives the following equation of the second kind:

$$v(\lambda, y) + \int_0^y G(\lambda, y, \eta) v(\lambda, \eta) d\eta = \tilde{f}_2(\lambda, y), \qquad (3)$$

$$G(\cdot) = \frac{1}{\pi g_0(y, y)} \int_\eta^y g_{0y} \cos(\lambda\varphi) - \lambda g_0 \varphi_y \sin(\lambda\varphi) \sqrt{\frac{y - \eta}{\xi - \eta}} d\xi,$$

$$\tilde{f}_1(\lambda, y) = \frac{1}{\pi g_0(y, y)} \frac{\partial}{\partial y} \int_0^y \frac{1}{\sqrt{y - \eta}} \tilde{f}(\lambda, \eta) d\eta.$$

The solution of equation (3) can be represented in the form

$$v(\lambda, y) = \tilde{f}_1(\lambda, y) + \int_0^y R(\lambda, y, \eta) \tilde{f}_1(\lambda, \eta) d\eta. \qquad (4)$$

In general, the norm of the integral operator in equations (3) increases without bound when λ increases, hence the problem of solving the family of equations (3), and hence equation (1), is ill-posed.

We carry out a complete investigation of the stability of equations (1) and (3) in one simplest case.

Let in equation (1)

$$\varphi_0(y, \eta) = g_0(y, \eta) = 1.$$

Then it takes the form

$$\int_0^y [u(x - \sqrt{y - \eta}, \eta) + u(x + \sqrt{y - \eta}, \eta)] \frac{d\eta}{\sqrt{y - \eta}} = f(x, y). \qquad (5)$$

8.4. Volterra-Type Problems of Integral Geometry

After Fourier transformation equation (5) becomes

$$\int_0^y \cos(\lambda \sqrt{y - \eta}) v(\lambda, \eta) \frac{d\eta}{\sqrt{y - \eta}} = \tilde{f}(\lambda, y). \tag{6}$$

We will show that the solution to equation (6) is given by

$$v(\lambda, y) = \frac{1}{\pi} \frac{\partial}{\partial y} \int_0^y \cosh(\lambda \sqrt{y - \eta}) \tilde{f}(\lambda, \eta) \frac{d\eta}{\sqrt{y - \eta}}. \tag{7}$$

Indeed, applying to equation (6) the Volterra operator with the kernel

$$\cosh(\lambda \sqrt{z - y}) \frac{1}{\sqrt{z - y}},$$

we get

$$\int_0^z \cosh(\lambda \sqrt{z - y}) \tilde{f}(\lambda, y) \frac{dy}{\sqrt{z - y}}$$

$$= \int_0^z \int_\eta^z \cos(\lambda \sqrt{y - \eta}) \cosh(\lambda \sqrt{z - y}) \frac{dy}{\sqrt{(y - \eta)(z - y)}} v(\lambda, \eta) d\eta,$$

whence, in view of the well-known formula

$$\int_0^1 \cos(p \sqrt{t}) \cosh(p \sqrt{1 - t}) \frac{dt}{\sqrt{t(1 - t)}} = \pi,$$

it follows that the solution (7) is valid.

The formula (7) shows that the instability of the solution of equation (5) is of the same type as that in the Cauchy problem for Laplace's equation. For the general equation (1) and the associated equation (2), similar estimates of conditional stability can be obtained. Specifically, it can be demonstrated that the function in equation (4) satisfies the inequality

$$|R(\lambda, y, \eta)| \leq a e^{b\sqrt{y - \eta}\lambda} \frac{1}{\sqrt{y - n}},$$

where a and b are constants.

Method 2

Let us treat the function $u(x, y)$ in equation (1) at fixed y as an element of a Hilbert space W,

$$u(x, y) = w(y) \in W.$$

Chapter 8. Problems of Integral Geometry

Then equation (1) may be treated as a Volterra operator equation, as defined in Section 6.8:

$$\int_0^y A(y, \eta) w(\eta) d\eta = \psi(y). \tag{8}$$

Here the family of operators $A(y, \eta)$ is defined as follows:

$$A(y, \eta) w(\eta) = g(y, \eta)[u(x - \varphi, \eta) + u(x + \varphi, \eta)].$$

It is not hard to show that the Volterra operator equation (8) defined in this way can be reduced to an equation satisfying the conditions of Theorem 1 in Section 6.8. For instance, in this theorem, as the operator B we can take the following integral operator:

$$Bw(y) = \int_{-l}^{l} p(x, \xi) u(\xi, y) d\xi,$$

$$p(x, \xi) = \frac{1}{2l} (l + \xi)(l - x) \xi \leq x, \quad \frac{1}{2l} (l - \xi)(l + x) x \leq \xi.$$

The same results can be obtained for integral geometry problems in spaces of higher dimension.

Supplement

Inversion Formulas in Inverse Problems

A.L. Bukhgeim

INTRODUCTION

This supplement consists of two chapters. Chapter 1 presents a new approach to problems of tomography, which has been set forth by the author [4] and developed in recent papers [5, 6]. This approach treats tomographic problems as boundary-value problems for elliptic equations with operator coefficients. The classical Radon problem on the plane becomes equivalent to the Cauchy problem for a Beltrami-type operator equation with Cauchy data on a closed contour. The Radon problem with incomplete data in a fan formulation reduces to a Cauchy problem for an operator equation of Laplace type $\Delta_A u = 0$. The Dirichlet problem for the operator Δ_A leads to a principally new class of inversion formulas for Radon's transformation, which requires, roughly speaking, only a half of the initial information.

Chapter 2 presents inversion formulas for the problem of reconstruction of a phonon spectrum of a crystal from its heat capacity data. This chapter is based on a paper of S.G. Kazantsev and the author [7]. It is worth noting that this problem was approached by many authors (see, e.g. [13, 21] and references cited therein), but no inversion formula for it has been derived.

The author would like to acknowledge the assistance of S.G. Kazantsev and V.B. Kardakov in writing this supplement. Specifically, Sections 1.2.2 and 1.2.3 and the entire Chapter 2 were written in coauthorship with S.G. Kazantsev, and Sector 1.4 was coauthored with V.B. Kardakov.

Chapter 1. TOMOGRAPHY AND THE THEORY OF A-ANALYTIC FUNCTIONS

1.1. Radon's Problem: a Complex Function Approach

As known, solving the Radon transform in a bounded domain $\Omega \in \mathbb{R}^2$ with a smooth boundary $\partial\Omega$ consists in the reconstruction of a function $a(x, y) \in \mathbb{C}(\overline{\Omega})$ from its integrals over all possible straight lines intersecting Ω. In a differential statement this problem reduces to the evaluation of the right-hand side of the simplest transport equation

$$\cos\alpha\, u_x + \sin\alpha\, u_y = a(x, y), \quad (x, y) \in \Omega, \quad \alpha \in [-\pi, \pi], \qquad (1)$$

by the values of the solution $u(x, y, \alpha)$ on the boundary $\Sigma = \partial\Omega \times [-\pi, \pi]$

$$u|_\Sigma = f.$$

Here u is a real-valued smooth function 2π-periodic in α. Let us identify $\mathbb{R}^2$ with the complex plane $\mathbb{C}$ and set

$$z = x + iy,$$
$$u(x, y, \alpha) = u(z, \alpha), \quad a(x, y) = a(z),$$
$$\partial u = \frac{\partial u}{\partial z} = \frac{1}{2}\left(\frac{\partial u}{\partial x} + \frac{1}{i}\frac{\partial u}{\partial y}\right),$$
$$\overline{\partial} u = \frac{\partial u}{\partial \overline{z}} = \frac{1}{2}\left(\frac{\partial u}{\partial x} - \frac{1}{i}\frac{\partial u}{\partial y}\right).$$

In view of Euler's formula,

$$\cos\alpha = \frac{e^{i\alpha} + e^{-i\alpha}}{2}, \quad \sin\alpha = \frac{e^{i\alpha} - e^{-i\alpha}}{2i},$$

equation (1) can be recast in complex form:

$$e^{i\alpha} u_z + e^{-i\alpha} u_{\overline{z}} = a(z),$$

or equivalently,

$$u_{\overline{z}} + e^{2i\alpha} u_z = a(z) e^{i\alpha}. \qquad (2)$$

Supplement

Expanding u in a Fourier series in α,

$$u(z, \alpha) \sum_{n=-\infty}^{\infty} u_n(z) e^{-in\alpha},$$

and substituting it into the equation we obtain

$$u_{n\bar{z}} + u_{n+2\,z} = 0, \quad n \neq -1, \tag{3}$$

$$u_{-1\bar{z}} + u_{1z} = a(z). \tag{4}$$

Since u is real-valued, the Fourier coefficients u_n will be complex-conjugate quantities

$$u_n = \bar{u}_{-n}. \tag{5}$$

This implies that to determine u uniquely, it will suffice to find the complex-valued vector

$$\mathbf{u} = (u_0, u_1, u_2, \ldots),$$

moreover, by virtue of (3), the vector-valued function $\mathbf{u}$ satisfies the Beltrami-type equation

$$\mathbf{u}_{\bar{z}} - A\mathbf{u}_z = 0, \quad z \in \Omega, \tag{6}$$

with an operator-valued coefficient A:

$$A: \mathbf{u} = (u_0, u_1, u_2, \ldots) \to -(u_2, u_3, \ldots). \tag{7}$$

If we determine the vector-valued function $\mathbf{u}$ uniquely from equation (6) and the boundary condition

$$\mathbf{u}|_{\partial\Omega} = \mathbf{f}, \tag{8}$$

where $\mathbf{f} = (f_0, f_1, \ldots)$ is constituted by the Fourier coefficients f_n, then in view of (4), (5) the function a will be defined by

$$a(z) = 2(\operatorname{Re} \mathbf{u}_z)_1. \tag{9}$$

Since for $A = 0$, the solution to the Cauchy problem (6), (8), if it exists, will be given by the Cauchy integral, we need an analog of Cauchy's formula for functions satisfying equation (6). For brevity, we will refer to such functions as A-analytic. An analog of Cauchy's formula for the system (6) will be obtained in the forthcoming section. Combined with equation (9), it yields a complex inversion formula for the Radon transform.

For elliptic systems of the type (6), in the finite-dimensional case, the

Cauchy problem on a closed contour is known to be overdetermined, while the boundary-value problem is well-posed. Therefore, it is natural to expect that $a(z)$ can be reconstructed only from the real or imaginary part of **f**.

To realize this idea, we let $\mathbf{u} = v + iw$, where w and v are real-valued vectors, and separate the real and imaginary parts in equation (6). As a result, we arrive at an analog of the Cauchy-Riemann system:

$$(I - A) v_x = (I + A) w_y, \qquad (10)$$
$$(I + A) v_y = -(I - A) w_x.$$

Applying the operator $(I - A) \partial_x$ to the first equation, $(I + A) \partial_y$ to the second equation, and adding the results we obtain

$$\Delta_A v \equiv (I - A)^2 v_{xx} + (I + A)^2 v_{yy} = 0.$$

The equation for the imaginary part $\Delta_A w = 0$ is obtained in a similar way. Hence, the problem (6), (8) is equivalent to solving two Dirichlet problems for the operator Δ_A:

$$\Delta_A v = 0, \quad v|_{\partial\Omega} = g,$$
$$\Delta_A w = 0, \quad w|_{\partial\Omega} = h, \qquad (11)$$

where g and h are the real and imaginary parts of the vector **f**, respectively.

We intend to demonstrate that a can be recovered only from v or w. Indeed, assuming the operator $(I + A)^{-1}$ exists, from equation (10) we get

$$w_y = (I + A)^{-1} (I - A) v_x,$$

and, consequently,

$$2\text{Re } \mathbf{u}_z = v_x + w_y = (I + A)^{-1} (I + A) v_x$$
$$+ (I + A)^{-1} (I - A) v_x = 2(I + A)^{-1} v_x.$$

Analogously, if the operator $(I - A)^{-1}$ exists,

$$2\text{Re } \mathbf{u}_2 = 2(I - A)^{-1} w_y.$$

Comparing these relations with (9), we find

$$a(z) = (2(I + A)^{-1} v_x)_1,$$
$$a(z) = (2(I - A)^{-1} w_y)_1.$$

Supplement

We will call functions obeying the equation $\Delta_A v = 0$ A-harmonic. Like for classical harmonic functions, it turns out that under the appropriate conditions on A the solution of the Dirichlet problem (11) for a wide class of domains Ω can be written explicitly in a form similar to Poisson's formula.

The basic theory of A-harmonic and A-analytic functions is outlined in two ensuing sections. The main difficulty of this theory is that in functional spaces, most interesting for applications, the norm and the spectral radius $\rho(A)$ of A are equal to unity, whereas in the classical theory of Beltrami equations $\|A\| < 1$.

We conclude this section by exposing some simplest properties of the operator A (7), which will serve us as a guide in constructing the general theory.

Let us consider A first in the scale of Hilbert spaces l_2^m for complex sequences $u = (u_0, u_1, \ldots)$ endowed with the norm

$$\|u\|_m^2 = \sum_{j=0}^{\infty} (1 + j)^{2m} |u_j|^2, \quad m \in \mathbb{R}.$$

For brevity we set $l_2^0 = l_2$.

Let U be a one-sided translation operator such that

$$U: u = (u_0, u_1, \ldots) \to (0, u_1, u_2, \ldots).$$

Then its adjoint U^* in l_2 is defined by

$$U^* u = (u_1, u_2, \ldots), \tag{12}$$

so that $A = -(U^*)^2$.

Since the equations for even and odd harmonics u_m (equations (3)) are mutually independent, and $a(z)$ is expressed in terms of the first harmonic u_1 or v_1, then going to the subspace of odd harmonics we can reduce the problem to the case of $A = -U^*$ which will be the focus of our further consideration. Clearly, in l_2^m for $m \geq 0$, $\|U^*\| = \rho(U^*) = 1$.

First we will show that the point spectrum $\Sigma_p(U^*)$ of U^* coincides with the interior of the unit disk, that is,

$$\Sigma_p(U^*) = \{\lambda \in \mathbb{C} : |\lambda| < 1\}. \tag{13}$$

Indeed, if $U^* u = \lambda u$, then from equation (12), $u_{j+1} = \lambda u_j$, $j = 0, 1, \ldots$. Normalizing the vector u by the condition $u_0 = 1$, we obtain $u_j = \lambda^j$.

Since
$$(1, \lambda, \lambda^2, \ldots) \in l_2^m \Leftrightarrow |\lambda| < 1,$$

the identity (13) is established. It implies that the resolvent[1] $R(\lambda) = (A - \lambda)^{-1}$ of A, which is bounded in l_2^m for $|\lambda| > 1$, can be extended to the unit circle $|\lambda| = 1$, but already as an operator unbounded in l_2^m.

Now we show that $R(\lambda)$, considered as an operator from l_2^{m+1}.) to l_2^m, is bounded for $|\lambda| = 1$, $m > -1/2$. This assertion is equivalent to the following estimate for all $u \in l_2^{m+1}$

$$\| (I - \lambda U^*)^{-1} u \|_m \leq C_m \| u \|_{m+1}. \tag{14}$$

Let
$$(I - \lambda U^*) v = u,$$
then
$$v_j - \lambda v_{j+1} = u_j, \quad j = 0, 1, 2, \ldots,$$
hence,
$$v_j = u_j + \lambda v_{j+1} = u_j + \lambda(u_{j+1} + \lambda u_{j+2}) = \ldots = \sum_{k=0}^{\infty} \lambda^k u_{j+k}, \tag{15}$$

and since $|\lambda| = 1$,
$$|v_j| \leq \sum_{k=0}^{\infty} |u_{j+k}| = \sum_{k=j}^{\infty} |u_k|. \tag{16}$$

Now we resort to the continuous analog of Minkowski's inequality:

$$\left\{ \int_1^{\infty} \left(\int_1^{\infty} \varphi(s, t) \, ds \right)^2 dt \right\}^{1/2} \leq \int_1^{\infty} \left\{ \int_1^{\infty} |\varphi(s, t)|^2 dt \right\}^{1/2} ds,$$

in which we set
$$\varphi(s, t) = t^{m+1} |u_{[st]-2}|.$$

Here $[x]$ denotes the integer part of x, and by definition $u_{-1} = 0$.

---

[1] This definition of the resolvent differs from the one given in Part One by sign.

Supplement

We have

$$\int_1^\infty \varphi(s,t)\,ds = \int_1^\infty t^{m+1} |u_{[st]-2}|\,ds = \int_t^\infty t^m |u_{[x]-2}|\,ds$$

$$= t^m \sum_{n=[t]-1}^\infty |u_n| \geq t^m |v_{[t]-1}|.$$

In deriving this inequality we made use of the definition of φ and the bound (16).

From the last inequality we have

$$\left\{ \int_1^\infty \left(\int_1^\infty \varphi(s,t)\,ds \right)^2 dt \right\}^{1/2} \geq \left\{ \int_1^\infty t^{2m} |v_{[t]-1}|^2 dt \right\}^{1/2}$$

$$= \left\{ \int_1^2 t^{2m} |v_0|^2 dt + \int_2^3 t^{2m} |v_1|^2 dt \ldots + \int_{j+1}^{j+2} t^{2m} |v_j|^2 dt + \ldots \right\}^{1/2}$$

$$\geq \left\{ \sum_{j=0}^\infty (1+j)^{2m} |v_j|^2 \right\}^{1/2} = \|v\|_m. \quad (17)$$

On the other hand,

$$\left\{ \int_1^\infty |\varphi(s,t)|^2 dt \right\}^{1/2} = \left\{ \int_1^\infty t^{2(m+1)} |u_{[st]-2}|^2 dt \right\}^{1/2}$$

$$= s^{-(m+3/2)} \left\{ \int_s^\infty x^{2(m+1)} |u_{[x]-2}|^2 dx \right\}^{1/2}$$

$$\leq s^{-(m+3/2)} \left\{ \int_1^\infty x^{2(m+1)} |u_{[x]-2}|^2 dx \right\}^{1/2}$$

$$= s^{-(m+3/2)} \left\{ \sum_{j=0}^\infty \int_{j+1}^{j+2} x^{2(m+1)} |u_{[x]-2}|^2 dx \right\}^{1/2}$$

$$= s^{-m-3/2} \left\{ \sum_{j=0}^\infty \int_{j+1}^{j+2} x^{2(m+1)} |u_{j-1}|^2 dx \right\}^{1/2}$$

$$\leq s^{-m-3/2} \left\{ \sum_{j=1}^\infty (j+2)^{2(m+1)} |u_{j-1}|^2 \right\}^{1/2}$$

$$\leq s^{-m-3/2} \left[\max_{k \geq 0} \left(\frac{k+3}{k+1} \right)^{m+1} \right] \left\{ \sum_{k=0}^\infty (1+k)^{2(m+1)} |u_k|^2 \right\}^{1/2}$$

$$\leq 3^{m+1} s^{-m-3/2} \|u\|_{m+1}.$$

Hence,

$$\int_1^\infty \left\{ \int_1^\infty |\varphi(s, t)|^2 \, dt \right\}^{1/2} ds \leq 3^{m+1} \|u\|_{m+1} \int_1^\infty s^{-m-3/2} \, ds$$

$$= 3^{m+1} \frac{1}{m + 1/2} \|u\|_m, \quad (18)$$

provided $m + 1/2 > 0$, that is, $m > -1/2$.

Recalling that $v = (I - \lambda U^*)^{-1} u$, from the estimates (17), (18) we get the bound (14) with the constant

$$C_m = 3^{m+1} \frac{1}{m + 1/2}, \quad m > -1/2.$$

Now we examine when the vector-valued function $v(\lambda) = (I - \lambda U^*)^{-1}$ is continuous in λ in l_2^m for $|\lambda| = 1$. We have

$$\|v(\lambda) - v(\mu)\|_m^2 = \sum_{j=0}^{N} (1 + j)^{2m} |v_j(\lambda) - v_j(\mu)|^2$$

$$+ \sum_{j=N+1}^{\infty} (1 + j)^{2m} |v_j(\lambda) - v_j(\mu)|^2 = S_1 + S_2,$$

where S_1 and S_2 stand for the first and second summand, respectively. Since

$$|a - b|^2 \leq 2(|a|^2 + |b|^2),$$

then

$$S_2 \leq 2 \sum_{j=N+1}^{\infty} (1 + j)^{2m} [|v_j(\lambda)|^2 + |v_j(\mu)|^2]$$

$$= 2 \|P_{N+1} v(\lambda)\|_m^2 + 2 \|P_{N+1} v(\mu)\|_m^2,$$

where

$$P_{N+1} v = (\underbrace{0, 0, \ldots, 0}_{N+1}, v_{N+1}, v_{N+2}, \ldots).$$

In view of equation (15)

$$v_j(\lambda) = \sum_{p=j}^{\infty} \lambda^{p-j} u_p,$$

Supplement

we have
$$P_{N+1} v = P_{N+1}(I - \lambda U^*)^{-1} u = P_{N+1}(I - \lambda U^*)^{-1} P_{N+1} u,$$

and since $\|P_{N+1}\| = 1$ in l_2^m,
$$\|P_{N+1} v\|_m \leq \|(I - \lambda U^*)^{-1} P_{N+1} u\|_m \leq C_m \|P_{N+1} u\|_{m+1},$$

whence
$$S_2 \leq 4 C_m \|P_{N+1} u\|_{m+1} \to 0$$

as $N \to \infty$, whenever $u \in l_2^{m+1}$.

For $j = 0, 1, ..., N$,
$$|v_j(\lambda) - v(\mu)| \leq \sum_{k=1}^{\infty} |\lambda^k - \mu^k||u_{j+k}|$$
$$= |\lambda - \mu| \sum_{k=1}^{\infty} |\lambda^{k-1} + \lambda^{k-2}\mu + ... + \mu^{k-1}||u_{j+k}|$$
$$\leq |\lambda - \mu| \sum_{k=1}^{\infty} k |u_{j+k}|,$$

because $|\lambda| = |\mu| = 1$.

Writing
$$k|u_{j+k}| = k^{-\alpha}(k^{1+\alpha}|u_{j+k}|),$$

and using Cauchy-Buniakovski's inequality, we find
$$\sum_{k=1}^{\infty} k|u_{j+k}| \leq \left\{\sum_{k=1}^{\infty} k^{-2\alpha}\right\}^{1/2} \left\{\sum_{k=1}^{\infty} k^{2(1+\alpha)}|u_{j+k}|^2\right\}^{1/2} \leq C_\alpha \|u\|_{1+\alpha},$$

whenever $\alpha > 1/2$. Thus,
$$|v_j(\lambda) - v_j(\mu)| \leq |\lambda - \mu| C_\alpha \|u\|_{1+\alpha}, \quad \alpha > 1/2.$$

And finally,
$$\|v(\lambda) - v(\mu)\|_m^2 \leq 4 C_m^2 \|P_{N+1} u\|_{m+1}^2 + |\lambda - \mu|^2 C_\alpha^2 C \|u\|_{1+\alpha}^2,$$
$$C = \sum_{j=0}^{N} (1+j)^{2m}.$$

Choosing N sufficiently large, and subsequently λ close to μ, we see that

$$\|v(\lambda) - v(\mu)\|_m \to 0 \text{ as } \lambda \to \mu,$$

whenever $u \in l_2^{m+1} \cap l_2^{\alpha+1}$, $\alpha > 1/2$. In particular, $v(\lambda)$ is continuous in l_2^m for $m > 1/2$.

Our investigation of the resolvent of $A = -U^*$ is summarized in the following

PROPOSITION 1. *The resolvent $R(\lambda, A) = (A - \lambda)^{-1}$ is extended from the exterior of the unit disk, $|\lambda| > 1$ to the unit circle as a linear bounded operator from l_2^{m+1} into l_2^{m+1} $m > -1/2$. Furthermore, for $m > 1/2$ it is strongly continuous in λ.*

In some cases, the condition that $R(\lambda, A)$ be unbounded in l_2^m is too restrictive, making one choose spaces in which the resolvent $R(\lambda, A)$ is bounded. In this connection we consider for $s \in (0, 1)$ the following scale of Hilbert spaces $l_{2,s}$:

$$\|u\|_s^2 = \sum_{j=0}^{\infty} s^{-2j} |u_j|^2.$$

Then

$$\|A\|_s^2 = \sum_{j=0}^{\infty} s^{-2j} |u_{j+1}|^2 s^2 s^{-2} \le s \|u\|_s^2,$$

consequently, $\|A\| \le s < 1$ in $l_{2,s}$. It is not hard to see that indeed

$$\|A\| = s, \quad \rho(A) = s.$$

Thus, in $l_{2,s}$, $R(\lambda, A)$ is a bounded operator for $|\lambda| > s$, and specifically for $|\lambda| = 1$ we have

$$\|R(\lambda, A)\| \le \frac{1}{(1-s)}.$$

1.2. A-Analytic Functions

1.2.1. The Cauchy Formula

Let X be a complex Banach space, $\mathcal{L}(X)$ be the algebra of bounded linear operators in X, and $A \in \mathcal{L}(X)$ be an operator with spectral radius $\rho(A) \le 1$. Then the resolvent $R(\lambda) = (A - \lambda)^{-1}$ of A is defined for all $\lambda \in \mathbb{C}$, $|\lambda| > 1$, it belongs to $\mathcal{L}(X)$ and is an analytic operator-valued function. We assume that for any integer $m \ge 0$,

Supplement

there is a chain of Banach spaces X^m embedded in X and dense there:

$$X^{m+1} \subseteq X^m \subseteq X, \quad X^0 = X,$$
$$\|u\|_{X^m} = \|u\|_m \leq \|u\|_{m+1}, \quad \forall u \in X^{m+1},$$
$$\|u\|_0 = \|u\|,$$

such that $A \in \mathcal{L}(X^m)$, $\rho(A) = 1$ in X^m, the resolvent $R(\lambda) \in \mathcal{L}(X^m)$, $|\lambda| > 1$, is extended by strong continuity in X^m to the circle $|\lambda| = 1$ as an operator from $\mathcal{L}(X^{m+1}, X^m)$.

From Hilbert's identity,

$$R(\lambda_1) - R(\lambda_2) = (\lambda_1 - \lambda_2) R(\lambda_1) R(\lambda_2),$$

we have that $R(\lambda_1)$ and $R(\lambda_2)$ commute for $|\lambda_j| > 1$, $j = 1, 2$, so by continuity, they will commute for $|\lambda_j| = 1$, more precisely

$$R(\lambda_1) R(\lambda_2) u = R(\lambda_2) R(\lambda_1) u, \quad u \in X^{m+2}.$$

Besides, for $u \in X^{m+2}$, as $\lambda_1 \to \lambda_2 = \lambda$ there exists a strong limit of the relation

$$\lim_{\lambda_1 \to \lambda_2} \frac{R(\lambda_1) - R(\lambda_2)}{(\lambda_1 - \lambda_2)} u = \frac{dR}{d\lambda} u = R^2(\lambda) u,$$

thus, $R(\lambda)$ has a strongly continuous derivative $R'(\lambda) \in \mathcal{L}(X^{m+2}, X^m)$. Moreover, $R'(\lambda) = R^2(\lambda)$, $|\lambda| \geq 1$. Hence, by induction, we find that the operator

$$R^{(n)}(\lambda) = n! R^{n+1}(\lambda) \qquad (1)$$

belongs to $\mathcal{L}(X^{m+n+1}, X^m)$ and is strongly continuous for $|\lambda| \geq 1$.

Having introduced the operator A, we are in a position to define A-analytic functions.

Let Ω be an open set in the complex plane $\mathbb{C}$ which we will identify with $\mathbb{R}^2$. If (x, y) are the real coordinates of a point $z = x + iy$, then, as usual, we assume

$$\partial u = \frac{\partial u}{\partial z} = \frac{1}{2}\left(\frac{\partial u}{\partial x} + \frac{1}{i}\frac{\partial u}{\partial y}\right),$$

$$\bar{\partial} u = \frac{\partial u}{\partial \bar{z}} = \frac{1}{2}\left(\frac{\partial u}{\partial x} - \frac{1}{i}\frac{\partial u}{\partial y}\right),$$

$$\partial_A = \partial - A\bar{\partial}, \quad \bar{\partial}_A = \bar{\partial} - A\partial.$$

Here $u: \Omega \to X$ belongs to the class $C^1(\Omega; X)$, where in the general case, $C^k(\Omega; X)$ is a space of k-times strongly continuously differentiable functions with values in X.

DEFINITION 1. *A function* $u \in C^1(\Omega; X)$ *is said to be A-analytic in* Ω *if for all* $z \in \Omega$,

$$\overline{\partial}_A u = 0.$$

The set of all A-analytic functions $u: \Omega \to X^m$ *will be denoted by* $A(\Omega; X^m)$.

Examples of A-analytic functions follow.

EXAMPLE 1. $u(z) = (z + \overline{z}A)^n e \in A(\mathbb{C}, X^m)$, where e is a fixed vector from X^m, $m = 0, 1$, etc. Since $A(\Omega; X^m)$ is a linear space, the function

$$u(z) = \sum_{k=0}^{n} (z + \overline{z}A)^k e_k, \quad e_k \in X^m, \quad k = 0, 1, \ldots, n,$$

also belongs to $A(\mathbb{C}; X^m)$.

EXAMPLE 2. For $z \neq 0$, define the operator

$$B(z) = (z + \overline{z}A)^{-1} = \overline{z}^{-1} R(-z/\overline{z}) \qquad (2)$$

and consider the function

$$u(z) = B(z)e, \quad e \in X^{m+2}.$$

From (2),

$$\partial B = -B^2, \quad \overline{\partial} B = -AB^2,$$

consequently, $u \in A(\mathbb{C} \setminus \{0\}; X^m)$. Analogously, for $e_k \in X^{m+k+1}$ ($k = 1, 2, \ldots, n$), $e_0 \in X^m$, the function

$$u(z) = \sum_{k=0}^{n} B^k(z) e_k = \sum_{k=0}^{n} (z + \overline{z}A)^{-k} e_k$$

also belongs to $A(\mathbb{C} \setminus \{0\}; X^m)$. Indeed,

$$\partial B^k = -kB^{k+1}, \quad \overline{\partial} B^k = -kAB^{k+1},$$

hence $B^k e_k \in A(C \setminus \{0\}; X^m)$.

Now let us obtain an analog of Cauchy's formula. Let Ω be a bounded open set on a plane, and let its boundary $\partial \Omega$ consist of a finite number of

Supplement

smooth Jordan's curves. Consider the differential form

$$\varphi = u(z)dz + Au(z)d\bar{z} = (dz + Ad\bar{z})u,$$

where $u \in C^1(\overline{\Omega}; X^m)$.

Since

$$d\varphi = \bar{\partial}_A u d\bar{z} \wedge dz,$$

then the Stokes formula

$$\int_{\partial \Omega} \varphi = \int_\Omega d\varphi$$

leads us to the identity

$$\int_{\partial \Omega} (dz + Ad\bar{z})u(z) = \int_\Omega \bar{\partial}_A u d\bar{z} \wedge dz. \tag{3}$$

This formula results in the following

THEOREM 1. *Let* $u \in C^1(\overline{\Omega}; X^m) \cap A(\Omega; X^m)$, *then*

$$\int_{\partial \Omega} (dz + Ad\bar{z})u(z) = 0.$$

Now let

$$\Omega_\varepsilon = \{z : z \in \Omega, \ |z - \xi| > \varepsilon\},$$

with $0 < \varepsilon < \rho$, where ρ is the distance from $\xi \in \Omega$ to $\mathbb{C} \setminus \Omega$.

Let us apply the formula (3) to the function

$$v(z) = B(z - \xi) u(z),$$

where the operator B is defined by equation (2). Taking into account Example 2, we have for $u \in C^1(\Omega; X^{m+2})$

$$v_{\bar{z}} - Av_z = B(z - \xi)(u_{\bar{z}} - Au_z). \tag{4}$$

Moreover, since X^{m+2} is dense in X^{m+1}, by continuity this formula remains valid for any function $u \in C^1(\overline{\Omega}_\varepsilon; X^{m+1})$ and $v \in C^1(\overline{\Omega}_\varepsilon; X^m)$. Substituting (4) into (3), with Ω_ε in place of Ω we have

$$\int_{\Omega_\varepsilon} (v_{\bar{z}} - Av_z) d\bar{z} \wedge dz = \int_{\Omega_\varepsilon} B(z - \xi)(u_{\bar{z}} - Au_z) d\bar{z} \wedge dz$$

$$= \int_{\partial \Omega_\varepsilon} (dz + Ad\bar{z}) B(z - \xi) u(z) = \int_{\partial \Omega} (dz + Ad\bar{z}) B(z - \xi) u(z)$$

$$- \int_{|z-\xi|=\varepsilon} (dz + Ad\bar{z}) B(z - \xi) u(z). \tag{5}$$

Denote the last integral by I_ε and set $z = \xi + \varepsilon e^{i\theta}$, $\theta \in [0, 2\pi]$. Then
$$dz = i\varepsilon e^{i\theta} d\theta,$$
$$d\bar{z} = -i\varepsilon e^{-i\theta} d\theta,$$
$$dz + A d\bar{z} = i\varepsilon(e^{i\theta} - e^{-i\theta} A) d\theta,$$
$$B(z - \xi)\big|_{|z-\xi|=\varepsilon} = \varepsilon^{-1}(e^{i\theta} + e^{-i\theta} A)^{-1},$$

hence,
$$I_\varepsilon = i \int_0^{2\pi} (e^{i\theta} - e^{-i\theta} A)(e^{i\theta} + e^{-i\theta} A)^{-1} u(\xi + \varepsilon e^{i\theta}) d\theta$$
$$= i \int_0^{2\pi} (e^{i\theta} - e^{-i\theta} A)(e^{i\theta} + e^{-i\theta} A)^{-1} u(\xi) d\theta$$
$$+ i \int_0^{2\pi} (e^{i\theta} - e^{-i\theta} A)(e^{i\theta} + e^{-i\theta} A)^{-1} [u(\xi + \varepsilon e^{i\theta}) - u(\xi)] d\theta$$
$$= I_1 + I_2.$$

It is clear that
$$\|I_2\|_m \leq C(1 + \|A\|_{\mathscr{L}(X^m)}) \int_0^{2\pi} \|u(\xi + \varepsilon e^{i\theta}) - u(\xi)\|_{m+1} d\theta \to 0$$

as $\varepsilon \to 0$, provided that $u \in C(\Omega; X^{m+1})$. Here we made use of the fact that for $|\lambda| = 1$ the operators $R(\lambda) \in \mathscr{L}(X^{m+1}, X^m)$ are strongly continuous in λ and so by the Banach-Steinhaus theorem there exists a constant $C = C_m$ such that
$$\|R(\lambda) u\|_m \leq C_m \|u\|_{m+1}, \quad u \in X^{m+1}.$$

To compute I_1, we notice that
$$\int_0^{2\pi} \frac{e^{i\theta} - e^{-i\theta}\lambda}{e^{i\theta} + e^{-i\theta}\lambda} d\theta = 2\pi, \quad \lambda \in \mathbb{C}, \ |\lambda| < 1. \tag{6}$$

Consequently, if we introduce the function
$$\varphi_r(\lambda, \theta) = \frac{e^{i\theta} - e^{-i\theta} \lambda r}{e^{i\theta} + e^{-i\theta} \lambda r}, \quad 0 \leq r < 1,$$

then in X the operator $\varphi_r(A, \theta) \in \mathscr{L}(X)$ will be properly posed together with the operator
$$\psi_r(A) = \int_0^{2\pi} \varphi_r(A, \theta) d\theta,$$

Supplement

which, in view of equation (6), is equal to $2\pi E$. Thus,

$$\psi_r(A)u = 2\pi u, \quad u \in X. \tag{7}$$

Consequently, there exists a strong limit in X as $r \to 1$:

$$\psi(A)u = \lim \psi_r(A)u,$$

and, by equation (7), $\psi(A)u = 2\pi u$. This proves that

$$I_1 = 2\pi i u(\xi). \tag{8}$$

Passing in equation (5) to the limit by letting $\varepsilon \to 0$ and observing (8), leads us to

THEOREM 2. *Let for some $m \geq 0$, $u \in C^1(\overline{\Omega}; X^{m+1})$. Then for all $\xi \in \Omega$,*

$$u(\xi) = (2\pi i)^{-1} \left\{ \int_{\partial\Omega} (dz + Ad\bar{z})B(z - \xi)u(z) \right.$$

$$\left. + \int_{\Omega} B(z - \xi)\bar{\partial}_A u \, dz \wedge d\bar{z} \right\},$$

which gives at once

THEOREM 3. *Let $u \in A(\Omega; X^m) \cap C(\overline{\Omega}; X^{m+1})$ and $\xi \in \Omega$. Then*

$$u(\xi) = (2\pi i)^{-1} \int_{\partial\Omega} (dz + Ad\bar{z})B(z - \xi)u(z).$$

1.2.2. Limit Values of Cauchy-Type Integrals

Now we examine the boundary properties of Cauchy-type integrals. Let Γ be a closed smooth curve, $\Gamma \in C^1$, $f(t)$ be a function with a domain in Γ and a range in the Banach space X^{m+2}, $m \geq 0$, $f(t)$ belongs to the Hölder space $C^{0,\alpha}(\Gamma; X^{m+2})$, $0 < \alpha \leq 1$, and Ω^+, Ω^- be the interior and exterior of Γ, respectively. The Cauchy-type integral given by

$$u(z) = (2\pi i)^{-1} \int (dt + Ad\bar{t})B(t - z)f(t), \quad z \notin \Gamma, \tag{1}$$

is an A-analytic function everywhere except Γ, $u \in A(\mathbb{C} \setminus \Gamma; X^m)$. If $z = t_0 \in \Gamma$ and the integral (1) exists in the sense of the Cauchy principal value, then we will speak about a singular integral:

$$S_\Gamma f(t_0) = (2\pi i)^{-1} \int_\Gamma (dt + Ad\bar{t})B(t - t_0)f(t). \tag{2}$$

THEOREM 1. *If $f(t) \in C^{0,\alpha}(\Gamma; X^{m+1})$, $m \geq 0$, $0 < \alpha \leq 1$, and $\Gamma \in C^1$ is a smooth closed contour, then the singular integral (2) exists.*

PROOF. Consider first the case of the density $f(t) = h$, where h is a constant vector from X^{m+1}.

Let $l = \{z: |z - t_0| < \delta\} \cap \Gamma$, $t_0 \in \Gamma$. If $t = t(s)$, $0 \leq s \leq d$, is an equation of Γ, $t(0) = t(d) = t_0$, then, for $t \in \Gamma \setminus l$,

$$t(s) = t_0 + r(t_0, s)e^{i\varphi(t_0, s)} \equiv t_0 + re^{i\varphi},$$

$$\frac{dt}{(t-t_0)} = \left(\frac{r'}{r} + i\varphi'\right)ds,$$

$$\frac{d\bar{t}}{(t-t_0)} = e^{-2i\varphi}\left(\frac{r'}{r} - i\varphi'\right)ds.$$

Using the identity $(I + \gamma A)^{-1} = I - \gamma A(I + \gamma A)^{-1}$, $\gamma = e^{-2i\varphi}$, we obtain

$$\int_{\Gamma \setminus l}(dt + Ad\bar{t})B(t-t_0)h = \int_{\Gamma \setminus l}\frac{h}{t-t_0}dt + \int_{\Gamma \setminus l}\gamma' A(I + \gamma A)^{-1}hds$$

$$= \int_{\Gamma \setminus l}\frac{h}{t-t_0}dt + \int_{\Gamma \setminus l}[\ln(I + \gamma A)]'_s hds = \int_{\Gamma \setminus l}\frac{h}{t-t_0}dt$$

$$+ [\ln(I + \gamma(s_2)A) - \ln(I + \gamma(s_1)A]h \to \pi i h,^2 \quad \delta \to 0.$$

Here $t(s_1)$, $t(s_2)$ are the end points of the arc l, $0 < s_1 < s_2 < d$,

$$\lim_{s_1 \to 0}\gamma(s_1) = \lim_{s_2 \to 0}\gamma(s_2) = \bar{t}'(0)/t'(0).$$

In the general case, we have

$$\int_{\Gamma \setminus l}(dt + Ad\bar{t})B(t-t_0)f(t)$$

$$= \int_{\Gamma \setminus l}(dt + Ad\bar{t})B(t-t_0)(f(t) - f(t_0))$$

$$+ \int_{\Gamma \setminus l}(dt + Ad\bar{t})B(t-t_0)f(t_0). \quad (3)$$

---

[2] For Hilbert spaces X^m, a correct definition of $\ln(I + \gamma A)$ was given in Sect. 1.2.1. For Banach spaces, this can be done proceeding from the properties of the resolvent $R(\lambda)$.

Supplement

Passing to the limit in (3), we get

$$S_\Gamma f(t_0) = \frac{1}{2\pi i} \int_\Gamma (dt + Ad\bar{t}) B(t - t_0)(f(t) - f(t_0))$$
$$+ \frac{f(t_0)}{2}. \blacksquare$$

THEOREM 2. *If $f(t) \in C^{0,\alpha}(\Gamma; X^{m+2})$, $0 < \alpha \leq 1$, $m \geq 0$, and $\Gamma \in C^1$, then the Cauchy-type integral (1) has limit values on Γ in the norm of X^m:*

$$u^\pm(t_0) = \lim_{z \to t_0} u(z), \quad z \in \Omega^\pm, \quad t_0 \in \Gamma,$$

and Sokhozki's formulas

$$u^\pm(t_0) = S_\Gamma f(t_0) \pm \frac{f(t_0)}{2}, \quad t_0 \in \Gamma \qquad (4)$$

hold true.

PROOF. Consider the function

$$\psi(z) = (2\pi i)^{-1} \int_\Gamma (dt + Ad\bar{t}) B(t - z)[f(t) - f(t_0)], \quad t_0 \in \Gamma.$$

We shall show that it is continuous at t_0. For this purpose, we examine the difference

$$\psi(z) - \psi(t_0) = (2\pi i)^{-1} \int_\Gamma (dt + Ad\bar{t})[B(t - z) - B(t - t_0)][f(t) - f(t_0)], \qquad (5)$$

and estimate the norm of the operator $B(t - z) - B(t - t_0)$, $t \neq t_0$. We have

$$B(t - z) - B(t - t_0) = ((z - t_0) + (\bar{z} - \bar{t}_0)A) B(t - z) B(t - t_0),$$

therefore,

$$\|B(t - z) - B(t - t_0)\|_{\mathscr{L}(X^{m+2}, X^m)} \leq \frac{\text{const} \cdot |z - t_0|}{|t - z| \cdot |t - t_0|}.$$

This inequality yields an estimate of the norm of the difference (5). To do this one may resort, for instance, to the scheme proposed in

[9]. Using the continuity of $\psi(z)$ at t_0 and the identities

$$\psi(z) = u(z) - f(t_0), \quad z \in \Omega^+,$$
$$\psi(z) = u(z), \quad z \in \Omega^-,$$
$$\psi(t_0) = S_\Gamma f(t_0) - \frac{f(t_0)}{2}, \quad t_0 \in \Gamma,$$

we obtain the Sokhozki formulas (4), as required. ∎

The following theorem is about the limit values of the derivative of a Cauchy-type integral.

THEOREM 3. *Let*

$$f(t) \in C^{1,\alpha}(\Gamma; X^{m+4}), \quad \Gamma \in C^{1,\alpha} \quad \text{and} \quad m \geq 0.$$

Then the first-order derivative $\partial u / \partial z$ of the integral (1) *has limit values on* Γ, *and these values satisfy the Sokhozki-type formulas*

$$\left[\frac{\partial u}{\partial z}\right]^\pm (t) = S_\Gamma g(t) \pm g(t)/2, \quad t \in \Gamma, \tag{6}$$

where

$$g(t) = B(t') f_s'(t),$$
$$f_s'(t) = \frac{d}{ds}(f(t(s))).$$

PROOF. Using formula for integration by parts, the derivative $\partial u/\partial z$ can be represented as a Cauchy-type integral

$$\frac{\partial u}{\partial z}(z) = (2\pi i)^{-1} \int_\Gamma (dt + Ad\bar{t}) B(t-z) g(t),$$

where $g(t) = B(t') f_s'(t) \in C^{0,\alpha}$ $(\Gamma; X^{m+2})$. Now Theorem 2 yields (6). ∎

THEOREM 4. *Let a function $f(t)$ from the class $C^{0,\alpha}(\Gamma; X^{m+2})$, $m \geq 0$, be given on a smooth closed contour* Γ. *The condition*

$$\frac{f(t_0)}{2} = \frac{1}{2\pi i} \int_\Gamma (dt + Ad\bar{t}) B(t-t_0) f(t), \quad t_0 \in \Gamma$$

is necessary and sufficient for $f(t)$ to be a boundary value of the function $u(z) \in A(\Omega^+; X^m) \cap C(\overline{\Omega}^+; X^m)$, where $u(z)$ is the Cauchy-type integral (1).

Supplement

The proof follows from Theorem 2.

To conclude this section we formulate an analog of the Morera theorem.

THEOREM 5. *Let* $u(z) \in C(\Omega; X^{m+3})$, $m \geq 0$. *Then* $u(z) \in A(\Omega; X^m)$ *if and only if*

$$\int_{\partial R} (dz + Ad\bar{z})u(z) = 0, \tag{7}$$

where R is any rectangle in Ω whose sides are parallel to the coordinate axes.

PROOF. If $u(z) \in A(\Omega; X^m)$ and R is a rectangle in Ω, then

$$u(z) \in A(R; X^m) \cap C^1(\bar{R}; X^m),$$

and by Theorem 1 of Sect. 1.2.1, the condition (7) is fulfilled.

Conversely, let $u(z) \in C(\Omega; X^{m+3})$ and the condition (7) be satisfied. We shall show that $u(z)$ is A-analytic in the neighborhood of $z_0 \in \Omega$. Let us consider a disc $D = \{z : |z - z_0| < d\} \subset \Omega$ in which the function

$$v(z) = \int_{z_0}^{z} (dt + Ad\bar{t})u(t)$$

is well defined. Here the integration is along the two sides of the rectangle $R \subset D$ which join the points z_0 and z.

Since $v_z = u$, $v_{\bar{z}} = Au$, $v_{\bar{z}} - Av_z = 0$, then $v \in A(D; X^{m+3}) \cap C^1(\bar{D}; X^{m+3})$. From Cauchy's formula (see Theorem 3 in Sect. 1.2.1) for the function v in D it follows that $v \in C^2(D, X^m)$, thus $u_{\bar{z}} = v_{z\bar{z}}$, $u_z = v_{zz}$.

Finally,

$$u_{\bar{z}} - Au_z = v_{z\bar{z}} - Av_{zz} = v_{z\bar{z}} - (Av_z)_z = v_{z\bar{z}} - v_{\bar{z}z} = 0,$$

where $u \in A(D; X^m)$, as asserted. ∎

1.2.3. Illustrative Examples

Now let us consider the integral geometry problem in a domain $\Omega \in \mathbb{C}$ when the family of curves is constituted by curves of constant curvature $k \geq 0$. Set in a differential form, this problem reduces to evaluating the real-valued right-hand side of the equation

$$e^{i\varphi} u_z + e^{-i\varphi} u_{\bar{z}} - k u_\varphi = a(z), \qquad (1)$$

given the values of the solution $u(z, \varphi)$ on a surface Σ,

$$u(t, \varphi)|_\Sigma = f(t, \varphi), \quad \Sigma = \partial\Omega \times [-\pi, \pi]. \qquad (2)$$

Let first $k = 0$, which corresponds to the Radon problem. In the integral formulation of the problem, we may assume that

$$u(z, \varphi) = \frac{1}{2} (\int_{L_2} a(\eta) |d\eta| - \int_{L_1} a(\xi) |d\xi|), \qquad (3)$$

where L_1, L_2 are the segments of the straight line L containing the points $t_1, t_2 \in \partial\Omega$:

$$L_1 = \{\xi: \xi(s) = t_1 + s(z - t_1), \ s \in [0, 1]\},$$
$$L_2 = \{\eta: \eta(s) = t_2 + s(z - t_2), \ s \in [0, 1]\},$$
$$\varphi = \arg(z - t_2) = \arg(t_1 - z) = \arg(t_1 - t_2), \ z \in L.$$

Then the function $f = u|_\Sigma$ satisfies the following additional conditions:

$$f(t, \varphi) = -f(t, \varphi + \pi), \quad t \in \partial\Omega, \qquad (4)$$
$$f(t_1, \Phi(t_1, t_2)) = -f(t_2, \Phi(t_1, t_2)), \quad t_1, t_2 \in \partial\Omega, \qquad (5)$$

where $\Phi(t_1, t_2) = \arg(t_1 - t_2)$.

The representation (3) implies that $u(z, \varphi)$ can be expanded in a Fourier series in terms of odd harmonics:

$$u(z, \varphi) = 2 \operatorname{Re} \sum_{n=0}^{\infty} u_n(z) e^{-i(2n+1)\varphi}.$$

Substituting this formula into (1), for $u(z) = (u_0(z), u_1(z), \ldots)$, we obtain the equation $\bar{\partial}_A u = 0$ with the operator $A = -U^*$, $U^*(u_0, u_1, \ldots) = (u_1, u_2, \ldots)$.

From Proposition 1 of Sect. 1.1, it follows that for $X^m = l_2^{m+2}$ all the conditions imposed on A in Sect. 1.2.1 are fulfilled.

Now we formulate the conditions under which the inverse problem (1), (2), (4), (5) is resolvable.

THEOREM 1. *Let Ω be a bounded, strictly convex domain in $\mathbb{C}$,*

Supplement

$\partial\Omega \in C^{1,\alpha}$; a function $f(t)$ belong to the class $C^{1,\alpha}(\partial\Omega; l_2^{m+4})$, $m \geq 1$, and satisfy the conditions

(i) $f(t_0)/2 = (2\pi i)^{-1} \int_{\partial\Omega} (dt - U^* d\bar{t})[(t - t_0)$
$\qquad\qquad\qquad - (\bar{t} - \bar{t}_0) U^*]^{-1} f(t), \quad t_0 \in \partial\Omega;$

(ii) $\operatorname{Re} \sum_{n=0}^{\infty} [f_n(t_1) + f_n(t_2)] \exp[-i(2n+1)\Phi(t_1, t_2)] = 0,$
$\qquad\qquad\qquad\qquad\qquad\qquad\qquad t_1, t_2 \in \partial\Omega.$

Then $a(z) = 2\operatorname{Re}(u_0)_z \in C(\overline{\Omega}) \cap C^{m+2}(\Omega)$, and $a(z)$ is the right-hand side of equation (1), whose solution satisfies the boundary condition (2), where

$f(t, \varphi) = 2 \operatorname{Re} \sum_{n=0}^{\infty} f_n(t) e^{-i(2n+1)\varphi} \in C^{m+3}([-\pi, \pi])$
$\qquad\qquad\qquad\qquad \cap W_2^{m+4}([-\pi, \pi]), \quad t \in \partial\Omega,$

and $u(z) = (u_0(z), u_1(z), \ldots)$ is representable as a Cauchy integral:

$u(z) = (2\pi i)^{-1} \int_{\partial\Omega} (dt - U^* d\bar{t})[(t - z) - (\bar{t} - \bar{z}) U^*]^{-1} f(t).$

PROOF. Making use of the Theorems 2, 3 of Sect. 1.2.2, we conclude that $u(z) \in C^1(\overline{\Omega}; l_2^m) \cap C(\overline{\Omega}; l_2^{m+2})$, and $u(t)|_{\partial\Omega} = f(t)$.

Consider the segment L_2 and the function $u(\eta)$ on it. We denote $\beta = z - t_2$, and evaluate the derivative of $u(\eta(s))$ with respect to s:

$u' = u_s'(\eta(s)) = \beta(I - \bar{\beta}\beta^{-1} U^*) u_\eta.$

Hence,
$u_\eta = (1/\beta)(I - \bar{\beta}\beta^{-1} U^*)^{-1} u',$

$\int_{L_2} u_\eta |d\eta| = \int_0^1 \frac{|\beta|}{\beta} (I - \bar{\beta}, \beta^{-1} U^*)^{-1} u' ds$

$\qquad\qquad = \frac{|\beta|}{\beta} (I - \bar{\beta}\beta^{-1} U^*)^{-1} [u(z) - u(t_2)].$

In a similar way, for the segment L_1, we have

$\int_{L_2} u_\xi |d\xi| = -|\beta|\beta^{-1}(I - \bar{\beta}\beta^{-1} U^*)^{-1} (u(z) - u(t_1)).$

Then, from the condition (ii), we have

$$u(z, \varphi) = \frac{1}{2}(\int_{L_2} a(\eta)|d\eta| - \int_{L_1} a(\xi)|d\xi|)$$

$$= \text{Re}\{\int_{L_2} u_\eta|d\eta| - \int_{L_1} u_\xi|d\xi|\}_0$$

$$= \text{Re}\left\{\frac{|\beta|}{\beta}\left(I - \frac{\overline{\beta}}{\beta}U^*\right)^{-1}(-u(t_1) - u(t_2) + 2u(z))\right\}_0$$

$$= 2\,\text{Re}\left\{\frac{|\beta|}{\beta}\left(I - \frac{\overline{\beta}}{\beta}U^*\right)^{-1}u(z)\right\}_0, \quad \varphi = \arg\beta.$$

Since $u(z) \in C(\overline{\Omega}; l_2^{m+2})$, then $u(z, \varphi) \in C(\overline{\Omega})$ for all φ, consequently,

$$u(t, \varphi)|_\Sigma = 2\,\text{Re}\left\{\frac{|\beta|}{\beta}\left(I - \frac{\overline{\beta}}{\beta}U^*\right)^{-1}f(t)\right\}_0 = f(t, \varphi);$$

that is, the function $u(z, \varphi)$ is a solution to the problem (1), (2) with the right-hand side $a(z) = 2\text{Re}\,(u_0)_z$. The conditions (3), (4) will be fulfilled automatically. ∎

Consider now the case $k = 1/R > 0$. We represent the solution to equation (1) in terms of the sought function $a(z)$:

$$u(z, \varphi) = \int_{L(z, \varphi)} a(\eta)|d\eta|,$$

where

$$L(z, \varphi) = \{\eta\colon \eta(\sigma) = z - iRe^{i\varphi} + iRe^{i(\varphi+\sigma)}, \quad \sigma \in [0, l/R],$$

$\eta(l/R) \in \partial\Omega$, and l is the length of the arc $L(z, \varphi)$. Hence, the function $u|_\Sigma = f(z, \varphi)$ has the property

$$f(t_2, \Phi(t_1, t_2)) = 0, \quad t_1, t_2 \in \partial\Omega,$$

where

$$\Phi(t_1, t_2) = \arg\left[\left(\sqrt{1 - \frac{|t_2 - t_1|^2}{4R^2}} + i\frac{|t_2 - t_1|}{2R}\right)\frac{t_1 - t_2}{|t_2 - t_1|}\right].$$

Supplement

Expanding the solution $u(z, \varphi)$ into a Fourier series,

$$u(z, \varphi) = 2\operatorname{Re}\sum_{n=0}^{\infty} u_n(z)e^{-in\varphi} - u_0(z),$$

we arrive at the equation with operator-valued coefficients

$$u_{\bar{z}} + U^*U^*u_z = -kidu, \tag{6}$$

where $u = (u_0, u_1, \ldots)$, and $D(u_0, u_1, u_2, \ldots) = (u_1, 2u_2, 3u_3, \ldots)$. For equation (6), we introduce a continuous scale of Banach spaces $\{X_s\}$, where $X_s = l_{2,s}$ is the Hilbert space of infinite sequences of complex numbers $(u_0, u_1, \ldots) = u$ endowed with the norm

$$\|u\|_{X_s}^2 = \sum_{j=0}^{\infty} |u_j|^2 e^{-2js}, \quad 0 < s < \infty.$$

For equation (6) and the scale of Banach spaces $\{X_s\}$, the corresponding analogs of the theorems of Sects. 1.2.1 and 1.2.2 hold, and the operator-valued kernel in the Cauchy-type integral is the operator $K(z) \in \mathcal{L}(X_s, X_{s'})$, where

$$s'' - s > 0, \quad s'' > s_0, \quad s'' - s' - \ln 2 > 0,$$
$$s_0 = \max[\ln 2, \ln(2k \operatorname{diam} \Omega), \ln(1 + k \operatorname{diam} \Omega)],$$
$$|z| \leq \operatorname{diam} \Omega,$$
$$K(z) = \exp(ki\bar{z}D)(I - ki\bar{z}U^*)(zI - kiz\bar{z}U^* - \bar{z}U^*U^*)^{-1}$$

The following theorem about the solvability of the integral geometry problem in a disc is valid.

THEOREM 2. Let $f(t) \in C^{1,\alpha}(\partial\Omega; X_s)$, where $\partial\Omega$ is a circle of radius $r < R = 1/k$, and $s > s_0$, satisfy the conditions:

(1) $f(t_0)/2 = (2\pi i)^{-1}\int_{\partial\Omega}(dt - U^*U^*\,d\bar{t})K(t - t_0)f(t),$

$$t_0 \in \partial\Omega,$$

(2) $2\operatorname{Re}\sum_{n=0}^{\infty} f_n(t_2)\exp[-in\phi(t_1, t_2)] = f_0(t_2), \quad t_1, t_2 \in \partial\Omega,$

(3) $\operatorname{Im} u_0(z) = 0, \quad z \in \bar{\Omega},$

where

$$u(z) = (2\pi i)^{-1}\int_{\partial\Omega}(dt - U^*U^*\,d\bar{t})K(t - z)f(t).$$

Then the function $a(z) = 2\text{Re}(u_1)_z \in C(\overline{\Omega}) \cap C^\infty(\Omega)$ is the right-hand side of equation (1), whose solution satisfies the boundary condition (2) with the function

$$f(t, \varphi) = 2\text{Re} \sum_{n=0}^{\infty} f_n(t) e^{-i\varphi n} - f_0(t),$$

which, for any fixed $t \in \partial\Omega$, is continued analytically to the horizontal strip $|\text{Im } \varphi| < \delta$ for some $\delta(s) > 0$.

This theorem, like Theorem 1, is proved by constructing a solution $u(z, \varphi)$ of the problem (1), (2) with the given right-hand side $a(z)$.

Let us consider the inverse problem for the transport equation in the domain $\Omega \times [-\pi, \pi]$, namely, determine the right-hand side of the equation

$$u_z e^{i\varphi} + u_{\bar{z}} e^{-i\varphi} + ku = b(z) \int_{-\pi}^{\pi} u(z, \theta) d\theta + a(z) \quad (7)$$

from the boundary values of the solution $u(z, \varphi)$ on the boundary $\Sigma = \partial\Omega \times [-\pi, \pi]$,

$$u(z, \varphi)|_\Sigma = f(z, \varphi). \quad (8)$$

In this case, expanding the solution $u(z, \varphi)$ into a Fourier series yields the equation

$$u_{\bar{z}} + U^* U^* u_z = -kU^* u,$$

with $u(z) = (u_0(z), u_1(z), \ldots)$, or, if we set $w(z) = \exp(k\bar{z} \ U^*) u(z)$, then

$$w_{\bar{z}} + U^* U^* w_z = 0.$$

Furthermore,

$$2\text{Re}(u_1)_z = b(z) \int_{-\pi}^{\pi} u(z, \theta) d\theta + a(z). \quad (9)$$

Thus, having evaluated the solution of the direct problem (7), (8),

$$u(z, \varphi) = 2\text{Re} \sum_{n=0}^{\infty} u_n(z) e^{-in\varphi} - u_0(z)$$

$$= 2\text{Re}[(I - \beta U^*)^{-1} u(z)]_0 - u_0(z),$$

where $u(z) = \exp(-k\bar{z}U^*)w(z)$, $w(z)$ is an A-analytic function in Ω, and $A = -U^*U^*$, we can determine with the aid of (9) the unknown function $a(z)$ for known $b(z)$, and vice versa.

1.3. A-Harmonic Functions

1.3.1. Green's Formulas

Let the Banach space X of Sect. 1.1 be the complexification of a real Banach space $Y: X = Y \oplus iY$. Then any one $u \in X$ is represented uniquely as $u = v + iw$; $v, w \in Y$. The elements v and w are the real and imaginary parts of u: $\operatorname{Re} u = v$, $\operatorname{Im} u = w$. Likewise, for an arbitrary operator $A \in \mathcal{L}(X)$ we let by definition,

$$(\operatorname{Re} A)u = \operatorname{Re}(Av) + i\operatorname{Re}(Aw),$$
$$(\operatorname{Im} A)u = \operatorname{Im}(Av) + i\operatorname{Im}(Aw).$$

It is not hard to see that $\operatorname{Re} A, \operatorname{Im} A \in \mathcal{L}(X) \cap \mathcal{L}(Y)$, and $A = \operatorname{Re} A + i\operatorname{Im} A$. If we assume in addition that the operators $\operatorname{Re} A$ and $\operatorname{Im} A$ commute, and u is an A-analytic function on $\Omega \subset \mathbb{R}^2$, then

$$\bar{\partial}_A u = \bar{\partial}u - A\partial u = 0 \Leftrightarrow \Delta_A v = 0, \quad \Delta_A w = 0,$$

where
$$\Delta_A v = [(I - \operatorname{Re} A)^2 + (\operatorname{Im} A)^2]v_{xx}$$
$$+ [(I + \operatorname{Re} A)^2 + (\operatorname{Im} A)^2]v_{yy} - 4\operatorname{Im} A v_{xy} = 0.$$

DEFINITION. *A function v of the class $C^2(\Omega; Y)$ or $C^2(\Omega; X)$, which satisfies the equation $\Delta_A v = 0$ is said to be A-harmonic.*

In what follows, to simplify presentation, we confine ourselves to the case of a real-valued A, $\operatorname{Im} A = 0$, and deem X to be a Hilbert space endowed with the inner product $\langle \cdot, \cdot \rangle$. Then, with vector notation for points in $\mathbb{R}^2$, $x = (x_1, x_2)$, we have

$$\Delta_A u = (I - A)^2 u_{x_1 x_1} + (I + A)^2 u_{x_2 x_2}.$$

Let Ω be a bounded domain with the boundary $\partial\Omega$ of class C^1, and ν be the unit outward normal to $\partial\Omega$. For an arbitrary vector field $w = (w_1, w_2)$, $(w_1, w_2 \in X)$ of class $C^1(\overline{\Omega}; X)$, we have the Gauss-Ostrogradski formula

$$\int_\Omega \operatorname{div} w(x) dx = \int_{\partial\Omega} \nu \cdot w ds, \tag{1}$$

where ds is an elementary arc length on $\partial\Omega$, and
$$v \cdot w = v_1 \cdot w_1 + v_2 \cdot w_2.$$

For functions u, v of the class $C^2(\Omega; X) \cap C^1(\overline{\Omega}; X)$ we put
$$\nabla_A u = ((I - A)^2 u_{x_1}, (I + A)^2 u_{x_2}), \quad \nabla = \nabla_0,$$
$$\frac{\partial u}{\partial v_A} = v \cdot \nabla_A u,$$
$$w(x) = (\langle v, (I - A)^2 u_{x_1}\rangle, \langle v, (I + A)^2 u_{x_2}\rangle).$$

Then
$$\operatorname{div} w = \langle v, \Delta_A u\rangle + \langle \nabla v, \nabla_A u\rangle,$$

where the inner product of the vector fields ∇v and $\nabla_A u$ is defined in a natural way:
$$\langle \nabla v, \nabla_A u\rangle = \langle (\nabla v)_1, (\nabla_A u)_1\rangle + \langle (\nabla v)_2, (\nabla_A u)_2\rangle.$$

Thus, the Gauss-Ostrogradski formula leads to the identity
$$\int_\Omega \langle v, \Delta_A u\rangle\, dx + \int_\Omega \langle \nabla v, \nabla_A u\rangle\, dx = \int_{\partial\Omega} \left\langle v, \frac{\partial u}{\partial v_A}\right\rangle ds. \qquad (2)$$

Similarly, if
$$w(x) = \langle \nabla_{A^*} v, u\rangle \stackrel{\text{def}}{=} (\langle (I - A^*)^2 v_{x_1}, u\rangle, \langle (I + A^*)^2 v_{x_2}, u\rangle),$$

where A^* is the adjoint of A in X, then
$$\int_\Omega \langle \Delta_{A^*} v, u\rangle\, dx + \int_\Omega \langle \nabla_{A^*} v, \nabla u\rangle\, dx = \int_{\partial\Omega} \left\langle \frac{\partial v}{\partial v_{A^*}}, v\right\rangle ds. \qquad (3)$$

Since $\langle \nabla_{A^*} v, \nabla u\rangle = \langle \nabla v, \nabla_A u\rangle$, subtracting (3) from the identity (2), we have
$$\int_\Omega (\langle v, \Delta_A u\rangle - \langle \nabla_{A^*} v, u\rangle)\, dx$$
$$= \int_{\partial\Omega} \left(\left\langle v, \frac{\partial u}{\partial v_A}\right\rangle - \left\langle \frac{\partial v}{\partial v_{A^*}}, u\right\rangle\right) ds. \qquad (4)$$

Supplement

The formulas (2), (4) will be called the first and second Green's identities in the scalar representation. Let an operator-valued function $V(x) \in C^2(\Omega; \mathcal{L}(X)) \cap C^1(\overline{\Omega}; \mathcal{L}(X))$. We put in equation (1)

$$w(x) = V(x)\nabla_A u = (V(x)(I - A)^2 u_{x_1}, V(x)(I + A)^2 u_{x_2}).$$

Then div $w = V(x)\Delta_A u + \nabla V \cdot \nabla_A u$, and so

$$\int_\Omega V\Delta_A u\, dx + \int_\Omega \nabla V \cdot \nabla_A u\, dx = \int_{\partial\Omega} V \frac{\partial u}{\partial v_A}\, ds. \tag{5}$$

On the other hand, if

$$w = (\nabla_A V)u = ((I - A)^2 V_{x_1} u, (I + A)^2 V_{x_2} u),$$

then div $w = (\Delta_A V)u + \nabla_A V \cdot \nabla u$, and consequently,

$$\int_\Omega (\Delta_A V)u\, dx + \int_\Omega \nabla_A V \cdot \nabla u\, dx = \int_{\partial\Omega} \frac{\partial V}{\partial v_A} u\, ds. \tag{6}$$

In particular, if for all $x \in \Omega$, the operator $V(x)$ commutes with A, then $\nabla_A V \cdot \nabla u = \nabla V \cdot \nabla_A u$, and from (5) and (6) we obtain the vector formula of the second Green's identity:

$$\int_\Omega (V\Delta_A u - \Delta_A Vu)\, dx = \int_{\partial\Omega} \left(V \frac{\partial u}{\partial v_A} - \frac{\partial V}{\partial v_A} u \right) ds. \tag{7}$$

It leads us to

THEOREM 1. *If u is an A-harmonic function of the class $C^2(\Omega; X) \cap C^1(\overline{\Omega}; X)$ in Ω, then*

$$\int_{\partial\Omega} \frac{\partial u}{\partial v_A}\, ds = 0.$$

To prove it, it suffices to put $V(x) = I$ in (7).

Now let us proceed to constructing the fundamental solution of the operator Δ_A. For an arbitrary point $x \in \mathbb{R}^2$, we introduce the operator-valued function

$$\Gamma(x) = \frac{1}{2\pi}(I - A^2)^{-1}\left\{\ln|x| - \ln(I + A) + \frac{1}{2}\ln(I - \mu(x)A) + \frac{1}{2}\ln(I - \overline{\mu}(x)A)\right\}, \quad (8)$$

$$\mu(x) = -q + i\sqrt{1 - q^2}, \quad q = \frac{(x_1^2 - x_2^2)}{|x|^2}, \quad |\mu| = 1. \quad (9)$$

To make this definition correct, we need some additional assumptions about the operator A.

Henceforth we will assume that there exists a scale of Hilbert spaces X^m dense in X:

$$X^{m+1} \subseteq X^m \subseteq \ldots \subseteq X = X^0,$$
$$\|u\|_m \leq \|u\|_{m+1}$$

such that, for all m, $A \in \mathcal{L}(X^m)$, $\|A\| \leq 1$ in X^m, and the resolvent of A is extended by strong continuity from the exterior of a unit circle $|\lambda|$ the boundary $|\lambda| = 1$ as an operator from $\mathcal{L}(X^{m+1}, X^m)$. Since in each X^{m+1} the operator A is a contraction, that is, $\|A\| \leq 1$, we can define functions of A using the functional calculus developed in [19].

Let H^∞ be a class of functions $f(\lambda)$ analytic in the open unit disc $D = \{\lambda : |\lambda| < 1\}$ and such that

$$|f|_\infty = \sup_{x \in D} |f(\lambda)| < \infty.$$

It is known that if $f \in H^\infty$, then, for almost all t on the unit circle, there exists the radial limit

$$f(e^{it}) = \lim_{r \to 1-0} f(r \cdot e^{it}).$$

Furthermore, $f(e^{it})$ exists almost everywhere as a nontangential limit, that is, the limit for λ tending to e^{it} inside the angle formed by two chords of the unit circle emanating from the point e^{it}. Denote by H_0^∞ a set of functions, form H^∞, whose nontangential limit exists for all t on the unit circle. Clearly, H_0^∞ is an algebra.

The following theorem is a special case of Theorem 2.3 in Ref. [19].

THEOREM 2. *Let A be a contraction of the Hilbert space X.*

Supplement

Then the mapping $f \to f(A)$ of the class H_0^∞,0) into $\mathcal{L}(X)$, defined by the condition

$$f(A) = \lim_{r \to 1-0} \sum_{k=0}^{\infty} r^k C_k A^k$$

for

$$f(\lambda) = \sum_{k=0}^{\infty} C_k \lambda^k \in H_0^\infty,$$

is a homeomorphism of the algebra H_0^∞ into $\mathcal{L}(X)$, posessing the following properties:

(a) $f(A) = \begin{cases} I \text{ for } f(\lambda) = 1, \\ A \text{ for } f(\lambda) = \lambda, \end{cases}$

(b) $\|f(A)\| \leq |f|_\infty$,

(c) $f_n(A) \to f(A)$ in the norm in $\mathcal{L}(X)$, if $f_n(\lambda) \to f(\lambda)$ uniformly in D.

With reference to this theorem, we give a correct definition of $\Gamma(x)$. For this purpose, in the complex plane with a cut along the negative part of the real axis, we pick a fixed branch of the logarithm by the formula

$$\ln z = \ln|z| + i \arg z, \quad -\pi < \arg z \leq \pi.$$

Since $f_\mu(\lambda) = (1 - \mu\lambda)\ln(1 - \mu\lambda) \in H_0^\infty$ for all μ such that $|\mu| = 1$, then by Theorem 2 the operator $f_\mu(A) \in \mathcal{L}(X^m)$, thus

$$\ln(I - \mu A) \stackrel{\text{def}}{=} (I - \mu A)^{-1} f_\mu(A) \in \mathcal{L}(X^{m+1}, X^m)$$

in view of the conditions imposed on the resolvent of A. Thus, the formula (8) defines the operator $\Gamma(x) \in \mathcal{L}(X^{m+3}, X^m)$, $x \neq 0$. Recalling the known property of the logarithm and invoking Theorem 2, the formula (8) may be recast in a more compact form:

$$\Gamma(x) = \frac{1}{2\pi}(I - A^2)^{-1}$$

$$\times \left\{ -\ln(I + A) + \frac{1}{2}\ln[x_1^2(I + A)^2 + x_2^2(I - A)^2] \right\}. \quad (10)$$

Denoting for brevity $(4\pi)^{-1}(I - A^2)^{-1} = C$, and $x_1^2(I + A)^2$

$+ x_2^2 (I - A)^2 = M(x)$, we have

$$\Gamma_{x_j} = CM^{-1}(x) \cdot M_{x_j}, \qquad (11)$$

$$\Gamma_{x_j x_j} = CM^{-2}(x)[MM_{x_j x_j} - M_{x_j}^2],$$

whence

$$\Delta_A \Gamma = CM^{-2}\{(I - A)^2 M_1 + (I + A)^2 M_2\}, \qquad (12)$$

where $M_j = MM_{x_j x_j} - Mx_{x_j}^2$. Since

$$M_{x_1} = 2x_1(I + A)^2, \quad M_{x_1 x_1} = 2(I + A)^2, \qquad (13)$$

and

$$M_{x_2} = 2x_2(I - A)^2, \quad M_{x_2 x_2} = 2(I - A)^2, \qquad (14)$$

we have

$$M_1 = [x_1^2(I + A)^2 + x_2^2(I - A)^2] \cdot 2(I + A)^2 - 4x_1^2(I + A)^4,$$
$$M_2 = [x_1^2(I + A)^2 + x_2^2(I - A)^2] \cdot 2(I - A)^2 - 4x_2^2(I - A)^4.$$

Substituting these expressions in (12) yields

$$\Delta_A \Gamma(x) = 0, \quad x \neq 0. \qquad (15)$$

Since M can be factorized as

$$M(x) = |x|^2(A - \mu(x))(A - \bar{\mu}(x)),$$

from (11), (13), and (14), we have

$$\frac{\partial \Gamma(x)}{\partial v_A} = v \cdot \nabla_A \Gamma$$

$$= v_1(I - A)^2 C|x|^{-2}(A - \mu)^{-1}(A - \bar{\mu})^{-1} \cdot 2x_1(I + A)^2$$
$$+ v_2(I + A)^2 C|x|^{-2}(A - \mu)^{-1}(A - \bar{\mu})^{-1} \cdot 2x_2(I - A)^2$$
$$= 2(v \cdot x)|x|^{-2} C(A - \mu)^{-1}(A - \bar{\mu})^{-1}(I - A^2)^2.$$

Supplement

Recalling the formula for the operator C, we obtain finally

$$\frac{\partial \Gamma(x)}{\partial v_A} = (2\pi)^{-1} |x|^{-2} (v \cdot x)(I - A^2)(A - \mu)^{-1}(A - \bar\mu)^{-1}. \quad (16)$$

Let $y \in \Omega$, and $B_\rho(y)$ be a closed ball of a sufficiently small radius ρ about y, $B_\rho(y) \in \Omega$. Since, by (15), $\Delta_A \Gamma(x - y) = 0$ for $x \in \Omega \setminus B_\rho(y)$, where Δ_A is evaluated in x, then substituting $\Gamma(x - y)$ for $V(x)$ and $\Omega \setminus B_\rho$ for Ω in Green's formula (7) we have

$$\int_{\Omega \setminus B_\rho} \Gamma \Delta_A u \, dx = \int_{\partial\Omega} \left(\Gamma \frac{\partial u}{\partial v_A} - \frac{\partial \Gamma}{\partial v_A} u \right) ds + \int_{\partial B_\rho} \left(\Gamma \frac{\partial u}{\partial v_A} - \frac{\partial \Gamma}{\partial v_A} u \right) ds. \quad (17)$$

Now we pass over to the polar coordinates

$$x_1 = y_1 + \rho \cos \varphi, \quad x_2 = y_2 + \rho \sin \varphi, \quad ds = \rho \, d\varphi,$$

and evaluate the limit of the second summand as $\rho \to 0$. We have

$$I_1 = \int_{\partial B_\rho} \Gamma \frac{\partial u}{\partial v_A} ds = \rho \int_0^{2\pi} \Gamma(\rho \cos \varphi_1, \rho \sin \varphi) \frac{\partial u}{\partial v_A} d\varphi.$$

According to the formula (8),

$$\|\Gamma(x)v\|_m \leq C_m \ln |x| \cdot \|v\|_{m+3}, \quad v \in X^{m+3},$$

and, consequently, if $u \in C^1(\bar\Omega; X^{m+3})$, then

$$\|I_1\|_m \leq C_m \rho \ln \rho \to 0, \quad \text{as } \rho \to 0.$$

Further on,

$$I_2 = \int_{\partial B_\rho} \frac{\partial \Gamma(x - y)}{\partial v_A} u(x) \, ds = \int_{\partial B_\rho} \frac{\partial \Gamma(x - y)}{\partial v_A} u(y) \, ds$$

$$+ \int_{\partial B_\rho} \frac{\partial \Gamma(x - y)}{\partial v_A} (u(x) - u(y)) \, ds. \quad (18)$$

The outward normal to the boundary of $\Omega \setminus B_\rho$ on ∂B_ρ is $v = -|x - y|^{-1} \times (x - y)$, consequently, by (16),

$$\frac{\partial \Gamma(x - y)}{\partial v_A} = -(2\pi)^{-1} |x - y|^{-1} (I - A^2)(A - \mu)^{-1}(A - \bar\mu)^{-1}$$

$$\in \mathscr{L}(X^{m+2}, X^m).$$

Therefore the second summand in (18) is estimated by

$$C \int_0^{2\pi} \|u(x) - u(y)\|_{m+2} \, d\varphi \to 0 \quad \text{as} \quad \rho \to 0.$$

The integral in the first summand in (18) is found explicitly, like it was done in the proof of Theorem 2 in Sect. 1.2.1. Namely, employing the identity

$$\int_0^{2\pi} \frac{d\varphi}{\lambda^2 + 2\lambda \cos 2\varphi + 1} = (1 - \lambda^2)^{-1}, \quad |\lambda| < 1,$$

one can prove that

$$-\int_{\partial \beta_\rho} \frac{\partial \Gamma}{\partial \nu_A} ds = (I - A^2) \int_0^{2\pi} (A^2 - 2A \operatorname{Re} \mu + I)^{-1} d\varphi = I,$$

where $-2\operatorname{Re} \mu = \cos 2\varphi$.

Finally, by letting in (17) $\rho \to 0$, we obtain

$$u(y) = \int_{\partial \Omega} \left(\frac{\partial \Gamma}{\partial \nu_A} u - \Gamma \frac{\partial u}{\partial \nu_A} \right) ds + \int_\Omega \Gamma(x - y) \Delta_A u \, ds. \quad (19)$$

By this we have established

THEOREM 3. *Let for some* $m \geq 0$ *a function* $u \in C^2(\Omega; X^{m+3}) \cap C^1(\overline{\Omega}; X^{m+2})$. *Then the Green's representation* (19) *holds for all* $y \in \Omega$. *If, in addition, u is A-harmonic, then*

$$u(y) = \int_{\partial \Omega} \left(\frac{\partial \Gamma}{\partial \nu_A} u - \Gamma \frac{\partial u}{\partial \nu_A} \right) ds.$$

Now, suppose that the operator-valued function $V(x) \in C^2(\Omega; \mathcal{L}(X)) \cap C^1(\overline{\Omega}; \mathcal{L}(X))$, possibly depends on a parameter $y \in \Omega$, and satisfies the equation $\Delta_A V = 0$ in Ω. Then, using the vector second Green's identity (7) we get

$$0 = \int_{\partial \Omega} \left(\frac{\partial V}{\partial \nu_A} u - V \frac{\partial u}{\partial \nu_A} \right) ds + \int_\Omega V \Delta_A u \, dx.$$

Denoting $G = \Gamma(x - y) + V$, and adding this identity with (19), we

Supplement

obtain

$$u(y) = \int_{\partial\Omega} \left(\frac{\partial G}{\partial v_A} u - G \frac{\partial u}{\partial v_A} \right) ds + \int_{\Omega} G \Delta_A u \, dx.$$

If, in addition, $G = 0$ on $\partial\Omega$, then such an operator-valued function $G(x, y)$ is said to be the Green's function (of the Dirichlet problem) for the domain Ω. In this case

$$u(y) = \int_{\partial\Omega} \frac{\partial G}{\partial v_A} u \, ds + \int_{\Omega} G \Delta_A u \, dx, \tag{20}$$

and if u is A-harmonic in Ω, then

$$u(y) = \int_{\partial\Omega} \frac{\partial G}{\partial v_A} u \, ds. \tag{21}$$

The Green's functions for particular regions (disc, half plane) will be constructed in Sect. 1.3.2.

Now we formulate without proof an analog of the theorem about the mean.

THEOREM 4. *Let* $u \in C^2(\Omega; X^{m+2})$ *and* A-*harmonic in* Ω. *Then for any ball* $B_r(y) = \{x : |x - y| < r\}$ *in* Ω

$$u(y) = \frac{1}{2\pi r} \int_{\partial B_r} \{[I - 2r^{-2} q(x - y) A(I + A^2)^{-1}] u(x) \, ds$$

$$- \frac{2}{\pi} A(I + A^2)^{-1} \int_{B_r} \frac{q(x - y)}{|x - y|^2} \} u(x) \, dx, \tag{22}$$

$$u(y) = \frac{1}{\pi r^2} \int_{B_r} \left\{ I - 2r^2 A(I + A^2)^{-1} \frac{q(x - y)}{|x - y|^2} \right\} u(x) \, dx, \tag{23}$$

$$q(x) = \frac{x_1^2 - x_2^2}{|x|^2}, \quad x = (x_1, x_2).$$

NOTE. The terms with kernel q in (22), (23) are singular integrals; that they exist follows from the fact that q is a homogeneous function of degree zero and

$$\int_{|x|=1} q(x) \, ds = 0.$$

1.3.2. The Dirichlet Problem

First we elucidate uniqueness. As in the preceding section, we assume that there is a scale of Hilbert spaces X^m, $m \geqslant 0$, embedded densely into $X = X^0$, and such that $A \in \mathcal{L}(X^m)$, and $\|A\| \leqslant 1$ in X^m for all m. (In fact, as we will see below, it would suffice to assume $m = 0, 1$, because only the spaces X and $X^1 \subset X$ will be used.) The conditions imposed on the resolvent of A will be significantly weaker than in the preceding section. Namely, we assume that for any $v \in X^{m+1}$ there exists the strong limit

$$\lim_{r \to 1-0} (I \pm rA)^{-1} v = (I \pm A)^{-1} v,$$

which defines the operators $(I \pm A)^{-1} \in \mathcal{L}(X^{m+1}, X^m)$.

Let

$$\Lambda_1 = (I - A)(I + A)^{-1}, \quad \Lambda_2 = (I + A)(I - A)^{-1}.$$

Treating Λ_1 and Λ_2 as unbounded operators in X with the domain $D(\Lambda_j) = X^1$, we evaluate the operators Λ_1^*, Λ_2^*.. Assuming that

$$D(\Lambda_j^*) \supseteq X^1$$

we define the operators

$$\operatorname{Re} \Lambda_j = \frac{\Lambda_j + \Lambda_j^*}{2} \in \mathcal{L}(X^1, X),$$

and the subspaces

$$L_j = \operatorname{Ker} \operatorname{Re} \Lambda_j = \{u \in X_1;\ \operatorname{Re} \Lambda_j u = 0\}.$$

THEOREM 1. *Let a function u of the class $C^2(\Omega; X) \cap C^1(\overline{\Omega}; X^1)$ satisfy the conditions*

$$\Delta_A u = 0, \quad x \in \Omega, \tag{1}$$

$$u|_{\partial \Omega} = 0. \tag{2}$$

Then if $\operatorname{Re} \Lambda_j$, $j = 1, 2$, are self-adjoint as operators in X with the domain X^1, and

$$L_1 \cap L_2 = \{0\}, \tag{3}$$

then $u = 0$ in Ω.

Supplement

PROOF. Let $v = Bu$ in the identity (2) of Section 1.3.1, where

$$B = ((I - r^2 A^2)^{-1})^*, \quad r < 1.$$

In view of the conditions (1) and (2), we have

$$\int_\Omega \langle \nabla Bu, \nabla_A u \rangle \, dx = 0.$$

Passing the operator B over to the second cofactor, and letting $r \to 1$, we obtain

$$\int_\Omega (\langle u_{x_1}, \Lambda_1 u_{x_1} \rangle + \langle u_{x_2}, \Lambda_2 u_{x_2} \rangle) \, dx = 0. \tag{4}$$

By a familiar theorem (see [19]) the operators Λ_j are accretive, that is,

$$\text{Re} \langle \Lambda_j v, v \rangle \geq 0, \quad v \in X^1, \tag{5}$$

hence, by evaluating the real part of (4) with allowance for (5) we find

$$\langle \text{Re} \, \Lambda_j u_{x_j}, u_{x_j} \rangle = 0, \quad j = 1, 2.$$

Since the operators $\text{Re} \, \Lambda_j$ are self-adjoint, this implies $u_{x_j} \in L_j$. Let x be an arbitrary point in Ω. Let us draw straight lines parallel to the coordinate axes through it and denote by $(a(x_2), x_2)$ and $(x_1, b(x_1))$ the points of intersection of these lines with the boundary $\partial \Omega$ nearest to x. By virtue of the boundary condition (2) and the linearity of subspaces L_j, we have

$$u(x) = \int_{a(x_2)}^{x_1} u_{x_1}(s, x_2) \, ds \in L_1,$$

$$u(x) = \int_{b(x_1)}^{x_2} u_{x_2}(x_1, s) \, ds \in L_2,$$

and, consequently, $u \in L_1 \cap L_2$. By the condition (3), this implies $u = 0$ for all $x \in \Omega$. ∎

The following example demonstrates the significance of condition (3).

EXAMPLE. Let $X^1 = X = \mathbb{C}^2$, $\Omega = [0, \pi] \times [0, \pi]$,

$$A = \begin{bmatrix} 0 & 1 \\ -1 & 0 \end{bmatrix}, \quad u = \begin{bmatrix} u_1(x) \\ u_2(x) \end{bmatrix},$$

and $u_1 = u_2 = \cos 2(x_1 + x_2) - \cos 2(x_1 - x_2)$.

Then $u|_{\partial\Omega} = 0$,

$$\Delta_A u = \begin{bmatrix} \Box & u_1 \\ -\Box & u_2 \end{bmatrix} = 0, \quad \Box = \partial_1^2 - \partial_2^2, \quad \partial_j = \frac{\partial}{\partial x_j}.$$

In the problems that conclude this chapter we intend to prove the uniqueness theorem for the case $X = l_2$, $A = -U^*$ or $A = -(U^*)^2$.

Here we give a method to obtain explicit solutions to the Dirichlet problem for A-harmonic functions:

$$\Delta_A u = 0, \quad x \in \Omega, \tag{6}$$

$$u|_{\partial\Omega} = f.$$

To this end we assume first that $A = \lambda I$, where $\lambda \in (-1, 1)$. Denoting points in $\mathbb{R}^2$ by (x, y), we have

$$\Delta_A u = (1-\lambda)^2 u_{xx} + (1+\lambda)^2 u_{yy} = 0.$$

Letting

$$\xi = (1-\lambda)^{-1} x, \tag{7}$$
$$\eta = (1+\lambda)^{-1} y,$$

and

$$u(x, y) = u_\lambda(\xi, \eta), \tag{8}$$

we obtain

$$\Delta u_\lambda = 0, \quad (\xi, \eta) \in \Omega_\lambda, \tag{9}$$
$$u_\lambda|_{\partial\Omega_\lambda} = f_\lambda, \tag{10}$$

where Ω_λ is the image of Ω under the mapping (7), and f_λ is related to f by the formula (8). The solution to the Dirichlet problem (9), (10) in terms of complex variables z, ξ is given by the formula [12]

$$u_\lambda(z) = \int_{\partial\Omega_\lambda} u_\lambda(\xi) \frac{\partial}{\partial \nu} G(z, \xi) |d\xi|, \quad z \in \Omega_\lambda, \tag{11}$$

where

$$G(z, \xi) = \frac{1}{2\pi} \ln |w_\xi(z)|,$$

$$w_\xi(z) = \frac{w(z) - w(\xi)}{1 - w(z)\overline{w(\xi)}},$$

and $w(z)$ maps Ω_λ conformally into the disc $|w| < 1$. Returning to the initial variables by the formulas (7), and using the "closure" of the Riesz-Dunford functional calculus (with the same assumptions about the resolvent as in Section 1.2) or the functional calculus for contractions developed in Ref. [19], we obtain an explicit formula for solving the problem (6).

For instance, if Ω is the half plane $\text{Im } z \geq 0$, then

$$u(x, y) = \int_{-\infty}^{\infty} P_y(x - t) f(t) dt = P_y * f,$$

$$P_y = \frac{1}{\pi} \frac{y}{y^2 + t^2} (I - A^2)(\xi - A)^{-1}(\overline{\xi} - A)^{-1},$$

$$\xi = \frac{(y - it)}{(y + it)}.$$

If Ω is a disc $|z| < 1$, then $\partial \Omega_\lambda$ are ellipses.

In this case, it is known that the function $w = w(z)$ which maps Ω_λ conformally onto $|w| < 1$ is given by the formula

$$w(z) = k^{1/2} \text{sn}\left(\frac{K}{\pi} \sin^{-1} z\right).$$

Here sn is the elliptic Jacobian function, $k = [\theta_2(0, q)/\theta_1(0, q)]^2$, θ_1, θ_2 are the theta functions,

$$K = \int_0^{\pi/2} (1 - k^2 \sin^2 \varphi)^{-1/2} d\varphi,$$

and

$$q = \left(\frac{a - b}{a + b}\right)^2,$$

where $a(\lambda)$ and $b(\lambda)$ are the semiaxes of the ellipse $\partial \Omega$:

$$\frac{x^2}{a^2} + \frac{y^2}{b^2} = 1.$$

Substituting this function into (11), returning to the initial variables, and substituting the operator A for the number λ, we arrive at Poisson's formula for a disc. Putting $x = y = 0$ in this formula yields another theorem of the mean (in addition to Theorem 4 in Section 1.3.1).

When instead of the Dirichlet problem for an operator we have a Cauchy problem with data given over a part of the boundary we obtain an appreciably ill-posed formulation. A treatment of the uniqueness and stability of such problems will be outlined in Section 1.5. It is based on the method of logarithmic convexity. Another approach, based on the solvability of the dual problem in the scale of Banach spaces, will be illustrated in the forthcoming section.

1.4. The Radon Problem on a Ring with Weight

In this section we obtain a stability estimate for the Radon problem with an arbitrary analytic weight in the case of incomplete projection data, in other words, when a sounded body contains an opaque screen. Here, we do not require the function to be smooth, but assume only that it is of bounded variation in each variable, which seems to be more natural for applications. Now we proceed to exact formulation of the problem.

In polar coordinates, this problem reduces to solving the following integral equation of the first kind:

$$p(\alpha, r) = (Au)(\alpha, r)$$
$$\equiv \int_\Pi \delta(r - \rho\cos(\beta - \alpha))a(\beta, \alpha, r)u(\beta, \rho)\rho d\rho d\beta. \quad (1)$$

Here $u(\beta, \rho)$ is the desired function in the ring $K = \{\beta, \rho; \rho \in [\kappa, 1], \kappa > 0, \beta \in [-\pi, \pi]$ periodic in β; $a(\beta, \alpha, r)$ is a given weight function periodic in α and β; $\delta(x)$ is Dirac's delta function, and $\Pi = [-\pi, \pi] \times [\kappa, 1]$.

The problem is to determine $u(\beta, \rho)$ given $p(\alpha, r)$ in the ring K.

This problem obviously is ill-posed in the sense of Hadamard. Therefore to obtain a conditional stability bound we need to introduce a set of well-posedness M. Let M be a set of functions $u(\beta, \rho)$ given in the ring K and such that the total variation in each variable is bounded by a constant B uniformly with respect to the other variables. The main result is as follows.

THEOREM 1. *Let the function* $q(\alpha, r, t) = a\left(\alpha + \arctan\left(\dfrac{t}{r}\right), \alpha, r\right) \cdot r$ *be continuous in* $r \in [\kappa, 1]$, $\kappa > 0$, *analytic in* α *and* t *within a*

Supplement

complex domain D:

$$|\operatorname{Im}\alpha| \leq \sigma\kappa,\ |\operatorname{Im} t| \leq \sigma\kappa,\ |\operatorname{Re} t| \leq \sqrt{1-\kappa^2}+\sigma\kappa,\ \sigma \in (0,1),$$

and periodic in $\operatorname{Re}\alpha$ *with period* 2π. *Furthermore, there exist* h_1, h_2 *such that, for all* $(r,\alpha,t) \in [\kappa,1] \times D$,

$$|q(\alpha,r,t)| \leq h_1 < \infty,\quad |a(\alpha,\alpha,r)| \geq h_2 > 0. \tag{2}$$

Then for any $0 < \kappa < 1$ *the solution of equation* (1) *is unique in* $L_1(\Pi)$; *moreover, if* $u \in M$ *and* κ *satisfies the condition*

$$(1-\kappa)\frac{h_1+h_2}{\kappa\mu h_2}C_1 < \frac{1}{e}, \tag{3}$$

where e is the Euler number, and C_1 is a constant depending on σ, then

$$\|u\|_{L_1(\Pi)} \leq C_2 \ln^{-1/2}\frac{C_2}{\|p\|_{L_1(\Pi)}}, \tag{4}$$

$$C_2 = C_2(h_1, h_2, \kappa, \sigma, B).$$

PROOF. The operator A is extended by continuity to $L_1(\Pi)$. We note that in (1) the integration is along the curves $l(\alpha,r)$: $r = \rho\cos(\beta-\alpha)$ with fixed α and r. The operator A^* adjoint to A with respect to the inner product $\langle\cdot,\cdot\rangle$ in L_2 is of the form

$$(A^*v)(\beta,\rho) = \int_\Pi \delta(r-\rho\cos(\beta-\alpha))a(\beta,\alpha,r)v(\alpha,r)r\,dr\,d\alpha. \tag{5}$$

Actually, the integration here is along the curves $l^*(\beta,\rho)$: $r = \rho\cos(\beta-\alpha)$ with fixed β and ρ. Consider a function $b(\alpha,r,t)$ which is a solution to the Cauchy problem:

$$Pb = tD_r b + D_\alpha b - rD_t b = q(\alpha,r,t)v(\alpha,r), \tag{6}$$

$$b(\alpha,\kappa,t) = f(\alpha,t), \tag{7}$$

where D_x is the operator of differentiation with respect to x. Putting $t = dr/d\alpha$, we observe that along the curve $l^*(\beta,\rho)$

$$(Pb)\left(\alpha,r,\frac{dr}{d\alpha}\right) = D_\alpha b\left(\alpha,r,\frac{dr}{d\alpha}\right). \tag{8}$$

From (5) and (8) we have

$$(A^*v)(\beta, \rho) = f\left(\beta + \arccos\frac{\kappa}{\rho}, -\sqrt{\rho^2 - \kappa^2}\right)$$
$$- f\left(\beta - \arccos\frac{\kappa}{\rho}, \sqrt{\rho^2 - \kappa^2}\right) = Tf, \quad (9)$$

and so,

$$\langle Au, v \rangle = \langle u, A^*v \rangle = \langle u, Tf \rangle = \langle Tf, u \rangle = \langle f, T^*u \rangle, \quad (10)$$

where T^* is the adjoint of T.

Denoting

$$\chi_1(\alpha, t) = \begin{cases} 1, & t \in [0, \sqrt{1-\kappa^2}], \\ 0, & t \notin [0, \sqrt{1-\kappa^2}], \end{cases}$$

$$\chi_2(\alpha, t) = \begin{cases} 1, & t \in [-\sqrt{1-\kappa^2}, 0], \\ 0, & t \notin [-\sqrt{1-\kappa^2}, 0], \end{cases}$$

we obtain an explicit expression for T^*:

$$(T^*u)(\alpha, t)$$
$$= \sum_{j=1}^{2} (-1)^j \chi_j(\alpha, t) u\left(\alpha + \arcsin\frac{t}{\sqrt{\kappa^2 + t^2}}, \sqrt{\kappa^2 + t^2}\right) |t|. \quad (11)$$

From this relation it is not hard to derive

$$\|T^*u\|_{L_1(\Pi_1)} = 2\|u\|_{L_1(\Pi)}, \quad (12)$$

where $\Pi_1 \equiv [-\pi, \pi] \times [-\sqrt{1-\kappa^2}, \sqrt{1-\kappa^2}]$.

By virtue of (2), the problem (6), (7) is equivalent to the integrodifferential equation

$$b(\alpha, r, t) = f(\alpha, t) + (Vb)(\alpha, r, t), \quad (13)$$

Supplement

where

$$(Vb)(\alpha, r, t) = \int_\kappa^r \{[a(\alpha, \alpha, \eta)\eta]^{-1}[D_\alpha b(\alpha, \eta, 0)$$
$$- \eta D_t b(\alpha, \eta, 0)]Rq(\alpha, \eta, t) - R(D_\alpha b - \eta D_t b)(\alpha, \eta, t)\}d\eta,$$

$$Rq(\alpha, \eta, t) \equiv \frac{q(\alpha, \eta, t) - q(\alpha, \eta, 0)}{t}.$$

Let X_s be a Banach space of functions analytic in the complex plane,

$$\Pi_s = \{(\alpha, t) \mid |\operatorname{Im}\alpha| \leq \mu(1+s), |\operatorname{Im} t| \leq \mu(1+s), |\operatorname{Re} t| \leq \sqrt{1-\kappa^2}$$
$$+ \mu(1+s), s \in [0, 1], \mu \leq \sigma\kappa\}$$

and bounded in its closure with the norm

$$\|f(\alpha, t)\|_s = \sup |f(\alpha, t)|, \quad (\alpha, t) \in \Pi_s.$$

Then, by the theorem on the solvability of Volterra operator equations in scales of Banach spaces (see [4], Theorem 1.1), we obtain that for any $f(\alpha, t) \in X_1$ there exists a solution $b(\alpha, r, t) \in C([\kappa, 1], X_{1/2})$ of equation (13) provided the condition (3) is fulfilled. In addition,

$$\|b\|_{C([\kappa, 1], X_{1/2})} \leq C_4(h_1, h_2, \kappa)\|f\|_{C([\kappa, 1], X_1)}. \tag{14}$$

Setting

$$f(\alpha, t) = \exp(-i\alpha n - it\xi), \tag{15}$$

$$U(\alpha, t) = \begin{cases} -u\left(\alpha + \arcsin\dfrac{t}{\sqrt{\kappa^2+t^2}}, \sqrt{\kappa^2+t^2}\right), & t \in (-\sqrt{1-\kappa^2}, \sqrt{1-\kappa^2}) \\ 0 & t \notin (-\sqrt{1-\kappa^2}, \sqrt{1-\kappa^2}) \end{cases}$$

we introduce the notations:

$$\langle T^*u, f\rangle = \int_{\Pi_1} U(\alpha, t)\exp(-i\alpha n - it\xi)d\alpha\, dt = \hat{U}_n(\xi),$$

$$\hat{U}(\alpha, \xi) = \int_{-\infty}^{\infty} U(\alpha, t)\exp(-it\xi)dt,$$

$$U_n(t) = \frac{1}{\pi}\int_{-\pi}^{\pi} U(\alpha, t)\exp(-i\alpha n)d\alpha.$$

Assuming ($\|p\|_{L_1(\Pi)} \leq \varepsilon$, from (10), (14), (15) we obtain

$$|\hat{U}_n(\xi)| \leq C_5(h_1, h_2, \kappa) \exp(2\mu(|n| + |\xi|)) \cdot \varepsilon, \quad (16)$$

and for $u \in M$ we have

$$\sup_\alpha |\hat{U}(\alpha, \xi)| \leq \frac{C_6(B)}{\sqrt{1 - \kappa^2 |\xi|}}, \quad (17)$$

$$\sup_t |U_n(t)| \leq \frac{C_6(B)}{\pi |n|}. \quad (18)$$

Using the Parseval identity together with the estimates (16), (18), we get

$$\int_\Pi |U(\alpha, t)|^2 \, d\alpha \, dt = \sum_{|n| \leq N} \int_{|\xi| < N} |\hat{U}_n(\xi)|^2 \, d\xi$$

$$+ \sum_{|n| \leq N} \int_{|\xi| \geq N} |\hat{U}_n(\xi)|^2 \, d\xi + \sum_{|n| \geq N} \int_{-\infty}^{\infty} |\hat{U}_n(\xi)|^2 \, d\xi$$

$$\leq \sum_{|n| \leq N} \int_{|\xi| \leq N} |\hat{U}_n(\xi)|^2 \, d\xi + \int_{|\xi| > N} \sum_{|n| \geq 0} |\hat{U}_n(\xi)|^2 \, d\xi$$

$$+ \sum_{|n| \geq N} \int_{-\infty}^{\infty} |U_n(t)|^2 \, dt \leq C_5^2 \exp(8\mu N) \cdot 4 N^2 \varepsilon^2 + \frac{8 C_6(B)}{N}.$$

Observing that $\|V\|_{L_1(\Pi_1)} \leq 2\pi^{1/2}(1-\kappa^2) \|U\|_{L_2}$ and taking into account (12) completes the proof. ∎

1.5. Problems

1. Find a Hilbert space in which the spectral radius of the operator U^* is zero.

2. Examine the properties of the resolvent of U^* in the scale of spaces l_2^m for $m \leq 1/2$.

3. Using the scale of spaces $l_{2,s}$ constructed in Section 1.1, prove for A-analytic functions ($A = -U^*$) an analog of Taylor's theorem.

4. Extend the results of Section 1.3 to the case of $\text{Im } A \neq 0$.

5. Find explicit solutions to the Dirichlet problem in the following cases:
   (1) $\Omega = [a, b] \times [c, d]$;
   (2) $\Omega = \{x \in \mathbb{R}^2; |x| > 1\}$.

6. Investigate uniqueness of the Dirichlet problem for a unitary operator A.

7. Study other types of boundary-value problems for the operator Δ_A.

Supplement

8. This problem outlines the proof of uniqueness of the Dirichlet problem for the operator Δ_A, $A = -(U^*)^2$.

Let $\langle \cdot, \cdot \rangle$ be the inner product in $\mathbb{R}^2$, and $v = (\cos \alpha, \sin \alpha)$ with $\alpha \in [-\pi, \pi]$. Consider the problem of determining the right-hand side of the equation

$$Pu = \langle v, \nabla u \rangle = a(x, y), \quad (x, y) \in \Omega,$$
$$u|_\Sigma = 0, \quad \Sigma = \partial\Omega \times [-\pi, \pi].$$

Here $u(x, y, \alpha)$ is a real-valued smooth function 2π-periodic in α. Let $v^\perp = (-\sin \alpha, \cos \alpha)$. Prove the following identity due to Mukhometov [15]:

$$2\langle v^\perp, \nabla u \rangle \partial_\alpha \langle v, \nabla u \rangle = |\nabla u|^2 + \partial_\alpha \{\langle u^\perp, \nabla u \rangle \langle v, \nabla u \rangle\}$$
$$+ (u_y u_\alpha)_x - (u_x u_\alpha)_y.$$

Applying the Stokes formula, derive from this identity that

$$\int_\Omega \int_{-\pi}^{\pi} |\nabla u|^2 \, dx\, dy\, d\alpha = -\int_{\partial\Omega} \int_{-\pi}^{\pi} u_\alpha u_r \, d\alpha\, ds,$$

where $u_\tau = \langle \tau, \nabla u \rangle$, τ is the unit vector tangential to $\partial\Omega$, and ds is the elementary length along the curve $\partial\Omega$. Putting

$$u_\pm = [u(x, y, \alpha) \pm u(x, y, -\alpha)]/2,$$

prove for all $\varepsilon > 0$ the estimate

$$2 \int_\Omega \int_{-\pi}^{\pi} |\nabla u|^2 \, dx\, dy\, d\alpha \leq \varepsilon \int_{\partial\Omega} \int_{-\pi}^{\pi} (|u_{-\tau}|^2 + |u_{-\alpha}|^2) \, ds\, d\alpha$$
$$+ \frac{1}{\varepsilon} \int_{\partial\Omega} \int_{-\pi}^{\pi} (|u_{+\alpha}|^2 + |u_{+r}|^2) \, ds\, d\alpha. \quad (1)$$

Expanding the function u into a Fourier series and using this estimate, prove the uniqueness theorem for the Dirichlet problem with Δ_A, $A = -(U^*)^2$.

9. Investigate the uniqueness and stability of the solution to the Cauchy problem for the operator Δ_A. Use for this purpose the solution of the forthcoming problem.

10. An abstract Cauchy problem.

Let H be a Hilbert space, and $A(t)$, $t \in [0, T]$, be a linear operator with a domain D dense in H and idependent of t. It is assumed that $D(A^*) \supseteq D$, the commutator $[A^*, A]$ is extended by continuity from

$D([A^*, A])$ to D, and the operator $\operatorname{Re} A$ is weakly continuously differentiable. Let $\lambda_0 = \inf \langle \operatorname{Re} Av, v \rangle$, $\lambda_1 = \sup \langle \operatorname{Re} Av, v \rangle$, $v \in D$, and $\|v\| = 1$. We will say that A *is hyponormal in the main* if there exist functions $\alpha, \varphi \in C^2[0, T]$, $\alpha'(t) > 0$ for $t \in [0, T]$, such that for all $\lambda \in [\lambda_0, \lambda_1)$,

$$[A^*A] + 4(\operatorname{Re} A - \lambda)^2 + 2\alpha'\{(\operatorname{Re} A)_t - \alpha''(\alpha')^{-2}(\operatorname{Re} A + \varphi_t) + \varphi_{tt}\} \geq 0.$$

Here $\operatorname{Re} A = (A + A^*)/2$.

Let $u(t)$ be a solution of the equation $u_t = A(t)u$. Prove the following assertion.

THEOREM. *For any $t_0, t, t_1 \in [0, T]$, $0 \leq t_0 < t < t_1 \leq T$ and all $\varepsilon > 0$, a necessary and sufficient condition for the estimate*

$$\begin{aligned} \|u(t)\| &\leq \varepsilon \|u(t_1)\| + c\varepsilon^{-\gamma} \|u(t_0)\|, \\ \gamma &= (\alpha(t) - \alpha(t_0))(\alpha(t_1) - \alpha(t_0))^{-1}, \\ c &= \beta^\gamma (1 - \beta) \exp\{\varphi(t_1)\gamma + \varphi(t_0) - \gamma\beta^{-1}\varphi(t)\}, \\ \beta &= \gamma(1 - \beta), \end{aligned} \quad (2)$$

to hold is that the operator A be hyponormal in the main.

COROLLARY. *Let $(\operatorname{Re} A)_t = 0$; $\alpha = t$, $\varphi_{tt} = 0$. Then for the estimate (2) to be valid it is sufficient that the operator A be hyponormal (that is, $[A^*, A] \geq 0$), and it is necessary that A be hyponormal on the eigenfunctions of the operator $\operatorname{Re} A$.*

A similar stability criterion for the finite difference scheme $u_{j+1} = Au_j$, $A \in \mathcal{L}(H)$ and A independent of j, has the form

$$A^*[A^*, A]A + (A^*A - \lambda)^2 \geq 0.$$

It enables the construction of stable finite difference solvers for the ill-posed Cauchy problem.

HINT. To prove this theorem, resort to the method of logarithmic convexity.

Logarithmic convexity can also be used to prove the uniqueness of solutions to the inverse coefficient problems.

11. For a real-valued A, prove the formula

$$\Delta_A = 4\bar\partial_A \partial_A.$$

Supplement

Chapter 2. RECONSTRUCTION OF PHONON SPECTRA FROM THE HEAT CAPACITY DATA

2.1. Problem Statement and Theoretical Background

As will be recalled the problem of recovering phonon spectra of crystals from the known heat capacity reduces to solving an integral equation of the first kind

$$\int_0^\infty \frac{(\omega/t)^2 e^{\omega/t}}{(e^{\omega/t} - 1)^2} g(\omega) d\omega = h(t), \quad t \in (0, \infty). \tag{1}$$

Here t and ω are the temperature and frequency measured in the same energy units.

We introduce the function

$$k(x) = \frac{x^2 e^x}{(e^x - 1)^2}, \tag{2}$$

and denote the operator given by the left-hand side of equation (1) by A. Then

$$Ag = \int_0^\infty k(\omega/t) g(\omega) d\omega.$$

Since

$$k'(x) = \frac{xe^x [e^x(2 - x) - 2 - x]}{(e^x - 1)^3},$$

and, for all $x > 0$,

$$e^x(2 - x) - (2 + x) < 0,$$

then $k(x)$ is a nonnegative, monotonically decreasing function for $x > 0$.

The graph of the function $k_t(\omega) = k(\omega/t)$ is obtained by expanding the plot of $k(x)$ t times along the x-axis. In particular, from (2) we have

$$k_t(\omega) \to 1 \quad \text{as} \quad t \to \infty, \quad \omega \geq 0,$$
$$k_t(\omega) \to 0 \quad \text{as} \quad t \to 0, \quad \omega > 0.$$

From the physics of the problem we have

$$g(\omega) \geq 0 \quad \text{and} \quad \int_0^\infty g(\omega)\,d\omega = 1.$$

Now, in view of the properties of $k_t(\omega)$ we conclude that the right-hand side of the equation $Ag = h$ is a monotonically increasing function, analytic for $t > 0$:

$$h(0) = 0, \quad \lim_{t \to \infty} h(t) = 1.$$

In what follows we need the Riemann zeta function $\zeta(s)$, defined for $\mathrm{Re}\, s > 1$ by the series

$$\zeta(s) = \sum_{n=1}^{\infty} 1/n^s,$$

and the Möbius function $\mu(n)$, which is defined for any natural $n \in N$ by the three properties:
(1) $\mu(1) = 1$,
(2) $\mu(n) = (-1)^k$, if n is a product of k different prime numbers,
(3) $\mu(n) = 0$, n is divisible by a square other than unity.
In particular,

$$\mu(1) = \mu(6) = \mu(10) = 1,$$
$$\mu(2) = \mu(3) = \mu(5) = \mu(7) = -1,$$
$$\mu(4) = \mu(8) = \mu(9) = 0.$$

This function will be required in connection with the Möbius series inversion formula (see Ref. [23]):

$$(M_k u)(x) = \sum_{n=1}^{\infty} \frac{u(nx)}{n^k}.$$

It turns out that if two functions $u(x)$ and $v(x)$ are related by a Möbius series, $M_k u = v$, then

$$u(x) = (M_k^{-1} v)(x) = \sum_{n=1}^{\infty} \mu(n) \frac{v(nx)}{n^k}.$$

Here is another pair of useful formulas (see [23]): if $P(n)$ is a completely multiplicative arithmetic function, that is,

$$P(n \cdot m) = P(n) \cdot P(m), \quad n, m \in N,$$

Supplement

and $u(x)$ is a function defined for all real $x > 0$, then

$$v(x) = \sum_{n \leq x} P(n) u\left(\frac{x}{n}\right)$$

implies

$$u(x) = \sum_{n \leq x} \mu(n) P(n) v\left(\frac{x}{n}\right).$$

We will also employ the following operators: the Laplace operator,

$$(Lu)(p) = \int_0^\infty e^{-pt} u(t) \, dt,$$

the Laplace-Carson operator, $\mathscr{L}$,

$$(\mathscr{L}u)(p) = p \cdot (Lu)(p),$$

the inversion operator, Inv

$$(\text{Inv } u)(t) = u(1/t), \quad t > 0,$$

the operator of multiplication by an independent variable, Λ,

$$(\Lambda u)(t) = tu(t),$$

the operator of integration, $\mathscr{S}$,

$$(\mathscr{S}u)(t) = \int_0^t u(\tau) \, d\tau,$$

and the operator of differentiation, ∂,

$$(\partial u)(t) = \frac{du}{dt}.$$

Under the appropriate conditions, each of these operators has an inverse (the expression for L^{-1} is given by the well-known inversion formulas for the Laplace transformation [11]:

$$\mathscr{L}^{-1} = L^{-1} \Lambda^{-1},$$
$$(\Lambda^{-1} u) = t^{-1} u(t),$$
$$\text{Inv}^{-1} = \text{Inv},$$
$$\partial \mathscr{S} = I,$$

$\mathscr{S}\partial u = u$ for $u(0) = 0$, that is,

$$\mathscr{S}^{-1} = \partial$$

on the class of functions $C_0^1[0, T] = \{ u \in C^1[0, T] : u(0) = 0 \}$. Here I is the identity operator as will be shown below, the operator A defining our equation $Ag = h$, factors into a product of operators listed above.

2.2. Inversion Formulas

THEOREM 1. *The following formulas are valid*:

$$A = \Lambda^{-1} \operatorname{Inv} M_0 \mathscr{L} \Lambda^2, \tag{1}$$

$$A^{-1} = \Lambda^{-2} \mathscr{L}^{-1} M_0^{-1} \operatorname{Inv} \Lambda. \tag{2}$$

If $Ag = h$, *then*

$$g(\omega) = \omega^{-2} \mathscr{L}^{-1} \left\{ \sum_{n=1}^{\infty} \mu(n) \frac{1}{nt} h\left(\frac{1}{nt}\right) \right\}. \tag{3}$$

PROOF. The relations (1)–(3) are mutually equivalent, hence it will suffice to prove, for instance, the formula (1). For this purpose, we notice that

$$\frac{1}{e^{\omega/t} - 1} = \sum_{n=1}^{\infty} e^{-n\omega/t}.$$

Differentiating the both sides with respect to ω and dividing by $-t^{-1}$ yields

$$\frac{e^{\omega/t}}{(e^{\omega/t} - 1)^2} = \sum_{n=1}^{\infty} n e^{-n\omega/t},$$

consequently,

$$(Ag)(t) = \int_0^{\infty} \sum_{n=1}^{\infty} n e^{-n\omega/t} (\omega/t)^2 g(\omega) d\omega$$

$$= t^{-1} \sum_{n=1}^{\infty} \frac{n}{t} \int_0^{\infty} \omega^2 g(\omega) e^{-n\omega/t} d\omega$$

$$= \frac{1}{t} \sum_{n=1}^{\infty} (\mathscr{L}(\omega^2 g))\left(\frac{n}{t}\right) = \Lambda^{-1} \operatorname{Inv} M_0 \mathscr{L} \Lambda^2 g. \blacksquare$$

Supplement

There are factorizations of A other than (1). They can be obtained with the following lemma.

LEMMA 1. *The following relations hold true*:

$$\mathscr{L}\mathscr{S}^k = \Lambda^{-k}\mathscr{L}, \tag{4}$$

$$M_0 \Lambda^{-k} = \Lambda^{-k} M_k, \tag{5}$$

$$M_k \mathscr{L} = \mathscr{L} \operatorname{Inv} M_k \operatorname{Inv}, \tag{6}$$

$$\operatorname{Inv} \Lambda^{-1} = \Lambda \operatorname{Inv}, \quad \operatorname{Inv} \Lambda = \Lambda^{-1} \operatorname{Inv}. \tag{7}$$

PROOF. The formula (4) is well-known [11] so we prove (5):

$$(M_0 \Lambda^{-k} u)(t) = \sum_{n=1}^{\infty} \frac{1}{(nt)^k} u(nt) = \frac{1}{t^k} \sum_{n=1}^{\infty} \frac{1}{n^k} u(nt) = \Lambda^{-k} M_k u.$$

The proof of (6) is achieved along the same lines:

$$(M_k \mathscr{L} u)(t) = \sum_{n=1}^{\infty} \frac{1}{n^k} (nt) \int_0^{\infty} e^{-nts} u(s)\, ds$$

$$= \sum_{n=1}^{\infty} \frac{1}{n^k} t \int_0^{\infty} e^{-tx} u\left(\frac{x}{n}\right) dx = t \int_0^{\infty} e^{-tx} \left\{ \sum_{n=1}^{\infty} \frac{1}{n^k} u\left(\frac{x}{n}\right) \right\}.$$

Since $u(x/n) = (\operatorname{Inv} u)(n/x)$, the last bracket is $(\operatorname{Inv} M_k \operatorname{Inv} u)(x)$, consequently,

$$M_k \mathscr{L} u = \mathscr{L} \operatorname{Inv} M_k \operatorname{Inv} u.$$

The relations (7) are evident.

Let $g(\omega)$ be such that $\omega^2 g(\omega) \in C_0^k[0, \infty) = \{ u(\omega) \in C^k[0, \infty) : u^m(0) = 0, m = 0, 1, \ldots, k-1 \}$. For such functions, $\Lambda^2 g = \mathscr{S}^k \partial^k \Lambda^2 g$, thus, on this class of functions we can substitute $\mathscr{S}^k \partial^k \Lambda^2$ for Λ^2 in (1). In view of Lemma 1 we have

$$A = \Lambda^{-1} \operatorname{Inv} M_0 \mathscr{L} \Lambda^2 = \Lambda^{-1} \operatorname{Inv} M_0 \mathscr{L} \mathscr{S}^k \partial^k \Lambda^2 = \Lambda^{-1} \operatorname{Inv} M_0 \Lambda^{-k} \mathscr{L} \partial^k \Lambda^2$$

$$= \Lambda^{-1} \operatorname{Inv} \Lambda^{-k} M_k \mathscr{L} \partial^k \Lambda^2 = \Lambda^{k-1} \operatorname{Inv} M_k \mathscr{L} \partial^k \Lambda^2$$

$$= \Lambda^{k-1} \operatorname{Inv} \mathscr{L} \operatorname{Inv} M_k \operatorname{Inv} \partial^k \Lambda^2.$$

By this we have established

THEOREM 2. *For finite functions $g(\omega)$ such that $\omega^2 g(\omega) \in C_0^k[0, \infty)$*

the following factorization of A is valid:

$$A = \Lambda^{k-1} \text{Inv } \mathcal{L} \text{ Inv } M_k \text{ Inv } \partial^k \Lambda^2.$$

This factorization has two advantages over (1).

Firstly, it immediately suggests that the right-hand side of $Ag = h$ is endowed with the property that the function $(\text{Inv } \Lambda^{1-k} h)(t) = t^{k-1} h(1/t)$ belongs to the range of the operator $\mathcal{L}$; that is, there is a function $f_k(\omega)$ such that

$$\mathcal{L} f_k = t^{k-1} h(1/t).$$

In particular, for $k = 0$, $t^{-1} h(t^{-1}) \in R(\mathcal{L})$, the range of the transformation $\mathcal{L}$, that is, $\mathcal{L} f_0 = t^{-1} h(1/t)$.

Secondly, the evaluation of the operator

$$(\text{Inv } M_k \text{ Inv})^{-1} = \text{Inv } M_k^{-1} \text{ Inv}$$

is simplified by the fact that the operator M_k^{-1},

$$(M_k^{-1} u)(t) = \sum_{n=1}^{\infty} \mu(n) \frac{1}{n^k} u(nt),$$

acquires the factor n^{-k} accelerating convergence of the series.

If the functions

$$f_k = \mathcal{L}^{-1}(t^{k-1} h(t^{-1})), \quad \varphi = \mathcal{L}^{-1}(h(t)),$$

exist then they are known to be related by [11, p. 372]

$$f_k(\omega) = \omega^{\frac{2-k}{2}} \int_0^{\infty} \frac{\mathcal{F}_{2-k}(2\sqrt{\omega x})}{x^{(2-k)/2}} \varphi(x) dx, \quad 0 \leq k < 3.$$

In terms of the functions $f = f_0(\omega)$, $\varphi(\omega)$, the inversion formula (3) takes on the form

$$\omega^2 g(\omega) = \sum_{n=1}^{\infty} \mu(n) f\left(\frac{\omega}{n}\right), \quad k = 0, \tag{8}$$

$$g(\omega) = \sum_{n=1}^{\infty} \frac{\mu(n)}{n^2} \int_0^{\infty} \frac{\mathcal{F}_2(2\sqrt{\omega x/n})}{(\omega x/n)} \varphi(x) dx. \tag{9}$$

Now we demonstrate that, using the Chebyshev-Laguerre polynomials $L_k(z, \alpha) \equiv L_k^{\alpha}(z)$, these relations can be transformed into more explicit

Supplement

inversion formulas, which give $g(\omega)$ directly in terms of h and its derivatives.

Let

$$\mathscr{L}f = \frac{1}{p} h(p)$$

and $f(t)$ satisfy the condition

$$\int_0^\infty e^{-t} t^{-\lambda} |f(t)|^2 dt < \infty, \quad \lambda > -1.$$

Then the expansion of $f(t)$ in terms of $L_k(z, \alpha)$ is of the form [14]:

$$f(t) = t^\lambda \sum_{k=0}^\infty a_k \frac{k!}{\Gamma(k + \lambda + 1)} L_k(t, \lambda),$$

where

$$a_k = \frac{(-1)^k}{k!} \left\{ \frac{h(z)}{z^{\lambda-1}} \right\}^{(k)}_{z=1}.$$

Consequently, from (8) we obtain

$$\omega^2 g(\omega) = \sum_{k=0}^\infty \frac{(-1)^k \omega^\lambda}{\Gamma(k + \lambda + 1)} \left\{ \frac{h(z)}{z^{\lambda-1}} \right\}^{(k)} \left(\sum_{n=1}^\infty \frac{\mu(n)}{n^\lambda} L_k\left(\frac{\omega}{n}, \lambda\right) \right).$$

Similarly, since (see [20])

$$(za)^{-\alpha/2} \mathscr{S}_\alpha(2\sqrt{za}) = e^{-a} \sum_{k=0}^\infty \frac{a^k}{\Gamma(k + \alpha + 1)} L_k(z, \alpha), \quad (10)$$

then putting here $\alpha = 2$, $a = \omega x$, $z = 1/n$, and substituting (10) into (9), we find

$$g(\omega) = \sum_{k=0}^\infty \frac{(-1)^k \omega^k}{(k + 2)!} \left(\frac{h(\omega)}{\omega} \right)^{(k)} \left(\sum_{n=1}^\infty \frac{L_k(1/n, 2)}{n^2} \mu(n) \right). \quad (11)$$

Letting in (10) $\alpha = 2$, $a = xp$, $z = \omega/pn$, $p > 0$, we get in a similar way

$$g(\omega) = \sum_{k=0}^\infty \frac{(-1)^k p^k}{(k + 2)!} \left(\frac{h(p)}{p} \right)^{(k)} \left(\sum_{n=1}^\infty \frac{\mu(n)}{n^2} L_k\left(\frac{\omega}{pn}, 2\right) \right). \quad (12)$$

Finally, letting here $p \to 0$ and recalling the representation for $1/\zeta(s)$

$$\frac{1}{\zeta(s)} = \sum_{n=1}^{\infty} \mu(n) \frac{1}{n^s},$$

we find

$$g(\omega) = \sum_{k=0}^{\infty} \frac{\omega^k}{k!(k+2)!\zeta(k+2)} \left(\frac{h(t)}{t}\right)^{(k)}_{t=0}.$$

Omitting the proofs for brevity, we give two more forms of inversion formulas. In the first, the inversion formula reads

$$\sqrt{v}\, g(\sqrt{v}) = \lim_{n \to \infty} \frac{(-1)^n}{2\pi} \left(\frac{e}{n}\right)^{2n} [v^{2n} f(n)(v)]^{(n)}, \qquad (13)$$

where

$$f(v) = \sum_{n=1}^{\infty} \mu(n) \frac{h_1(n\sqrt{v}) - n\sqrt{v}}{\sqrt{v}},$$

$$h_1(t) = \int_0^t h(\tau)\, d\tau,$$

and the normalization $\int_0^{\infty} g(\omega)\, d\omega = 1$ has been taken into account.

In the second version, it is assumed that the function $g(\omega)$ is known in a neighborhood of zero (say, on an interval $[0, a]$, $a > 0$), and is sought for $\omega > a$. Since by a change of variables the case of $[0, a]$ reduces to $[0, 1]$, we present this inversion formula for $a = 1$:

$$\omega^2 g(\omega) = \sum_{n \le \omega} \mu(n) u(\omega/n), \quad \omega \ge 1, \qquad (14)$$

$$u(\omega) = \mathcal{L}^{-1} \left\{ t^{-1} h(t^{-1}) - t \int_0^1 \frac{e^{\omega t}}{(e^{\omega t} - 1)^2} g(\omega) \omega^2 \, d\omega \right\}.$$

2.3. Problems

1. Prove the inversion formulas (13), (14).
2. Prove the identity.

$$A(\omega^{\beta} e^{-q\omega})(t) = t^{\beta+1} \Gamma(\beta+3)[\zeta(\beta+2, 1+qt) - qt \cdot \zeta(\beta+3, 1+qt)].$$

Supplement

Here Γ is the gamma function, and

$$\zeta(s, a) = \sum_{n=0}^{\infty} 1/(a+n)^s.$$

What should be the values of parameters β, q for this formula to be valid?

Orthogonalize the family of polynomials $\{\omega^n\}$ on $(0, \infty)$ with the weight $\omega^2 e^{-q\omega}$ ($q > 0$), and construct on the basis of this formula the equation $Ag = h$.

3. Let $u_p(\omega) = \omega^p \theta(b - \omega)$, where $p \geq 0$, $b > 0$, and θ is the Heaviside function. Prove that for $t > b/2\pi$

$$(Au_p)(t) = (2t)^{p+1} \left\{ -\left(\frac{b}{2t}\right)^p \coth\left(\frac{b}{2t}\right) \right.$$

$$\left. + p \sum_{k=0}^{\infty} \frac{2^{2\kappa} B_{2\kappa}}{(p + 2\kappa - 1)(2\kappa)!} \frac{b^{p+2\kappa}}{t^{p+2\kappa-1}} \right\}.$$

Here B_{2k} are Bernoulli numbers.

4. Noting that

$$-t^{-1} \frac{d}{dt} t \int_0^{\infty} \frac{(\omega/t) g(\omega) d\omega}{e^{\omega/t} - 1} = t(Ag)(t),$$

obtain inversion formulas for the Planck integral equation (see [8]):

$$\nu^3 \int_0^{\infty} \frac{\varphi(t) dt}{e^{\hbar \nu/kt} - 1} = \psi(\nu), \quad \nu \geq 0.$$

Here $\hbar$ and k are given constants. Compare with the formula given in Ref. [8].

5. Construct an algorithm for solving the equation $Ag = h$ in the parametric class of functions

$$\omega^2 g(\omega) = a_0 \omega^\alpha \theta(c - \omega) + \sum_{k=1}^{N} a_\kappa \delta(\omega - \omega_\kappa),$$

where a_0, c, a_k, ω_κ, $k = 1, 2, ..., N$, and the number N itself are parameters to be determined, θ is the Heaviside function, δ the Dirac delta function, and $\alpha > 0$ is a given number.

Reference guide

To Section 1.1: The properties of the resolvent are well-known, see, e.g., [18].

To Sections 1.2 and 1.3: Results outlined in these sections may be found in [6] and [5], respectively. Almost all of them have been obtained by analogy with the familiar properties of usual analytic and harmonic functions, see, e.g., [9, 10, and 22].

To Section 1.4: The material of this section is based on an unpublished work of A.L. Bukhgeim and V.G. Kardakov.

To Chapter 2: Results presented in this chapter may be found in [7]. For closely related tomographic problem formulations in the plane, see [1–3 and 15–17], and the references cited therein.

REFERENCES

to Part One: Linear Operators

1. A.N. Kolmogorov, S.V. Fomin. *Elements of Function Theory and Functional Analysis* [in Russian], Nauka, Moscow (1989).
2. N. Dunford and J.T. Schwartz. *Linear Operators*, Wiley Interscience, New York, Vol. 1 (1958), Vol. 2 (1963), Vol. 3 (1971).
3. A.A. Kirillov and A.D. Gvishiani. *Theorems and Problems of Functional Analysis* [in Russian], Nauka, Moscow (1989).
4. K. Yošida. *Functional Analysis*, Springer Verlag, New York (1965).
5. W. Rudin. *Functional Analysis*, McGraw-Hill, New York (1973).
6. S. Helgason. *Groups and Geometric Analysis*, Academic Press, New York (1984).
7. T. Kato. *Perturbation Theory for Linear Operators*, Springer Verlag, Berlin (1966).
8. J. Weidmann. *Linear Operators in Hilbert Spaces*, Springer Verlag, Berlin (1980).
9. M. Reed and B. Simon. *Methods of Modern Mathematical Physics*, Vol. 1: Functional Analysis, Academic Press, New York (1972).
10. A.V. Balakrishnan. *Applied Functional Analysis*, Springer Verlag, Berlin (1976).
11. L.Ya. Savel'ev. *Differentiation of Vector Functions of Vector Variable* [in Russian], NGU Press, Novosibirsk (1991).
12. L.Ya. Savel'ev. *Lectures on Mathematical Analysis. Introduction and Parts One-Four.* [in Russian], NGU Press, Novosibirsk (1969–1975, pp. 159, 285, 292, 339, 396).

to Part Two: Ill-Posed Problems

13. A.N. Tikhonov. "Stability of inverse problems," *Dokl. Akad. Nauk SSSR*, **39**, No. 5, 195–198 (1943).
14. M.M. Lavrent'ev. *Some Improperly Posed Problems of Mathematical Physics*, Springer, New York (1967).
15. A.N. Tikhonov. "On the solution of ill-posed problems and the method of regularization," *Dokl. Akad. Nauk SSSR*, **151**, No. 3, 49–52 (1963).
16. R. Lattes and J.L. Liones. *Methode de Quasi-Reversibilite et Applications*, Dunod, Paris (1967); English transl.: North Holland, Amsterdam (1969).
17. V.K. Ivanov, V.V. Vasin, and V.P. Tanana. *Theory of Linear Ill-Posed Problems and Its Applications* [in Russian], Nauka, Moscow (1978).
18. A.N. Tikhonov and V.Ya. Arsenin. *Solution of Ill-Posed Problems*, Wiley, New York (1977).
19. M.M. Lavrent'ev, V.G. Romanov and S.P. Shishatskii. *Ill-Posed Problems of Mathematical Physics and Analysis* [in Russian], Nauka, Moscow (1980); English transl.: *Amer. Mathem. Soc.*, Providence Vol. 64 (1986), .
20. S.G. Krein. *Linear Differential Equations in Banach Space* [in Russian], Nauka, Moscow (1967); English transl.: *Amer. Mathem. Soc.*, Providence (1971).
21. R. Courant. *Methods of Mathematical Physics*, Vol. 2: Partial Differential Equations, Wiley Interscience, New York (1962).

to Supplement:

1. Yu.E. Anikonov. *Methods of Analysis of Multidimensional Inverse Problems for Differential Equations* [in Russian], Nauka, Novosibirsk (1978).
2. Yu.E. Anikonov and L.N. Pestov. *Formulas in Linear and Nonlinear Tomography Problems* [in Russian], NGU, Novosibirsk (1990).
3. A.L. Bukhgeim. *Special Operator Equations in Scales of Banach Spaces and Their Applications* [in Russian], Preprint No. 280. Computing Center, Novosibirsk (1981).
4. A.L. Bukhgeim. *Introduction into the Theory of Inverse Problems* [in Russian], Nauka, Novosibirsk, 1988.
5. A.L. Bukhgeim. "Tomography and A-harmonic functions," *Dokl. Akad. Nauk* (in print).
6. A.L. Bukhgeim and S.G. Kazantsev. "Beltrami type elliptic systems and problems of tomography," *Dokl. Akad. Nauk SSSR*, **351** (1), 15–19 (1990).
7. A.L. Bukhgeim and S.G. Kazantsev. "Reconstruction of the phonon spectrum of a crystal from its heat capacity," *Dokl. Akad. Nauk* (in print).
8. N. Wiener and R. Paley. *Fourier Transforms in the Complex Domain*, Amer. Mathem. Soc., New York (1934).
9. F.D. Gakhov. *Boundary Problems* [in Russian], Fizmatgiz, Moscow (1963).
10. D. Gilbary and N.S. Trudinger. *Elliptic Partial Differential Equations of Second Order*, 2nd ed. Springer Verlag, Heidelberg (1983).
11. V.A. Ditkin and A.G. Prudnikov. *Integral Transformations and Operational Calculus* [in Russian], Nauka, Moscow (1974).
12. M.A. Evgrafov. *Analytic Functions* [in Russian], Nauka, Moscow (1968).
13. V.A. Korshunov and V.A. Tanana. "Evaluation of phonon density states by the thermodynamic functions of a crystal," *Dokl. Akad. Nauk*, **231** (4), 845–848 (1972).
14. V.I. Krylov and N.S. Skoblya. *Methods of Approximate Fourier and Inverse Laplace Transformations* [in Russian], Nauka, Moscow (1974).
15. R.G. Mukhometov. "Reconstruction of two-dimensional Riemannian metric and integral geometry," *Dokl. Akad. Nauk SSSR*, **232** (11), 32–35 (1977).
16. R.G. Mukhometov. "Integral geometry in regions with a reflecting part of the boundary," *Dokl. Akad. Nauk SSSR*, **296** (2), 279–283 (1987).
17. R.G. Mukhometov. Stability estimator for one problem of computer tomography. In: *Well-Posedness of Analysis Problems* [in Russian], Institute of Mathematics, Novosibirsk (1989), 122–124.
18. S. Prössdorf. *Einige Klassen von singulärer Gleichungen*, Akademia, Berlin (1974).
19. B. Sz.-Nagy and C. Folias. *Analyse Harmonique des Opérateurs de l'Espace de Hilbert*. Academia Kiado, Szeged (1967).
20. P.K. Suetin. *Classical Orthogonal Polynomials* [in Russian], Nauka, Moscow (1979).
21. A.N. Tikhonov. On ill-posed problems. In: *Computing Methods and Programming* [in Russian], Vol. 8, 3–33 MGU, Moscow (1967).
22. L. Hörmander. *An Introduction to Complex Analysis in Several Variables*, Van Nostrand, Princeton, New Jersey (1966).
23. T.M. Apostol. *Introduction to Analytic Number Theory*, Springer Verlag, New York (1976).

Subject Index

Absolute value of operator 184
Approximating sequence 73

Base 128
    orthonormal 127
    trigonometric 128
Boundedness
    integral 69
    uniform 139
    weak 169

Coefficient
    Fourier 129
    functional 121
Components
    monotonic 73
    positive 73
Convergence
    almost everywhere 66
    in itself 4
    in terms of components 73
    in the mean 124, 145
    of a sequence 137
    to a point 4
Criterion
    integrability 84, 85

Criterion
    Riesz 147
    weak convergence 169
    Weyl 194

Decompositon
    Fourier 130
    orthogonal 119
    polar 184
Derivative 14
    directional 58
    generalized 156
    of distribution 155
    partial 17
    total 28
Differential 11
    partial 17
    total 17
Distance 133
Distribution 150
    finite 154
    tempered 152

Equation
    evolution 227, 283
    Fredholm 185
    functional 36

Equation
    operator 254, 280
    Parseval 132
    resolvent 190
ε-net 137

Form
    canonical (of operator) 199
    Hermitian 112
    quadratic 113
Formula
    Leibnitz's 45, 156
    Maclaurin's 44
    Taylor's 41, 141
Fourie series 129
Fourier series expansion 130
Function
    A-analytic 333
    A-harmonic 327, 347
    characteristic 161
    continuous 135, 138
    continuous at a point 136
    delta 140
    distribution 202
    finite 142
    generalized 150, 152
    Green's 349, 355
    integrable 75
    Lagrange 53
    locally integrable 150
    measurable 71
    monotonically integrable 75
    of rapid decrease 148
    of slow growth 153
    partial 92
    smooth 141
    uniformly continuous 136
Functional 65, 119
    continuous 119
    linear 119
    positive 65

Gradient 59

Hyperplane 120

Identity
    Green's 349
    polarization 115
    triangle 115
Increment 10
Inequality
    Bessel 128
    Cauchy 6, 113
    integral triangle 80
Infinitesimal generator 223
Intergal 76
    Cauchy-type 337
    double 92
    indefinite 108
    iterated 93
    with respect to projection-valued measure 102
Integral sum 63, 203
    with respect to projection-valued measure 203
Isometry 134

Kernel of an operator 119

Lemma
    Fatou's 82
    Levy's 69
    Riesz's 124
    Sobolev's 157

Measure 62
    Baire 151
    continuous 67
    countably additive 68
    countably semiadditive 68
    of slow growth 153
    projection-valued 200, 203, 208, 212
    spectral 217

Subject Index

Metric 133
Metric topology 136
Minimum
   local 50
   strict local 50
Multiindex 149

Norm 1
   Euclidean 114, 197
   graph 169
   operator 6

Operator
   adjoint 173, 174, 180
   bounded 5
   closed 168
   compact 171
   completely continuous 170
   degenerate 171
   diagonal 171
   differential 141
   Fredholm 175, 185
   Hermitian 182
   integral 171, 172
   isometric 183, 184
   Laplace 142
   linear 5
   matrix 171
   normal 182
   positive 183
   self-adjoint 179, 182
   symmetric 182
   unitary 183
Operator-valued exponential 224
Orthogonal projection 116
Orthonormal family 127

Parallelogram law 115
Problem
   Cauchy 291
   Dirichlet 356
   integral geometry 298

Problem
   interior 249
Product
   inner 113
   of measures 91
Polynomials
   Chebyshev 128
   Hermit 128
   Laquerre 128
   Legendre 128

Root of operator 184

Schmidt method (of orthogonalization) 127
Set
   Baire 151
   closed 137
   compact 138
   completely bounded 137
   equicontinuous 139, 147
   integrable 90
   measurable 89
   of measure zero 63
   open 137
   relatively compact 137
   resolvent 185
   simple 62
Smallness
   comparative 9
   local 8
Smooth surface 54
Space
   Banach 4
   basic 147
   Euclidean 114
   Frechet 142
   Hilbert 115
   Lebesque 143
   metric 133
   normed 2, 3

Subject Index

Space
  of basic functions 147
  of distributions 150
  of smooth functions 138
  reflexive 169
  separable 2, 127
  separated 133
  Sobolev 157
Spectrum 184
  continuous 186
  limit 194

Spectum
  of operator 186
    bounded 191
    closed 190
    compact 192
    compact Hermitian 198
    Hermitian 194
    positive Hermitian 197
    self-adjoint 193, 194
  point 186
  residual 186

Tangent 56
Theorem
  Alaoglu's
  Arzela's 140
  Ascoli's 139
  Baire's 165, 1
  Banach's 167
  Banach-Steinhaus 164
  Bochner's 161
  Borel's 137
  Cantor's 138
  Fatou's 81, 83
  Fredholm's 177, 178, 261
  Fubini's 93

Theorem
  Fubini-Tonelli 99
  Hahn-Banach 163
  Hilbert-Schmidt 198
  Lagrange's 23
  Lebesque's 86
  Levy's 80
  Mackey's 169
  Mazur's 170
  Picard's 264, 265, 274
  Plancherel's 161
  Phythagorean 116
  Radon-Nikodym 108, 109
  Riesz's 121, 140, 176
  Riesz-Fisher 132
  spectral 215
  Stone's 225
  Tonelli's 97
  Tychonoff's 267
  Weierstrass' 137
Theorem
  implicit function 36
  integral-derivative commutation 101
  integral-integral commutation 104
  integral-limit commutation 100
  inverse function 37
  on the change of variable 109
  orthogonal projection 116
  fixed point 39
Transform(ation)
  Fourier 87, 157 159
  Fourier-Plancherel 160
  Fourier-Stieltjes 161
  Laplace 223
  Radon 162

Well-posedness 178, 243, 264, 266